Recent Challenges and Advances in Geotechnical Earthquake Engineering

T.G. Sitharam
Indian Institute of Science, India

J. S. Vinod
University of Wollongong, Australia

A volume in the Advances in Environmental Engineering and Green Technologies (AEEGT) Book Series

IGI Global
DISSEMINATOR OF KNOWLEDGE

Published in the United States of America by
IGI Global
Engineering Science Reference (an imprint of IGI Global)
701 E. Chocolate Avenue
Hershey PA, USA 17033
Tel: 717-533-8845
Fax: 717-533-8661
E-mail: cust@igi-global.com
Web site: http://www.igi-global.com

Library of Congress Cataloging-in-Publication Data

Names: Sitharam, T. G., 1961- editor. | Vinod, Jayan, editor.
Title: Recent challenges and advances in geotechnical earthquake engineering / T.G. Sitharam and J.S. Vinod, editors.
Description: Hershey PA : Engineering Science Reference, [2019] | Includes bibliographical references.
Identifiers: LCCN 2018011672| ISBN 9781522569480 (h/c) | ISBN 9781522569497 (eISBN)
Subjects: LCSH: Earthquake engineering. | Geotechnical engineering.
Classification: LCC TA654.6 .R433 2019 | DDC 624.1/762--dc23 LC record available at https://lccn.loc.gov/2018011672

This book is published in the IGI Global book series Advances in Environmental Engineering and Green Technologies (AEEGT) (ISSN: 2326-9162; eISSN: 2326-9170)

British Cataloguing in Publication Data
A Cataloguing in Publication record for this book is available from the British Library.

Advances in Environmental Engineering and Green Technologies (AEEGT) Book Series

ISSN:2326-9162
EISSN:2326-9170

Editor-in-Chief: Sang-Bing Tsai, University of Electronic Science and Technology of China Zhongshan Institute, China & Ming-Lang Tseng, Lunghwa University of Science and Technology, Taiwan & Yuchi Wang, University of Electronic Science and Technology of China Zhongshan Institute, China

MISSION

Growing awareness and an increased focus on environmental issues such as climate change, energy use, and loss of non-renewable resources have brought about a greater need for research that provides potential solutions to these problems. Research in environmental science and engineering continues to play a vital role in uncovering new opportunities for a "green" future.

The **Advances in Environmental Engineering and Green Technologies (AEEGT)** book series is a mouthpiece for research in all aspects of environmental science, earth science, and green initiatives. This series supports the ongoing research in this field through publishing books that discuss topics within environmental engineering or that deal with the interdisciplinary field of green technologies.

COVERAGE

- Cleantech
- Water Supply and Treatment
- Waste Management
- Industrial Waste Management and Minimization
- Pollution Management
- Policies Involving Green Technologies and Environmental Engineering
- Biofilters and Biofiltration
- Sustainable Communities
- Contaminated Site Remediation
- Radioactive Waste Treatment

IGI Global is currently accepting manuscripts for publication within this series. To submit a proposal for a volume in this series, please contact our Acquisition Editors at Acquisitions@igi-global.com or visit: http://www.igi-global.com/publish/.

The Advances in Environmental Engineering and Green Technologies (AEEGT) Book Series (ISSN 2326-9162) is published by IGI Global, 701 E. Chocolate Avenue, Hershey, PA 17033-1240, USA, www.igi-global.com. This series is composed of titles available for purchase individually; each title is edited to be contextually exclusive from any other title within the series. For pricing and ordering information please visit http://www.igi-global.com/book-series/advances-environmental-engineering-green-technologies/73679. Postmaster: Send all address changes to above address.

Titles in this Series

For a list of additional titles in this series, please visit:
https://www.igi-global.com/book-series/advances-environmental-engineering-green-technologies/73679

Biostimulation Remediation Technologies for Groundwater Contaminants
Ashok K. Rathoure (Independent Researcher, India)
Engineering Science Reference • ©2018 • 355pp • H/C (ISBN: 9781522541622) • US $215.00

Innovations and Trends in Environmental and Agricultural Informatics
Petraq Papajorgji (University of New York Tirana, Albania) and Francois Pinet (Irstea Centre de Clermont-Ferrand, France)
Engineering Science Reference • ©2018 • 340pp • H/C (ISBN: 9781522559788) • US $205.00

Effective Solutions to Pollution Mitigation for Public Welfare
Ahmet Ozan Gezerman (Yildiz Technical University, Turkey) Burcu Didem Corbacioglu (Yildiz Technical University, Turkey) and Bhola R. Gurjar (Indian Institute of Technology Roorkee, India)
Engineering Science Reference • ©2018 • 274pp • H/C (ISBN: 9781522533795) • US $180.00

Handbook of Research on Microbial Tools for Environmental Waste Management
Vinay Mohan Pathak (Gurukul Kangri University, India) and Navneet (Gurukul Kangri University, India)
Engineering Science Reference • ©2018 • 496pp • H/C (ISBN: 9781522535409) • US $215.00

Food Systems Sustainability and Environmental Policies in Modern Economies
Abiodun Elijah Obayelu (Federal University of Agriculture – Abeokuta (FUNAAB), Nigeria)
Engineering Science Reference • ©2018 • 371pp • H/C (ISBN: 9781522536314) • US $195.00

Microbial Biotechnology in Environmental Monitoring and Cleanup
Pankaj (G. B. Pant University of Agriculture and Technology, India) and Anita Sharma (G. B. Pant University of Agriculture and Technology, India)
Engineering Science Reference • ©2018 • 427pp • H/C (ISBN: 9781522531265) • US $235.00

For an entire list of titles in this series, please visit:
https://www.igi-global.com/book-series/advances-environmental-engineering-green-technologies/73679

IGI Global
DISSEMINATOR OF KNOWLEDGE

701 East Chocolate Avenue, Hershey, PA 17033, USA
Tel: 717-533-8845 x100 • Fax: 717-533-8661
E-Mail: cust@igi-global.com • www.igi-global.com

Table of Contents

Detailed Table of Contents

Time-frequency representation and spectral features extraction from a digitally recorded ground motion time history of an earthquake is cornerstone in earthquake engineering signal processing and interpretation. Recently developed time-frequency analysis (TFA) techniques are one of the most suitable techniques for the spectral estimation of signals whose frequency content varies with time. The most often used TFA techniques are short-term Fourier transform, Gabor transform, wavelet transform, Wigner-Ville distribution, Choi-William distribution, and cone shape distribution. The spectrograms of TFA reveal better spectral estimation in time-frequency domain and hence recommended to estimate local frequencies, dominate frequency and their incident time. Moreover, the time of occurrence of frequency component corresponding to maximum energy burst as well as its variation with time can also be identified. Results obtained from TFA techniques shows better picture of the spectral content in the data than the other conventional techniques.

Design ground motions are usually developed by one of two approaches: by performing site-specific analyses or from provisions of building codes. Although contemporary codes consider the site effects to an extent, they provide more conservative estimates.

Hence, site-specific analysis, which involves both the seismic hazard analysis and ground response analysis, is a preferred approach to obtain design ground motions. This chapter presents a site-specific analysis for a site near Ahmedabad, Gujarat. The seismic hazard analysis was carried out by DSHA approach. The site is predominantly characterized by deep stiff sandy clay deposits. Extensive shear wave velocity measurement is used for site classification and ground response analysis. The ground response analysis was carried out by using two approaches: the equivalent linear approach using SHAKE2000 and the non-linear approach using FLAC2D. The deep-stiff-soil site is found to amplify the ground motion. The response from nonlinear analysis is found to be considerably higher than those obtained from the equivalent linear approach.

Srikanth Venkatesan, Victoria University, Australia
Abbas Rajabifard, University of Melbourne, Australia
Nelson Lam, University of Melbourne, Australia
Emad Gad, Swinburne University of Technology, Australia
Helen Goldsworthy, University of Melbourne, Australia
Ged Griffin, University of Melbourne, Australia

Disaster management is recognized as a complex task. Despite current facilities and advancements, each disaster continues to frustrate the government and the community. Well-known issues include lack of awareness, direction, preparation and planning, response, recovery, and policy vacuum, besides many other factors. Other issues include emerging drivers of change such as globalization, climate change, technology, social construct, global finance, and education. Despite the recognition of such complex issues, the pace of disaster management strategies seems to be inadequate. This is mainly due to the lack of an approach that can integrate the myriad of issues with the roles and responsibilities of stakeholders such as governments, practitioners, and the community. This chapter identifies a spatially enabled platform as a tool to overcome the aforementioned issues. A preliminary roadmap with "stakeholder position" as the central point of integration is presented. It is anticipated that the roadmap will provide governments and decision makers with the direction needed for future planning.

A. Murali Krishna, Indian Institute of Technology Guwahati, India

Ground improvement techniques are employed to mitigate liquefaction hazards. Granular piles are the most widely preferred alternative all over the world, due to several advantages associated with them. Different mechanisms operate in the function of stone columns/granular piles in liquefaction mitigation like drainage, storage, dilation, densification, and reinforcement. This chapter presents an overview of the use of granular piles as a liquefaction remedial measure for sand deposits. A brief description on the phenomenon of liquefaction and the associated features has been presented. A short discussion on various ground improvement methods available for liquefaction mitigation is presented in light of importance of granular piles. Different installation methods and design concepts for granular piles are presented. Various mechanisms of granular piles in mitigating the liquefaction potential of loose sand deposits are discussed and quantified in detail proving their effectiveness in hazard mitigation.

Seismically induced accelerations are greatly affected by the transmission through soil layers. The stiffness and damping of the material beneath a structure can therefore influence the seismic loads experienced in the structure. Recently, there has been interest in the use of very low shear stiffness materials in subsoil beneath structures in order to act as a filter to seismic motions, but there is significant skepticism as to its practicability. In this chapter, some notable contributions are discussed, and some work showing the physical modelling and numerical modelling of a soft-material scheme that showed initial promise is presented. This is seen to be readily modelled as a simple mass-spring system, which could be applied to other similar cases. Future challenges include development of a suitable material for the purpose, and reliably proving static stability.

Modeling a semi-infinite soil domain has always been a challenge, especially in problems where the interaction between the soil domain and structure domain is a dynamically responsive one, as in the case of seismic wave loading. Under such conditions, researchers used various strategies to simulate, both physically and numerically, the behavior of pile foundations under seismic loading. While describing the challenges and hitherto applied strategies, two comprehensive numerical studies are described: one at behavioral state in terms of inelastic response spectrum and the other at the limiting state of collapse.

Seismic behavior of landfills need for a better understanding of the dynamic properties of municipal solid waste (MSW) and site response of MSW landfill during seismic events. This chapter presents unit weight, shear wave velocity, strain-dependent normalized shear modulus reduction, and material damping ratio relationships for Mavallipura landfill based on field testing, laboratory measurements, and validated using semi-empirical methods. In addition, one-dimensional seismic response analysis by an equivalent linear method for Mavallipura landfill, Bangalore is done using software like SHAKE2000 and DEEPSOIL. Results indicated that the MSW landfill has less shear stiffness and more amplification due to the loose filling and damping, which need to be accounted for by seismically safe MSW landfill design in India. Also, results of seismic response analyses performed by the authors and other researchers are examined to assess the impact of stiffness and height of the landfill refuse fill on the overall response.

Discrete element method (DEM) provides insight into the fundamental physical principles leading to a better understanding of the complex behavior of granular materials under different loading conditions. In the reported studies adopting DEM, real particle shape is ignored, and the particles are modeled as spheres. Spherical shapes are preferred as they provide the simplest case with least computational effort in calculations and efficient algorithms. Hence, in this study, a comprehensive and comparative study on the mechanical behavior of assemblies consisting of particles with different aspect ratios is reported. The results indicate that the assemblies with lower aspect ratio lose its strength completely at less number of cycles when subjected to lower confining pressures. At higher confining stresses, the particles are not quick enough to rearrange themselves resulting in the reduction of average coordination number. This causes a drop of shear strength leading to lower resistance to liquefaction for non-spherical assemblies at high confining stresses.

Blasting is the most common method of rock excavation technique in mining and civil construction and infrastructure projects. General blast-induced damage was extensively studied globally, but the studies on repeated blast induced damage is not yet reported, quantitatively. This chapter deals with the research work carried on the effect of repeated dynamic loading imparted on the jointed rock mass from subsequent blasts in the vicinity, on the jointed rock mass at Lohari Nag Pala Hydroelectric Power Construction Project. The blast-induced damage was monitored by borehole extensometers, borehole camera inspection surveys and triaxial geophones installed at three test sites of different joint orientations at the Main Access Tunnel of power house. The study reveals that the repeated blast vibrations resulted in extra damage of 60%. The results of the study indicate that repeated vibrations, resulted in inducing damage even at 23-33% of the critical vibrations (Vc) levels.

Preface

This book contains important papers published recently in the *International Journal of Geotechnical Earthquake Engineering*. A complete insight into the state-of-the-art technologies in the field of geotechnical earthquake engineering is the key feature of this book. The technical chapters selected in this book offers an excellent mix of topics covering the concepts in geotechnical earthquake engineering with proper emphasis on seismology, liquefaction mitigation and rock engineering. This book will serve as a comprehensive reference volume for different problems in the area of geotechnical earthquake engineering and will greatly benefit engineers and researchers.

The chapter on "Advances in Extraction of Signal from Ground Motion Time Histories Using Time-Frequency Analysis" presents a powerful tool for analysing and processing of seismograms. The chapter summaries the methodologies to distinguish the frequencies in ground motion time history with their incident time at the recording station.

The chapter on "Site-Specific Ground Motion Studies for a Deep Soil Site Near Ahmedabad, Gujarat" highlights the evaluation of site specific design ground motion parameters for a seismically vulnerable site near Ahmedabad, Gujarat. The results of 1D ground response analysis carried out using an equivalent linear and a nonlinear approach and the ground are reported.

The chapter on "A Novel Meta-Integrative Platform for Effective Disaster Management" investigates the developments in the field of information and technology on the news and alerts on a natural disaster. An effective way of integrating the needs and requirements to effectively address the challenges in disaster management is presented in this chapter.

The chapter on "Mitigation of Liquefaction Hazard Using Granular Piles" presents an overview of the use of granular piles for mitigating the liquefaction of sand. The mechanisms of granular piles and its performance in mitigating liquefaction are discussed along with some case histories.

The chapter on "Mitigation of Seismic Accelerations by Soft Materials Embedded in the Ground" highlights the beneficial use of a soft layer, low-stiffness material, in reducing the structural acceleration. The effect of utilising the filtering effect of a soft layer in the ground, akin to liquefied soil related to a sliding surface, with soft trenches to prevent diffraction of shear waves around the horizontal layer is discussed.

The chapter on "Numerical Simulation of Seismic Interaction of Pile Foundation and Structure" reports the strategies for seismic soil-pile-structure interactions. This chapter demonstrates the inelastic response spectrum for a given pile foundation, and the lateral collapse load under liquefying soil conditions.

The chapter on "Seismic Behavior and Dynamic Site Response of Municipal Solid Waste Landfill in India" presents the strain dependent dynamic properties of municipal solid waste evaluated from laboratory and field testing. The chapter also discusses the seismic response analysis of Mavallipura landfill using the equivalent linear method.

The chapter on "DEM Studies on the Liquefaction Behavior of Particles With Different Aspect Ratios" investigates the liquefaction behaviour of assemblies of particles with different aspect ratios during undrained cyclic loading. The article also provides a discussion on the macroscopic and microscopic behaviour of the isotropically compressed samples during cyclic loading.

The chapter on "Effect of Repeated Blast Vibrations on Rock Mass Damage in Tunnels" presents the monitoring of The blast induced damage using borehole extensometers, borehole camera inspection surveys and triaxial geophones at three test sites of different joint orientations at the Main Access Tunnel of power house.

The contributors of this book are among the top researchers and academics in the area of geotechnical and earthquake engineering. All the chapters have been updated by the respective Authors according to publisher guidelines to maintain high standards. The assistance of the publisher during the entire period of editing and publishing is greatly acknowledged.

T. G. Sitharam
Indian Institute of Science, India

J. S. Vinod
University of Wollongong, Australia

Chapter 1
Advances in Extraction of Signal From Ground Motion Time Histories Using Time-Frequency Analysis

Vaneeta Devi
IIT Roorkee, India

M. L. Sharma
IIT Roorkee, India

ABSTRACT

Time-frequency representation and spectral features extraction from a digitally recorded ground motion time history of an earthquake is cornerstone in earthquake engineering signal processing and interpretation. Recently developed time-frequency analysis (TFA) techniques are one of the most suitable techniques for the spectral estimation of signals whose frequency content varies with time. The most often used TFA techniques are short-term Fourier transform, Gabor transform, wavelet transform, Wigner-Ville distribution, Choi-William distribution, and cone shape distribution. The spectrograms of TFA reveal better spectral estimation in time-frequency domain and hence recommended to estimate local frequencies, dominate frequency and their incident time. Moreover, the time of occurrence of frequency component corresponding to maximum energy burst as well as its variation with time can also be identified. Results obtained from TFA techniques shows better picture of the spectral content in the data than the other conventional techniques.

DOI: 10.4018/978-1-5225-6948-0.ch001

INTRODUCTION

A simple 1-D frequency transformation (viz., Fourier Transform) is not sufficient to reveal basic characteristics of ground motion time histories (seismogram) whose frequency content varies with time. Major research efforts have been dedicated in the past decades to seismological data processing to interpret the physical processes using various outcomes of data processing techniques. The extraction of time of the incoming seismic phases use automatic phase pickers which make use of time domain as well as frequency domain analyses. Most of the event detectors and phase pickers are based on the STA (Short Term Average) and LTA (Long Term Average) algorithm (Sharma et al 1990, Sharma, 1992a and b). However, due to non-stationary nature of seismograms, techniques like Time-Frequency Analysis (TFA) which includes time information for the frequencies present in signals are preferred. TFA techniques maps a one-dimensional signal into two-dimensional as a function of time and frequency i.e. Time- Frequency (T-F) domain and show how the spectral content of the signal changes with time. Time- Frequency domain representation a is better frame work then time domain or frequency domain to employ for feature extraction analysis such as the concentration of energy in certain frequency band or at certain time and existence of frequency components and respective localised time in different phases of the waves. In this concern, different types of TFA approaches have been developed for analysis and processing the non-stationary signals like seismograms. The Time-Frequency distributions consist of two categories which can be classified as linear time-frequency representation and quadratic time- frequency distributions (Cohen, 1995, Hlawatsch and Boundreaux-Bartels, 1992; Devi and Sharma, 2016). The linear time-frequency distributions satisfy the basic property of linearity principle/superposition principle which is the desired characteristic in any application involving multicomponent signals, whereas, quadratic time-frequency distribution violate this principle because of quadratic structure of time-frequency representation. The approaches which comes under the linear time-frequency distribution are Short-Time Fourier Transform (STFT), Gabor Transform (GT) and Wavelet Transform (WT). On the other hand Wigner-Ville Distribution (WVD), Choi–Williams Distribution (CWD) and Cone Shaped Distribution (CSD) approaches belongs to quadratic time-frequency distribution and generally known as the classes of Cohen's Distribution. Later two CWD and CSD are also called reduced interference distribution which is an extension of the WVD (Jeong and Williams, 1992).

The widely used Short-Time Fourier Transform (STFT) technique has been well documented by Nawad and Quatieri (1988) and Allen (1977). Huerta-Lopez et al., (2000) and Upergui-Botero et. al., (2012) applied STFT for characterisation of seismic/earthquake waves which is essential for understanding the wave propagation phenomena. The characteristics of waves include the variation of ground motion intensity and spectral content with time. Gabor Transform is a special case of STFT with application of Gaussian window function and the discrete Gabor expansion is the inverse of the discrete Gabor transform. Hence, the pair of the discrete Gabor transform, and discrete Gabor expansion provides a feasible bridge for converting an arbitrary signal from 1D the time domain into 2D the time-frequency domain, or vice versa (Qian et al., 1993, 1999 and 2003; Devi and Sharma, 2016b). The Gabor Transform and expansion has been implemented for the spectral enhancement and removal of culture noise from the doppler ultrasound signals and seismic signals (Zhang et al., 2005; Kumar et. al., 2015; Devi & Sharma, 2016b). STFT and GT both deals with uncertainty principle which precludes to simultaneously obtaining fine time-frequency resolution. The uncertainty principle is given by $\Delta t\Delta\omega \geq 1 / 4\pi$, where Δt is time resolution and $\Delta\omega$ is the frequency resolution and are not arbitrary parameters. In order to achieve good resolution depending upon the signal characteristics a trade off between is generally considered.

Wigner-Ville distribution (WVD) (Wigner, 1932) represents time-frequency components by using the reverse of the signal as an analysis window function. This distribution overcomes the limitation of time- frequency resolution of STFT and GT. WVD provides significant enhancement in time-frequency resolution i.e. good temporal resolution at high frequencies and good frequency resolution at a low-frequency (Cohen, 1989). However, WVD suffers from cross-terms interference when implemented on time varying frequency content signal such as seismological earthquake data. The presence of cross-term in the time-frequency plan makes it impossible to carry out a detailed characterisation/ identification of multicomponent signal data, hence impede its applications on time evolved frequency content signals (Boashash, 1988; Huerta-Lopez et. al., 2000; Tray et. al., 2014; Devi and Sharma, 2016 b). In order to suppress the cross term and overcome the limitation of WVD, few new Cohen's classes are introduced viz., Choi–Williams distribution (CWD) was devised by Hyung-Ill Choi and William J. Williams in 1989 and Cone Shaped Distribution (CSD) proposed by Zhao, Atlas and Marks (1990). CWD adopts exponential kernel, whereas CSD used twin cone shape of the distribution's kernel function to minimise the cross-term interference effect of WVD (Sejdic, 2009; Marks II, 2009). Steeghs and Drijkoningen (1996) applied the CWD to recognise the reflection pattern which reveals the remarkable information about the lithologic stratification. Because the exponential kernel function, the CWD does not suppress the cross-terms that two auto-terms with the same time center or frequency center generate whereas, the CSD suppresses the cross-terms that two auto-terms with the same frequency center generate because of having cone-shaped kernel function (Bradford, 2007).

Over the past two decades, wavelet-based techniques have been applied to time-frequency analysis of seismic data (Ahmed et al., 2007). A comprehensive foundation of the wavelet transform was developed by Daubechies, (1990), Chui, (1992) and Mallat, (1989). The Wavelet Transform (WT) provides a time-scale map, known as a Scalogram (Rioul and Vetterli, 1991), rather than a time-frequency spectrum. Because a scale represents a frequency band, Hlawatsch and Boudreaux-Bartels, (1992) choose their scale to be proportional to the centre frequency of the wavelet, which allowed them to transform their Scalogram into a time-frequency map. Chakraborty and Okaya (1995) and Devi & Sharma (2016 b) have compared wavelet-based with Fourier-based techniques for performing time-frequency analysis on ground motion time history and showed that the wavelet-based technique improves spectral resolution for low frequencies and time resolution for higher frequencies. In addition to this WT can also be used to identify the range of frequency components appearing in different phases of wave like p-wave phase and s-wave phase. WT using non-orthogonal Morlet wavelet function in reflection seismic data analysis has been computed and recommended in the geophysical data processing by Miao and Moon, (1999). Shokrollahi and Riahi, (2013) has applied STFT and CWT techniques in imaging the stratigraphic buried channel in the oil fields and concluded that CWT showed better results than STFT.

In view of above literature survey, TFA approach can also be used for analyses, and processing of ground motion time histories. The overall processing steps to demonstrate the use of TFA includes, (i) stacking the p-onset time histories of ground motion in Time-Frequency (T-F) domain using discrete Gabor Transform in order to suppress the culture noise and for further processing converted in time domain using Discrete Gabor Expansion technique, (ii) comparing the conventional FFT with t-f domain as well as comparison among the techniques of Time-Frequency analyses, (iii) identification of frequencies in p-wave and s-wave phase of the stacked ground time history and (iv) extraction of dominant frequency and study its variation pattern with time. The Advance Signal Processing toolkit of transform using LabVIEW software developed by National Instrument (NI) has been used for processing.

TIME-FREQUENCY ANALYSIS (TFA) TECHNIQUES

Signal processing techniques are classified as time domain, frequency domain, and time-frequency domain techniques. These techniques are complementary to each other and selection of suitable technique for specific objectives in observational seismology becomes a challenging task. Therefore, some of the widely used time-frequency domain signals processing techniques have been used for identification of frequency components; locations of energy concentration burst and examining the frequency pattern variation with time in non-stationary multi-components seismic signals obtained from an active seismic region in Himalayas through deployment of network of short period seismographs. Moreover, time-frequency analysis can give interesting information with regards to energy distribution over frequency bands. The Time-Frequency distributions consist of two categories which are classified as linear time-frequency representation and quadratic time- frequency distributions (Cohen, 1995, Hlawatsch & Boundreaux-Bartels, 1992; Devi & Sharma, 2016b).

LINEAR TIME-FREQUENCY REPRESENTATION

The linear time-frequency distributions satisfy the basic property of linearity i.e. superposition principle which is the desired characteristic in any application comprising multi-component signals. The approaches which comes under the linear time-frequency distribution are; Short-Time Fourier Transform (STFT), Gabor Transform (GT) and Wavelet Transform (WT). These approaches are briefly discussed in the following subsection.

- **Short Time Fourier Transforms (STFT):** The Fourier Transform represents an adequate analytic technique in many signals processing application, but it has a major drawback. It provides the frequency domain information but all the information belonging to time is lost. Thus, the Fourier Technique becomes less effective in case of analysing the non-stationary signals. To improve the Fourier Transform, Gabor introduced a windowing technique in 1946 known as the Short Time Fourier Transform (STFT). The concept in STFT algorithm is to divide the signal into small segments of some window function. Then, Fourier analysis is performed on each window segment to get the frequency present in each segment. Next, this window function is moved to a new segment of the signal and the above-mentioned process is repeated. In this way, time-frequency plane also called 'spectrogram' is formed with colour intensity showing amplitude coefficients as a function of time and frequency. The amplitude distribution is defined by the Short-Time Fourier Transform through linear integral,

$$STFT\left(\tau, f\right) = \int_{-\infty}^{+\infty} x\left(t\right)\gamma\left(\tau - t\right)e^{-j2\pi ft}dt \tag{1}$$

where, x(t) is the time domain seismogram, $\gamma(\tau - t)$ localised windowing function and $e^{-j2\pi ft}$ is Fourier kernel.

- **Gabor Transform (GT):** The Gabor Transform is a special case of Short Time Fourier Transforms (STFT) where window function is chosen as Gaussian function. Since the Gaussian function signals are more concentrated than the other function in frequency domain, the frequency resolution of the Gabor Transform (GT) behaves much better than Short Time Fourier Transforms (STFT).

Albeit, STFT and GT provide both the frequency spectrum and the time evolution of the signal, they also have major drawbacks. They have a fixed resolution with respect to the time window size at all frequencies. This is due to Heisenberg's Uncertainty Principle. If the length of the time window is increased, better frequency resolution can be obtained in the spectrogram. But at the same time, process loos the local time information. Conversely, shorting the time window improves the time resolution at the cost of inferior frequency resolution in the spectrogram.

- **Wavelet Transform (WT):** Wavelet Transform is a recently developed technique in signal analysis and processing. It overcomes the shortcomings of the other TFA techniques such as STFT and GT. Thus, enormous interest of WT technique has been observed in engineering applications. The WT techniques decompose a signal on the basis of wavelet which are scaled and shifted version of the so called mother wavelet. The WT allows exceptional localization both in time domain via translation and in frequency (scale) domain via dilations. Hence, WT techniques have their energy concentrated in time and are better suited to analysis of transient signals such as seismograms. A diagram which display the time as a function of frequency or scale with colour intensity showing amplitude coefficient is called as "Scalogram". The wavelet analysis used here employed application of the Continuous Wavelet Transform (CWT) using different standard type of wavelets. Wavelet Transform (WT) is defined as

$$X_{wT}\left(s,\tau\right)=\frac{1}{\sqrt{s}}\int x\left(t\right)\psi\left(\frac{x-\tau}{s}\right)dt \qquad (2)$$

where, x(t) is the signal to be analyzed, $\Psi(t)$ is the mother wavelet or the basic function, variables s and τ corresponds to the scaling and shifting parameters.

QUADRATICS TIME-FREQUENCY REPRESENTATIONS

Quadratic time-frequency analyses techniques violate the principle of superposition, because of non-linear structure decomposition of multi-component signal in time-frequency domain. Quadratic time-frequency distribution generally, known as the classes of Cohen's distribution. Wigner-Ville distribution (WVD), Choi–Williams distribution (CWD) and Cone Shaped Distribution (CSD) approaches belongs to quadratic time-frequency distribution and discussed in the subsequent section.

Wigner-Ville Distribution (WVD): Wigner-Ville Distribution was developed in the area of quantum mechanics by Wigner in 1932 and later it was introduced in area of signal processing and spectral analysis by Ville (1948). Wigner-Ville Distribution (WVD) is one of the quadratic time-frequency representation techniques. WVD is define as

$$W_s(\tau, f) = \frac{1}{2\pi} \int_{-\infty}^{+\infty} x\left(t + \tfrac{\tau}{2}\right) x^*\left(t - \tfrac{\tau}{2}\right) e^{-j2\pi f\tau} d\tau \tag{3}$$

where; $x\left(t + \tfrac{\tau}{2}\right) x^*\left(t - \tfrac{\tau}{2}\right)$ is time-dependent autocorrelations of signal x (t). 'τ' is time lag and '*'represent the complex conjugate of signal 'x(t)'. It preserves many useful properties like time marginal (equation 4) condition and frequency marginal condition (equation 5).

$$\int_{-\infty}^{+\infty} W_s(\tau, f)\, df = \left|x(t)\right|^2 \tag{4}$$

$$\int_{-\infty}^{+\infty} W_s(\tau, f)\, dt = \left|x(f)\right|^2 \tag{5}$$

However, signal components with a different central time or central frequency generate cross-term interference in the WVD. The cross-term interference reduces the readability of the time-frequency representation.

Choi-William Distribution (CWD)

The Choi-Williams distribution was recognised and described by Choi and Williams in 1989. This distribution is useful for reduction of amplitude of the cross-terms by multiplying basic WVD with an exponential kernel and is enhancing the readability of the time-frequency representation (Tseng, 2012; Huerta et. al. 2000). For signal x(t), CWD is expressed as

$$CWD_x\left(t,f\right)=\int_{\tau=-\infty}^{\infty} e^{-j2\pi f\tau}\int_{\mu=-\infty}^{\infty}\frac{1}{\sqrt{\frac{4\pi\tau^2}{\sigma}}}e^{\frac{-(\mu-t)^2}{4\tau^2/\sigma}}\,x\left(\mu+\frac{\tau}{2}\right)x^*\left(\mu-\frac{\tau}{2}\right)d\mu d\tau \tag{6}$$

where, t is time interval, f is the frequency interval, ' τ ' is time lag, σ is the positive valued scaling factor and * indicates the complex conjugate.

Although, CWD has better readability than WVD but poor time-frequency resolution, CWD alleviates the cross-terms interference between two signal components that have different frequency centres or time centres. However, CWD does not suppress the cross-term interference between two signal components that have the same central time or the same central frequency. Therefore, for large values of the signal length a huge computation time and more memory are required (NI Labview).

Cone Shape Distribution (CSD)

It was first proposed by Yunxin Zhao, Les E. Atlas, and Robert J. Marks II in 1990. The distribution's name stems from the twin cone shape of its kernel function (which is two dimensional functions) on the time-frequency representation. The cone shape distribution of bilinear time-frequency distribution is given by

$$CSD_x=\iiint\frac{1}{4\pi^2}2\zeta\left(\tau\right)\frac{\sin\left(\theta\tau/2\right)}{\theta}e^{\left(-2\pi\alpha\tau^2\right)}x\left(\mu+\frac{\tau}{2}\right)x^*\left(\mu\frac{\tau}{2}\right)e^{-j\theta t-j2\pi\tau-j\theta\mu}d\mu d\tau d\theta \tag{7}$$

where, t is time interval, f is the frequency interval, ' τ ' is time lag, σ is the positive valued scaling factor and * indicates the complex conjugate.

Unlike CWD, it can completely suppress the cross-term between two components having the same center frequency. However, cross-term results from components with the same time center cannot be completely removed (Bradford, 2007).

A CASE STUDY FROM GARHWAL HIMALAYA REGION

To demonstrate the advantages of TFA over classical approaches, a data set consisting of earthquake records from local microearthquake was considered. The recorded data from the Garhwal Himalaya region in the central lower Himalayan Seismic gap which falls in Seismic Zone IV of the seismic zoning map of India (IS code 1893-2001) was taken as an example. The region has experienced moderate size earthquakes viz., 1991, Uttarkashi earthquake (M-6.6) and 1996, Chamoli earthquake (M 6.3) in recent past. These earthquakes occurred within approximately 500 km epicentral distance range from location of the Tehri Dam. Kumaon-Garhwal Himalayan region is bounded by number of thrust/ faults namely: Munsiari Thrust (MT), Vaikrita Thrust (VT), Srinagar fault, Berimag Thrust (BT), Ramgarh Thrust (RT) etc. exhibiting significant seismic potential in this region. An array of twelve short period stations (model CMG 40 T-1, sampling rate 1-100Hz) was deployed to monitor the seismicity of this area in 1995. These twelve stations array is distributed over an area of 100 km x 50 km around the Tehri dam site in Uttarakhand (India). The name of location of these stations are; Ayarchali (AYR), Chandrabadni (CHN), New Tehri Town (NTT), Srikot (SRT), Surkanda (SUR), Chentarbagi (CNT), Khurmola (KHU), Partapnagar (PRT), Sirala (SRL), Giyanja (GYN), Rajgadi (RAJ) and Vinakkhal (VIN) and shown in Figure1a (Sharma et al., 2014). All these stations are monitored by Department of Earthquake Engineering, IIT Roorkee (India). The vertical component seismograms for an event of magnitude 2.08 recorded at different stations namely, Ayarchali, Vinakkhal, Chentarbagi, Surkanda, Rajgadi, New Tehri Town and Vinakkhal on dated 25.08.2009 has been shown in Figure 1b for illustration. All seismograms traces are sampled at equal sampling rate of 100sps. The background culture noise attributes are almost similar in all seismograms traces, because all signals (with few exceptions like, local ground heterogeneity) pass through the same electronic component fitted seismometer before being recorded.

The dataset for the study has been compiled from the vast data recorded by the array from the year 2009 to 2011. The stipulated epicentral distance and earthquake magnitude (M) has been kept between 70 to 80 km and M=2.0-2.25, respectively for making a data set for the study. Total nineteen events of 60 sec duration were acquired to the aforesaid range and summarised in Table 1. The vertical seismograms time series listed in Table 1 are aligned with respect to first arriving p-wave i.e., p- onset at the stations. To minimise effect of background culture noise and to emphasize the effect of the focal process, stacking of 19 events in time-frequency domain has been applied using Discrete Gabor Transform and time series has been converted using Discrete Gabor Expansion techniques for further processing. The Time-Frequency Analysis has been performed on the noise suppressed converted time history of the recorded vertical components of ground motion time histories shown in Figure 2 and thereafter obtained results has been processed and discussed.

TIME-FREQUENCY ANALYSIS (TFA) OF NOISY SEISMOGRAM: AN EXAMPLE

The TFA techniques have been applied on the compiled vertical component of noisy seismograms from events having aforementioned magnitude size range and epicentral distance range. Noisy seismogram is the time series having cultural noise relatively more hindering the interpretation of signals (consisting various phases) from seismic sources. The cultural noise includes the manmade disturbances and the amplitudes from sources other than the seismic sources. The noise suppressed average earthquake time history (i.e. final average Seismogram) has been acquired by stacking of all 19 events time series listed in Table 1 in Time-Frequency (T-F) domain with respect to their first arriving p-wave at the stations and added in the distance range 70-80 km from the epicentre. The resulted time history has been shown in Figure 2 and Fourier transform (FT) of the same is shown in Figure 3 a & b. Figure 3a represents Fourier spectrum on the log map, while Figure 3b only frequency is on log scale so the frequencies in the average earthquake time history can be easily identified. Few local frequency components including dominant frequency having maxi amplitude coefficient in Fourier spectrum and frequency component with respect to highest amplitude in time domain has been annotated in figure 3b. The Fourier spectrum (Figure 3) reveals the frequency content information in the seismogram (Figure 2); however the identification of temporal location of these frequencies as well as the change of these frequencies in the time series is not divulged in the spectrum. Since, Seismograms exhibit clearly multi-component characteristics hence, FT is not enough for the accurate interpretation of seismological data. Therefore, now the seismogram has been examined in Time-Frequency (T-F) domain with Time-frequency analysis techniques viz., STFT, GT, WT, WVD, CWD and CSD. The former two TFA techniques (SFTF and GT) have been studied for various window length/ size (WL) and window types with constant time-frequency sampling information (N), middle one i.e. WT has been examined with various mother wavelet types on same scale and latte three (WVD, CWD,CSD) have been analysed only for different time-frequency sampling information (N).

Figure 1. (a) Array of twelve stations deployed around Tehri Dam in Uttarakhand (India) (Sharma et al., 2014). (b) Example of typical seismograms of an event at different stations recorded by the array

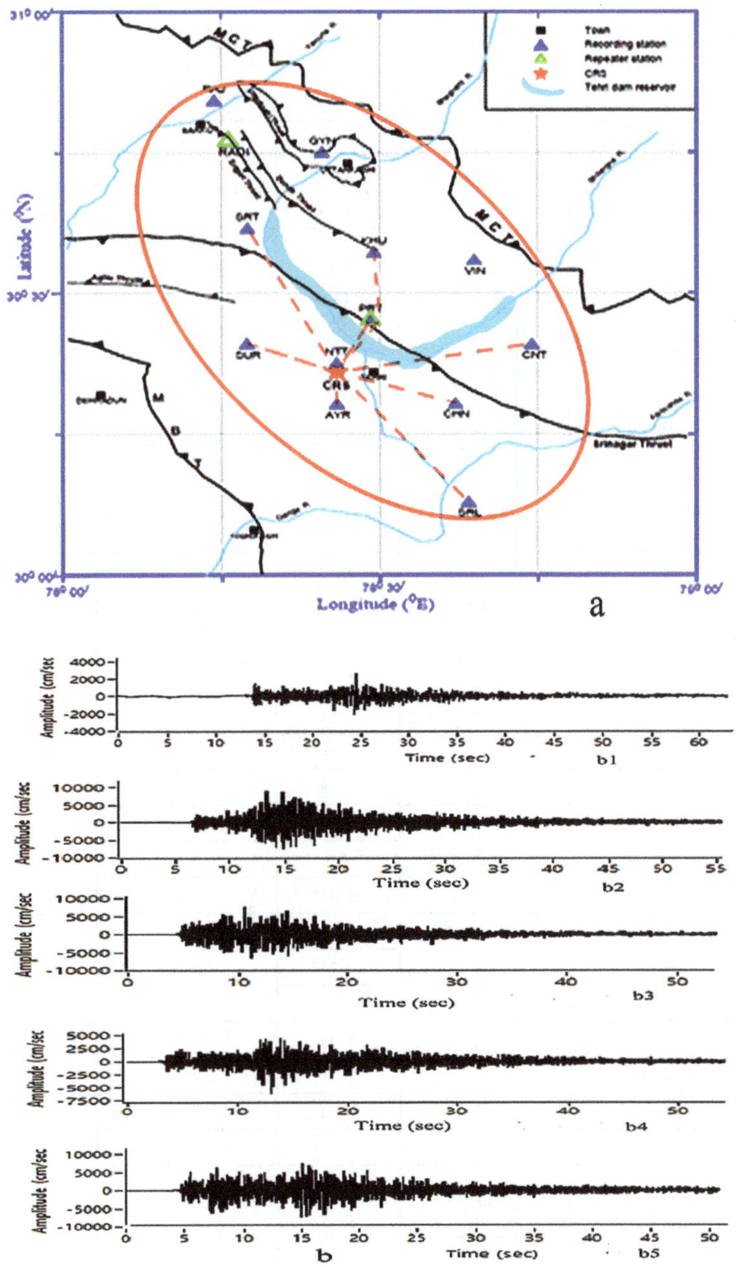

Table 1. Summary of events time series considered for the study

S. No.	Julian Date	Origin Time Hr: Min	Sec	Epicentral Distance D, (km)	Magnitude (M)
1	2009361	01:28	57.18	79.48	2.04
2	2011327	13:28	43.40	76.62	2.12
			43.62	74.92	
			43.63	74.92	
3	2009003	13:10	09.40	76.08	2.19
			08.08	76.45	
			08.79	73.46	
4	2009261	14:17	59.40	77.80	2.02
			59.49	76.51	
			59.87	74.94	
5	2009322	19:53	47.05	76.37	2
6	2009061	14:49	16.01	78.68	2.05
			16.11	77.41	
			16.39	75.85	
7	2010287	08:45	25.09	76.73	2.03
			26.10	75.04	
8	2010283	15:06	41.50	79.74	2.1
			40.39	70.49	
			41.30	77.74	
9	2011147	02:50	21.14	72.68	2.12
			22.88	79.46	
			22.28	73.09	
10	2010029	01:16	19.28	74.77	2.17
11	2011171	06:27	31.87	74.89	2.06
			32.39	75.41	
			31.47	71.02	
13	2009135	19:11	10.13	77.07	2.1
			10.79	78.50	
			10.16	76.20	
14	2009255	22:54	27.16	77.68	2.06
			27.77	78.97	
			27.62	76.18	
16	2009255	22:37	31.92	76.86	2.1
17	2010334	13:52	37.86	70.60	2.07
			38.95	77.67	
18	2010003	18:10	12.54	76.55	2.2
19	2009056	04:04	36.25	72.97	2.01

Devi & Sharma, 2016b.

The time-frequency sampling information and window length/size in terms of Samples points are denoted here by N and WL, respectively. The Time-Frequency (T-F) representation i.e. Spectrogram obtained from STFT depends on the length/size and type of sliding window. The STFT is computed hereby using different type of window includes Hanning Window, Flat Top Window and Low Side-Lobe Window, the time window information is generally taken with 99% overlapping. GT requires only length/size of the sliding time window to construct the spectrogram for given information (Devi & Sharma, 2016b). The spectrogram computed using quadratic TFA techniques namely WVD, CWD and CSD are independent of sliding window and hence, has been examined only for different N values (Devi & Sharma, 2016b). WT is computed here using Continuous Wavelet Transform (CWT) with different standard types of mother wavelets or wavelet function viz., Morlet wavelet, db 04, Coilet 01, Mexican Hat etc. Time-scale (frequency) resolution of given seismogram using CWT is depends on the scale of dilated wavelet 'a' being used in the analysis of spectral decomposition i.e., where signal has low frequency, wavelet window scale will be longer scale, while signal has high frequency, wavelet window scale will be shorter. The specified maximum number of scales of the dilated wavelet chosen in this study is 64 in term of samples. Thus, Scalogram gives amplitude coefficients at different scales for particular time. The distinct quasi-relationship between scale of wavelet and frequency in the signal is expressed in equation (6). The frequency in signal is inversely related to the scale of the wavelet. Frequency (f) in signal corresponds to wavelet dilated scale is represented as

$$f = f_c / \left(a \times dt\right) \tag{6}$$

where, f_c is the centre frequency of the wavelet, 'a' is scale of dilated wavelet and 'dt' is the sampling period.

The colour intensity contours pattern/scale of t-f domain plan/ Spectrogram/ Scalogram represents the energy concentration in various frequency components and corresponding time information. The intense red colour contours represents the stronger amplitudes coefficient, while less density colour contours have weaker amplitude coefficients, of the spectrogram/ Scalogram and exhibits the temporal information of the local frequency component corresponding to the maximum energy concentration.

Figure 2. Time-Frequency (T-F) domain stacked average vertical component seismogram at stipulated epicentral distance

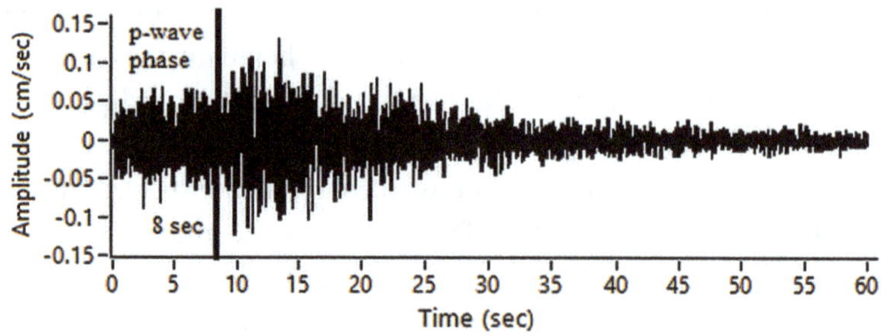

Figure 3. Fourier spectrum of T-F domain stacked average vertical component seismogram; (a) on log graph and (b) only frequency on Log scale

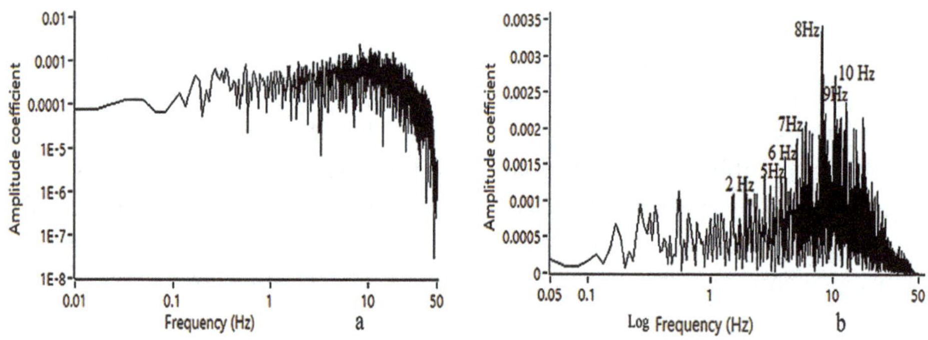

PERFORMANCE OF TIME FREQUENCY ANALYSIS TECHNIQUES

The time-frequency domain stacked seismogram as shown in Figure 2 has been analysed using STFT, GT, WT, WVD, CWD and CSD TFA techniques as discussed in earlier section. Time-Frequency (T-F) domain results thus obtained from various Time-frequency techniques are shown in Figure 4, 5, 6, 7, 8 & 9. Time-Frequency (T-F) domain deciphered that the considered seismogram of magnitude range M=2.0-2.25 consist frequency content in the range of 2 Hz to 12 Hz at the stipulated epicentral distance range 70 to 80 km. Few spectral components have been extracted and annotated in Fourier spectral on log map and only frequency on log scale as shown in Figure 3a and Figure 3b respectively. The Fourier spectrum on log scale in Figure 3a shows the corner frequency band present in the analysed seismogram is around 1-10 Hz. This range is also verified by scaling law of frequency vs magnitude curve which decipher the frequency content in the recorded ground motion time history for magnitude less than 3 (M<3) exists upto10 Hz. Hence, the content of background culture noise present in the stacked seismogram in 2-12 Hz frequency band is acceptable. In addition to this Figure 3 a & b elucidate that frequency content has a sharp decay in its spectral amplitude coefficient at about 0.1 Hz and then amplitude increase progressively upto 10 Hz with small drift in spectral amplitude coefficient between 0.1 Hz to 1.0 Hz. This shows the noise suppression upto some extent i.e. less than 0.1 Hz due to averaging of events having same magnitude range in time-frequency domain. The Fourier spectrum (Figure 3 b) of seismogram revealed that this seismogram has one dominant frequency component of 8 Hz which has maximum spectral amplitude coefficient. However, time information of this frequency has lost completely in Figure 3 b. Fourier spectrum also has not revealed clearly the frequency component corresponding highest amplitude in time domain of seismogram. In addition to this, Fourier spectrum did not describe the distribution of energy in frequency bands with time. All disadvantages of Fourier spectrum can be overcome by the time-frequency analysis techniques. Figure 4 presents the time-frequency representation for STFT using various sliding window length with 99% overlapping and windows type. Figure 4 shows STFT spectrogram for three types of windows namely Hanning window (Figure a1, a2, a3), flat top window (Figure b1, b2, b3) and low side-lobe window (Figure c1, c2, c3) for constant value N=512. The subscripts in a, b and c signifies the WL considered as 128, 256 and 512 respectively. Figure 4 reveals that Hanning window portrays better TFA resolution than other types of windows for same given information. It has been also observed that with increase of the length/size of the analysis sliding window, time information leads to smear, whereas localised frequency information become prominent for a given constant time-frequency sampling points. This is because of the inverse relationship between the sampling rate in the frequency domain and size of analysis window followed by Heisenberg's Uncertainty Principle. It is very difficult to make accurate adjustment between time and frequency resolution. Based on the analysis, it is concluded that a good time-frequency resolution can be achieved when selected sliding windows length/size is taken as greater than and equal to half of the provided time-frequency sampling information. Figure 5 shows GT spectrogram for different window length (Figure 5(a1) for WL=128, Figure 5(a2) for WL=256 and Figure 5(a3) for WL=512. The time-frequency information exhibited in spectrogram portrayed by GT technique shown in Figure 5 is comparatively superior to STFT technique as shown in Figure 4.

The spectrogram of WVD for different time-frequency sampling (N) data is shown by Figure 6 (Figure 6(a1) for N=128, Figure 6(a2) for N=256 and Figure 6(a3) for N=512). Although, a significant time-frequency resolution has been obtained, exhibited information is blurred with some interference cross-terms. These interference cross-terms have no physical meaning and their effect increases

Figure 4. STFT spectrograms of noisy seismogram with 99% window overlapping

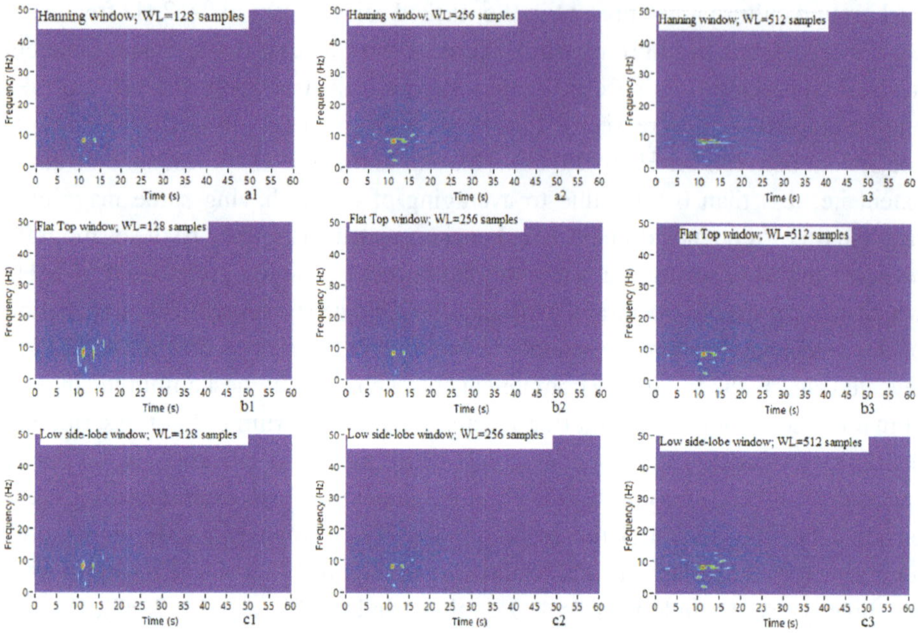

Figure 5. GT Spectrograms of noisy seismogram with 99% window overlapping

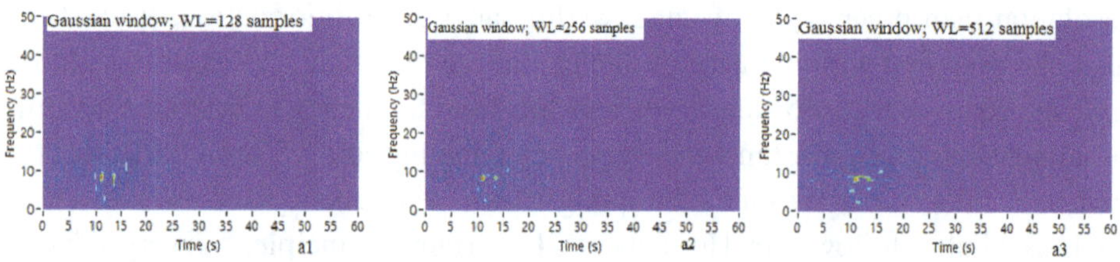

Figure 6. Wigner-Ville Distribution of noisy seismogram
Devi & Sharma, 2016b.

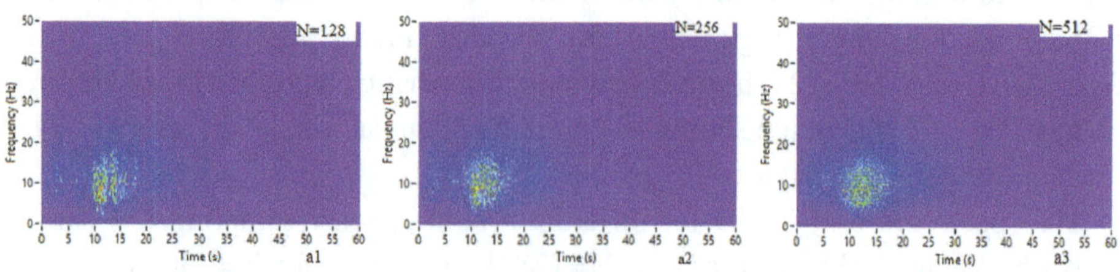

with increase in the time-frequency sampling information (N). Figure 6 a1, a2, & a3 explicitly describes that the interference cross-term effect increases horizontally as well as vertically with increase in the time-frequency sampling rate. Thus, WVD technique has been observed to be not inferring back the original information of the seismogram.

Time-scale distribution (Scalogram) by Continuous Wavelet Transform (CWT) with different mother wavelet is shown in Figure 7 a, b, c & d. Figure 7a shows CWT using Morlet wavelet of 0.75Hz center frequency (f_c), Figure 7b shows CWT using db04 wavelet of 0.71Hz center frequency (f_c), Figure 7c shows CWT using Coiflet01 wavelet 0.725Hz center frequency and figure 7d shows CWT using Mexican Hat of f_c=0.25Hz. Results have manifested that the information produced with CWT using Morlet wavelet is superior to other wavelets being used in the analysis. It is because of non-orthogonal characteristics of Morlet wavelet and hence, no phase distortion occurs during transform (Morlet, 1982 & Daubechies, 1990). The signal energy at equally spaced time interval over duration is represented by the amplitude coefficients of the transform in a particular frequency component. The decrease in maximum amplitude coefficients toward the higher scale value 'a' due to frequency widening of the basic wavelet and wavelet energy conservation can be clearly visualised in Scalogram of WT. Hence, transfer of energy from one frequency band to another frequency band in time is clearly demonstrated with CWT scalogram (Figure 7a). Here, WT scalogram delineates the transfer of ground motion time history energy initially from higher frequencies to lower frequencies in p-wave phase, latterly lower to higher frequencies in s-wave phase.

The spectrograms computed using CWD and CSD are shown by Figure 8 (a1, a2 and a3) and figure 9 (a1, a2 and a3) for different time-frequency sampling (N=128, 256 and 512) information and constant alpha ($\alpha = 0.01$) because of blurring of time-frequency resolution with increase in alpha value. Time-Frequency sampling information and alpha values both are specify the relationship between cross-terms interference and time-frequency resolution. CWD and CSD distribution shows that many of cross-terms have been alleviated. Moreover time- frequency resolution of CWD approach is better than CSD approach, thus CWD can be used to identify the frequency component and its occurrence time corresponding to the maximum amplitude in time domain.

All TFA techniques have revealed almost similar pattern of frequency contents of given stacked seismogram as shown in Figure 2. The Spectrogram /Scalogram obtained from different TFA techniques have been exhibited that the frequency content of the acquired seismogram is presented in the range of 2 Hz to 10 Hz and few identify frequency components are annotated in the Fourier Spectrum (Figure 3b), STFT's Spectrogram (Figure 10a) and GT's Spectrogram (Figure 10b). It has observed that the seismogram have two major frequency components 8 Hz and 9

Figure 7. Continuous wavelets transform Scalogram using different mother wavelets

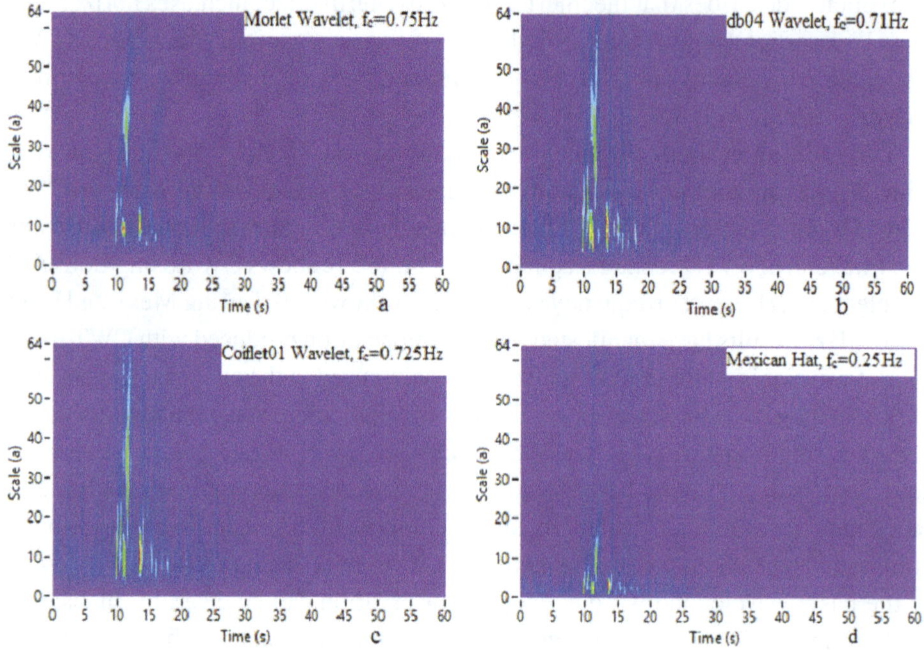

Figure 8. Choi-William Distribution (CWD) of noisy seismogram for (a1) N=128 (a2) N=256 and (a3) N=512

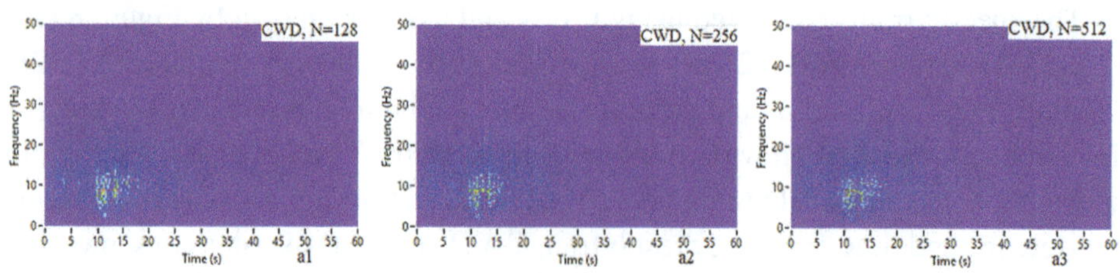

Figure 9. Cone-shaped Distribution (CSD) of noisy seismogram for (a1) N=128 (a2) N=256 and (a3) N=512

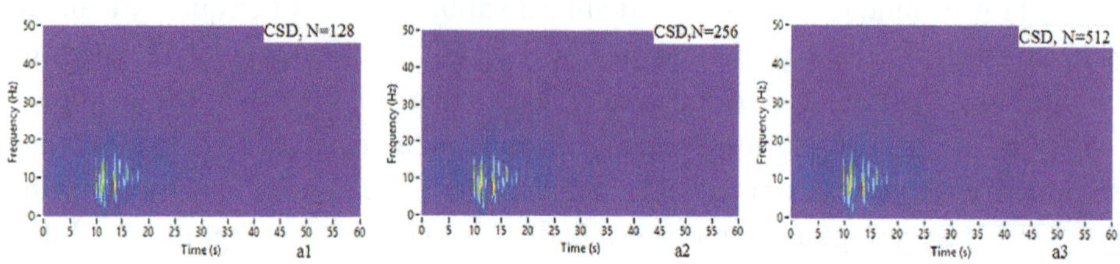

Hz which describes the characteristic feature of seismogram time history, 8 Hz frequency component is corresponding to maximum spectral amplitude coefficient in frequency domain called dominant frequency and 9 Hz corresponding to highest amplitude in time domain. The frequency corresponding to the highest amplitude of seismogram in time domain has been clearly depicted by STFT's Spectrogram (Figure 10a), while the maximum frequency amplitude coefficient in frequency domain has been represented by GT's Spectrogram (Figure 10b). Results of TFA techniques shows that the time of occurrence 8 Hz dominant frequency is about around 10.5 sec whereas for 9 Hz frequency is around 11.4 sec. The concentration/release of maximum energy of the seismogram is marked in the GT's Spectrogram (Figure 10b) and is attributed to the frequency components between 7.5 to 9.5 Hz which covers the frequencies corresponding to highest amplitude in time domain and maximum amplitude coefficient in frequency domain. Hence, the seismogram may be released its maximum energy around 10.5 to 11.5 sec. The STFT and GT's spectrogram manifests maximum energy burst around 11 sec. whereas WT's Scalogram has depicted two significant energy bursts; one of 9 Hz at 11sec and another of 8 Hz at 13sec as shown in Figure 11. WT's scalogram also elucidate that the lowest frequency (known as fundamental frequency) of 2 Hz in the seismogram arrive at around 11.5 sec. The arrival time of s-wave phase can be easily inferred from the energy transfer pattern of WT. The arrival time of s-wave phase is around 10.5 sec to 11.0 sec followed by the lowest frequency 2 Hz in the seismogram.

The spectrogram of WVD depicts addition apparent concentration of energy burst near frequency component of 6 Hz in addition to aforementioned by STFT GT & CWT as shown in Figure 6a. This is because of the unlike decomposition process has involved in different TFA technologies and misleads the correct information of seismogram. The spectrogram of CWD has revealed the maximum energy released frequency component is 9 Hz as shown in Figure 8. While, the spectrogram obtained using CSD still evinced some ambiguous information as shown in Figure 9.

Wavelet Transform has decomposed the signal on the basis of the adjustable wavelet and its scale (flexible time frequency window) which is not an infinite sinusoidal whereas, STFT and GT both have decomposed the signal over a fixed window length of finite sinusoidal. Due to the fixed window size/length, The STFT and GT techniques analysis is inconsistent with different frequencies presented in the signal i.e., at low frequencies there are few oscillations lie within the window that lost localisation of the frequency, while at high frequencies there are number of oscillations laying within the window, hence, the time localization lost. This discrepancy of STFT and GT's TFA is overcome by WT. Time-frequency resolution of WT can be adjusted by choosing suitable mother wavelet (basic function) and scale.

Figure 10. (a) STFT's Spectrogram and (b) GT's Spectrogram; frequency component are annotated corresponding to highest amplitude in time domain, maxi coefficient amplitude in frequency domain and maxi energy released range of frequency presented in the seismogram

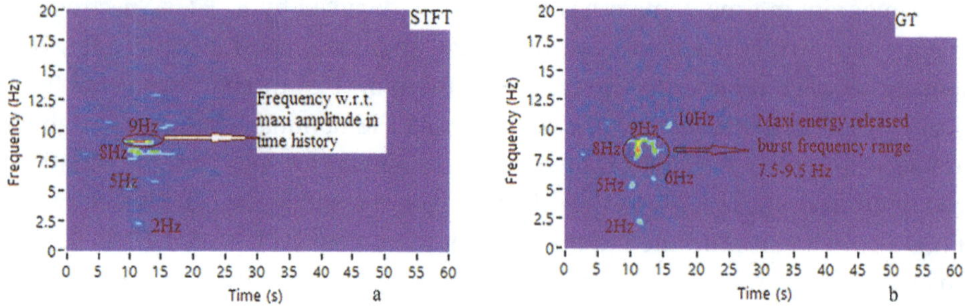

Figure 11. CWT using Morlet Wavelet, f_c=0.75Hz scalogram annotated for the maxi energy released range in the seismogram

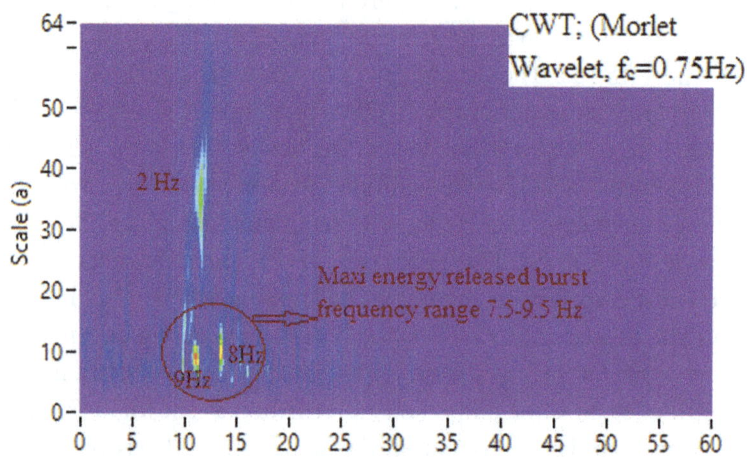

TIME FREQUENCY ANALYSIS OF p-WAVE PHASE

Time- Frequency Analysis of whole noisy seismogram as discussed in the earlier section has not demonstrated any information regarding the frequency components corresponding to the onset arrival of p-wave and frequency component present in p-wave phase of the seismogram. In this concern, the seismogram of 8.0 sec duration from onset of seismogram of figure 2 has been examined and shown in figure 12 and figure13. Figure 12 & 13 exhibits the time frequency planes of GT and CWT using Morlet wavelet with f_c=0.75 Hz respectively. It has observed that the arrival of p- waves reports with high frequencies at epicentral distance range of 70-80 km for magnitude range 2.0- 2.25. The spectrograms of 8 sec duration obtained using GT and CWT shows the existence of 7 Hz, 8 Hz, 9Hz &10 Hz frequencies component in the p-wave phase. Figure 12 and Figure 13 exhibits that the onset arrival of p-wave reported with the 7 Hz frequency at the stipulated epicentral distance. GT spectrogram of p-wave phase reveals that there are two energy released bursts, one for small energy which released around 2.0 to 3.0 sec near to dominant frequency of 8 Hz of seismogram and other for the highest energy released around 6 to 7 sec at frequency of 9 Hz corresponding to the highest amplitude in time domain. Furthermore, the p-wave phase analysis of seismogram does not show fundamental frequency however, the lowest frequency present in p-wave phase is about around 4 Hz. The results of TFA of the vertical component seismogram discussed here exhibit the consistent time-frequency distribution trend and maximum energy concentration in the seismogram time series. The different time frequency analysis techniques being considered here deciphered that the maximum energy released in the whole seismogram has delayed by about 5.0 sec from the maximum energy released by p- wave phase at aforementioned epicentral distance range.

VARIATION OF DOMINANT FREQUENCY (8Hz) WITH TIME

To explore the variation of the amplitude coefficients of frequency components with time, a time history trace corresponding to 8 Hz frequency has been filtered

Figure 12. GT Spectrogram of p-wave phase for duration of 8.0 sec

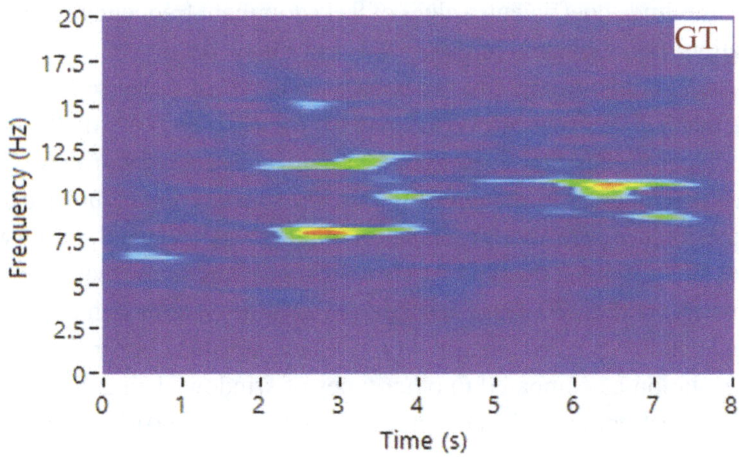

Figure 13. WT Spectrogram of P-wave phase for duration of 8.0 sec using Morlet wavelet with f_c=0.75 Hz

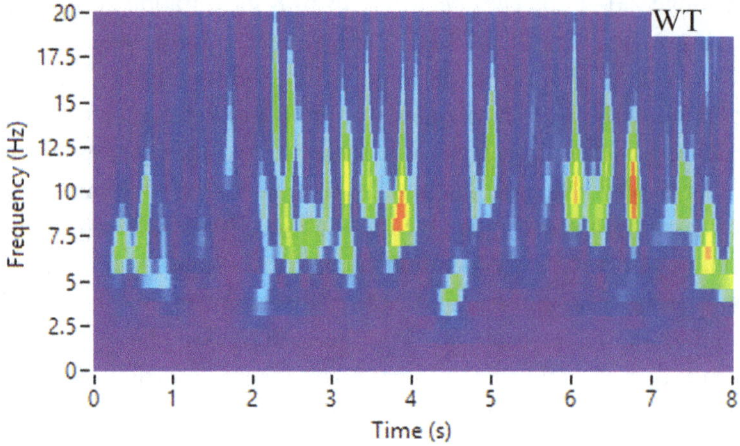

using FIR filter from the acquired stacked seismogram (Figure 2) and is shown in Figure 14a with its Fourier spectrum in Figure 14b. The spectrogram obtained from GT of the 8 Hz filter time trace has been used to extract the frequency amplitude coefficients along time. These amplitudes are called Gabor amplitude coefficient. The Gabor amplitude coefficient of this frequency has been normalised with its maximum coefficient and plotted against time of the time history. Figure 15 shows the 8 Hz dominant frequency variation with time. Frequency variation pattern is nearly similar to the Fourier spectrum of the filtered time trace. Figure 15 elucidates that the arrival time of dominant frequency is about around 10 sec after p-onset of the seismogram, as already discussed in TFA analysis for whole noisy seismogram. Furthermore, it is observed

Gabor amplitude coefficients values of 8 Hz dominant frequency have increased after filtration and elucidated in Figure 14b and Figure 15.

From the aforementioned time-frequency analyses of noisy seismogram, it is concluded that STFT (Figure 4), GT (Figure 5) are suitable to identify the frequency corresponding to highest amplitude in time domain and occurrence time of dominant frequencies in frequency domain constituted in the seismogram as shown in Figure 2. A good time-frequency resolution has been obtained when selected sliding windows length/size (WL) is greater than and equal to half of provided time-frequency sampling information (N) with STFT (Figure 4) and GT (Figure 5) approaches. In the STFT technique, Hanning Window Figure 4 (a, b, c) has resolved better time-frequency spectral resolution as compared to other types of windows being used here in the analysis. The spectrogram of CWD (Figure 8) and CSD (Figure9) has represented better

Figure 14. (a) Filtered time history trace of 8 Hz dominant frequency (b) Fourier spectrum of dominant frequency components at epicentral distance 70-80 km

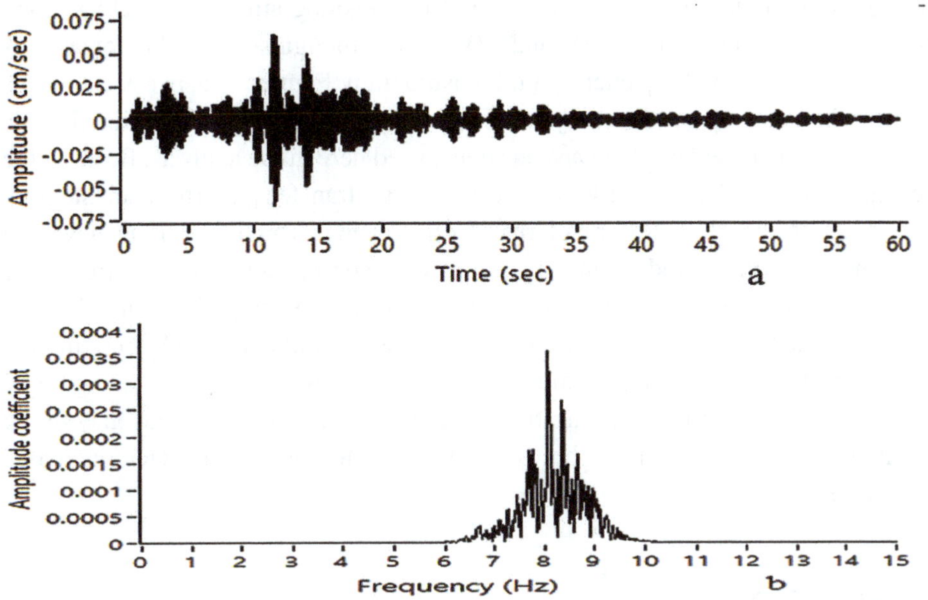

Figure 15. Normalised amplitude coefficients variation of 8 Hz dominant frequency components with time

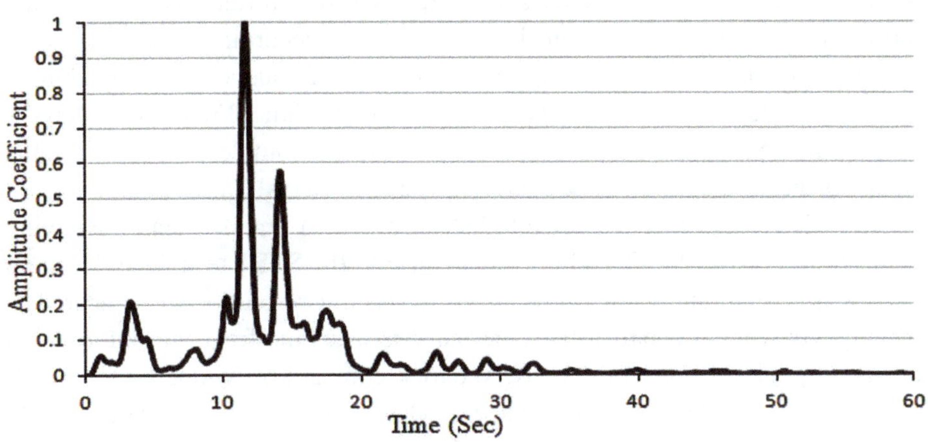

time-frequency distribution as compared to WVD (Figure 6). Furthermore, CWD has explicitly identified 9 Hz frequency and its respective time that corresponding to highest amplitude in the time domain of the seismogram. The CWD has good frequencies resolution than CSD and WVD, even though some ambiguousness is still present for some frequencies of the seismogram. Figures 7 (a, b, c & d) reveals that the information produced by CWT using Morlet wavelet with f_c=0.75 Hz is superior to other used mother wavelets being used here in the analysis. Besides, the scalogram of CWT has clearly shown the energy transfer pattern in seismogram over time. The energy transfer pattern of seismogram is from lower frequency band to higher frequency band (Figure 11). In p-wave phase, energy transfer frequency range is 7 Hz to 10 Hz and in case of the s-wave phase, frequency range is 2 Hz to 10Hz. The time variation of frequency has been examined here by plotting the Gabor amplitude coefficient obtained from GT's spectrogram with time. Figure 15 presents the time- frequency variation pattern of 8 Hz dominant frequency in the seismogram and clearly shows how the frequency amplitude coefficient increases and decreases with time.

CONCLUSION

The time-frequency techniques have been shown as a powerful tool for analyses and processing of seismograms. From the aforementioned discussions it is concluded that STFT and GT both are better suits to distinguish the frequencies in ground motion time history with their incident time at the recording station. Incident time of dominant frequency in frequency domain and frequency corresponding to highest amplitude in time domain can be elucidated in Time- Frequency Analysis, unlike Fourier Spectrum where time information is lost completely. An acceptable time-frequency resolution can be achieved when selected sliding windows length/ size (WL) is greater than and equal to half of provided time-frequency sampling information (N) with STFT and GT techniques. In the STFT technique, Hanning Window has been able to better resolved t-f spectral resolution as compared to other types of windows being used in the analyses. The time-frequency information portrayed by GT's spectrogram is observed to be comparatively superior to STFT's spectrogram due to less spectral leakage by Gaussian window. At window length (WL) equal to time-frequency sampling information (N) and equal to 512 samples, less leakage of spectral content has been observed during analysis process. The time-frequency representation of STFT and GT are subjected to some resolution limitations due to their fixed time window information. Besides this, spectrograms of STFT and GT no information could be inferred related to energy distribution pattern in the time histories of seismogram. All These limitation can be removed with

WT time-frequency technique having flexible time-frequency window framework and wavelets basis decomposition procedure which allow much explicit resolution to frequency varying signals. CWT using Morlet wavelet (Figure 7a) has resolved better time-frequency resolution. Moreover, WT is suitable to recognise the energy transfer pattern among the different frequency bands in the ground motion time history. The energy changing pattern of examined seismogram has been observed here from lower to higher frequencies. It is interesting to note that CWT clearly detects some low frequency components and their incident time at recorded stations which are not seen in other spectrograms. Thus, CWT has advantages over the GT and STFT in describing the basic characteristics features of non- stationary signals like seismogram whose frequency content varies with time. WVD has significant TFA resolution, information exhibited in it is blurred by interference cross-terms and hence, energy concentrations and changing pattern have not been possible to be observed, which limits its applications. As apparent concentration of energy at 5 Hz and 4.5 Hz has been observed that has not been observed in the STFT, GT and WT. However, some time-frequency distribution trends are possible to be interpreted. CWD and CSD TFA techniques suppress much of the cross-terms interference and thus, CWD has identified frequency component corresponding to highest amplitude in the time domain of the seismogram. The CWD has good frequency resolution than CSD, even though some ambiguity is still present for some frequencies of the seismogram. The time variation of frequency has been examined by plotting the Gabor amplitude coefficient obtained from GT's spectrogram with time.

The most important advantages of using STFT, GT, WT over WVD, CWD and CSD is that the former are reconstructive and reversible whereas latter are non-reconstructive and irreversible. Hence, STFT, GT and WT time-frequency techniques are good for the processing, analysis and interpretation of the signal whose frequency content varies with time. The time-frequency analysis of multi-component ground motion time history is non-unique, have different time-frequency techniques yield different time-frequency representations.

ACKNOWLEDGMENT

Authors would like to thank the Department of Earthquake Engineering IIT Roorkee, Uttarakhand, (India) for providing data support as well as laboratory facility to compute this work. The data used is from seismological array funded by THDC. Signal processing module of NI LabVIEW software has been used for processing and analysis of the data. Special thanks to developer National Instruments (NI). I also acknowledge QIP center IIT, Roorkee for providing financial support.

REFERENCES

Ahmed, A. M. L. S., & Wavelet, A. S. (2007). Based Automatic Phase Picking Algorithm for 3-Component Broadband Seismological Data. *Journal of Seismology and Earthquake Engineering*, *9*(1), 15–24.

Allen, J. B. (1977). Short-Time Spectral Analysis, Synthesis and Modifications by Discrete Fourier Transform. *IEEE Transactions on Acoustics, Speech, and Signal Processing*, *25*(3), 235–238. doi:10.1109/TASSP.1977.1162950

Auger, F., Flandrin, P., Goncalves, P., & Lemoine, O. (1996). *Time-Frequency Toolbox*. Rice University, CNRS France.

Boashash, B. (1988). Note on the Use of the Wigner Distribution for Time Frequency Signal Analysis. *IEEE Transactions on Acoustics, Speech, and Signal Processing*, *36*(9), 1518–1521. doi:10.1109/29.90380

Bracewell, R. N. (1978). *The Fourier Transform and its Applications*. New York: McGraw-Hill.

Bradford, S. C. (2007). *Time-frequency Analysis of Systems with Changing Dynamic Properties* (PhD dissertation). California Institute of Technology.

Chakraborty, A., & Okaya, D. (1995). Frequency-Time Decomposition of Seismic Data using Wavelet-Based Method. *Geophysics*, *60*(6), 1906–1916. doi:10.1190/1.1443922

Choi, H., & Williams, W. J. (1989). Improved Time-Frequency Representation of Multicomponent Signals using Exponential kernels. *IEEE Transactions on Acoustics, Speech, and Signal Processing*, *37*(6), 862–871. doi:10.1109/ASSP.1989.28057

Chui, C. K. (1992). Wavelet Transforms and Time-Frequency Analysis. In C. K. Chui (Ed.), *An Introduction to Wavelets* (pp. 49–74). Academic Press Inc.

Cohen, L. (1989). Time-Frequency Distributions: A Review. *Proceedings of the IEEE*, *77*(7), 94–981. doi:10.1109/5.30749

Cohen, L. (1995). *Time-Frequency Signal Analysis*. New York: Prentice Hall.

Daubechies, I. (1990). The Wavelet Transform, Time-Frequency Localization and Signal Analysis. *IEEE Transactions on Information Theory*, *36*(5), 395–457. doi:10.1109/18.57199

Devi, V., & Sharma, M.L. (2016a). Recent Spectral Decomposition Techniques and its Application in analysis of Seismological Data: A review. *IJIRSET, 5.*

Devi, V., & Sharma, M.L. (2016b). Spectral Estimation of Noisy Seismogram using Time-Frequency Analyses. *International Journal of Geotechnical Earthquake Engineering, 7*, 19-32.

Ding, J.J. (n.d.). *Time Frequency Analysis Tutorial Gabor Feature and its Applications*. Academic Press.

Heidari, A., & Salajegheh, E. (2008). *Wavelet Analysis for Processing of Earthquake Records. Asian Journal of Civil Engineering*, *9*(5), 513–524.

Hlawatsch, F. (1998). *Time-Frequency Analysis and Synthesis of Linear Signal Spaces*. Boston: Kluwer Academic Publishers; doi:10.1007/978-1-4757-2815-6

Hlawatsch, F., & Boudreaux-Bartels, G. F. (1992). Linear and Quadratic Time-Frequency Signal Representations. *IEEE SP Magazine*, *9*(2), 21–67. doi:10.1109/79.127284

Huerta, C. I., Shin, Y., Powers, E., & Roesset, J. (2000). Time-Frequency Analysis of Earthquake Records. Pro. of the 12th WCEE.

Iyama, J., & Kuwamura, H. (1999). Application of Wavelets to Analysis and Simulation of Earthquake Motions. *Earthquake Engineering & Structural Dynamics*, *28*(3), 255–272. doi:10.1002/(SICI)1096-9845(199903)28:3<255::AID-EQE815>3.0.CO;2-C

Jeong, J., & Williams, W. J. (1992). Kernel Design for Reduced Interference Distributions. *IEEE Transactions on Signal Processing, 40*(2), 402-412.

Kareem, A., & Gurley, K. (1999). Applications of Wavelet Transforms in Earthquake, Wind and Ocean Engineering. *Engineering Structures*, *26*(3).

Kumar, A., Gupta, S. C., Sharma, M. L., Pandey, A. D., Jindal, A. K., & Jain, S. K. Goswami, Neetu, (2010). Attributes of local seismicity of The Garhwal Himalaya. *Proceedings of the14th Symp. Earthquake Engineering*, 90-99.

Kumar, R., Sumathi, P., & Kumar, A. (2015). A Time–Frequency Approach for Generation of Synthetic Time-Histories of Earthquake Signals. *Acta Geodaetica et Geophysica*. doi:10.100740328-015-0115-7

Liu, G., Fomel, S., & Chen, X. (2011). Time-frequency Analysis of Seismic Data using Local Attributes. *Geophysics*, *76*(6), 23–34. doi:10.1190/geo2010-0185.1

Lockwood, O. G., & Kanamori, H. (2006). Wavelet Analysis of the Seismograms of the 2004 Sumatra-Andaman Earthquake and its Application to Tsunami Early Warning. *Geochemistry Geophysics Geosystems, 7*(9). doi:.10.1029/2006GC001272

Mallat, S. (1999). *A Wavelet Tour of Signal Processing*. New York: Academi Press.

Marks, J. R. II. (2009). *Handbook of Fourier Analysis & its Applications*. Oxford University Press.

Miao, X. G., & Moon, W. M. (1999). Application of Wavelet Transform in Reflection Seismic Data Analysis. *Geosciences Journal*, *3*(3), 171–179. doi:10.1007/BF02910273

Nawab, S. N., & Quatieri, T. F. (1988). *Short Time Fourier Transform, Chapter in Advanced Topics in Signal Processing* (J. S. Lim & A. V. Oppenheim, Eds.). Englewood Cliffs, NJ: Prentice Hall.

Qian, S., & Chen, D. (1993). Discrete Gabor Transform. IEEE Trans. *Signal Processing*, *41*, 2429–2438.

Qian, S., & Chen, D. (1994). Decomposition of the Wigner-Ville Distribution and Time-Frequency Distribution Series. *IEEE Transactions on Signal Processing*, *42*(10), 2836–2842. doi:10.1109/78.324750

Qian, S., & Chen, D. (1996). *Joint Time-Frequency Analysis, Methods and Applications*. Prentice Hall PTR.

Qian, S., & Chen, D. (1999). Understanding the Nature of the Signals whose Power Spectrum Change with Time Joint Analysis. *IEEE Signal Processing Magazine*, 53–67.

Qian, S., Hui, S., & Wei, J. (2003). *Discrete Gabor Expansion for Order Tracking. IEEE Trans. of Instrumentation and Measurements*.

Qiang, Z., Wang, Y., & Shen, G. (2011). Comparison and Application of Time-Frequency Analysis Methods for Non-stationary Signal Processing. In *Advanced Research on Computer Education, Simulation and Modeling* (Vol. 175, pp. 286–291). CCIS. doi:10.1007/978-3-642-21783-8_47

Rioul, O., & Vetterli, M. (1991). Wavelets and Signal Processing. *IEEE Signal Processing Magazine*, *14*(38).

Saurez, L. E., & Montejo, L. A. (2005). Generation of Artificial Earthquakes via the Wavelet Transform. *International Journal of Solids and Structures*, *21-22*(21-22), 5905–5919. doi:10.1016/j.ijsolstr.2005.03.025

Sejdic, E., Djurovic, I., & Jiang, J. (2009). Time-Frequency Feature Representation using Energy Concentration: An Overview of Recent Advances. *Digital Signal Processing*, *19*(1), 153–183. doi:10.1016/j.dsp.2007.12.004

Sharma, M. L. (1992a). *Sample Telemetered Digital Seismic Array in Garhwal Himalaya-Software and Data Management*. Geo. Soc. of India.

Sharma, M. L. (1992b). *Procedures for Analysis of Digital Telemetered Seismic Array Data* (Ph.D. thesis). EQD, Univ. of Roorkee, India.

Sharma, M. L., Kumar, A., Gupta, S.C., Jindal, A.K., Sen, A., Jain, S.K., … Vandana. (2014). Earthquake source parameters and focal mechanism of local earthquakes around Tehri region. *Proceedings of SEE '10*.

Sharma, M. L., Gupta, S. C., & Wason, H. R. (1990). An Automatic Phase Picker for Earthquakes. *Proc. International Conference on Automatic Robotics and Computer Vision*, 247-250.

Shokrollahi, E., Zargar, G., & Riahi, M.A. (2013). Using Continuous Wavelet Transform and Short Time Fourier Transform as Spectral Decomposition Methods to Detect of Stratigraphic Channel in One of the Iranian South-West Oil Fields. *Int. J Sci. Emerging Tech, 5*(5).

Sifuzzaman, M., Islam, M. R., & Ali, M. Z. (2009). Application of Wavelet Transform and its Advantages Compared to Fourier Transform. *The Journal of Physiological Sciences; JPS*, *13*, 121–134.

Simons, F. J., Dando, B., & Allen, R. (2006). Automatic Detection and Rapid Determination of Earthquake Magnitude by Wavelet Multi Scale Analysis of the Primary Arrival. *Earth and Planetary Science Letters*, *250*(1-2), 214–223. doi:10.1016/j.epsl.2006.07.039

Steeghs, P., & Drijkoningen, G. G. (1996). Time-Frequency Analysis of Seismic Reflection Signals. *Proceedings of the IEEE International Conference on Acoustics, Speech, and Signal Processing*, *5*, 2972–2975.

Tary, J. B., Herrera, R. H., Han, J., & Baan, M. (2014). Spectral Estimation—what is new? what is next? *Reviews of Geophysics*, *52*(4), 723–749. doi:10.1002/2014RG000461

The Time Frequency Toolbox Tutorial. (n.d.). Retrieved from http://tftb.nongnu.org/tutorial.pdf

Tseng, C. (2012). *Tutorial: Time-Frequency Analysis for Seismology*. R99942044.

Upegui-Botero, F. M., Huerta-Lopez, C. I., Caro-Cortes, J. A., Martinez-Cruzado, J. A., & Suarez-Colche, L. E. (2012). Joint Time- Frequency Analysis of Seismic Records. *Proceedings of the 15 WCEE*.

Ville, J. (1948). Theory and Applications of the Notion of the Analytic Signal. *Câbles et Transmissions*, *2*, 61–74.

Wigner, E. (1932). On the Quantum Correction for Thermodynamic Equilibrium. *Physical Review*, *40*(5), 749–759. doi:10.1103/PhysRev.40.749

Yilmaz, S., & Bayrak, Y. (2013). Time Frequency Analysis of Georgia Earthquake. *Proceedings of the 7th Congress of Balkan Geophysical Society*. doi:10.3997/2214-4609.20131753

Zabihi, N. E., & Siahkoohi, H. R. (2006). Single Frequency Seismic Attribute Based on Short Time Fourier Transform, Continuous Wavelet Transform and S Transform. *Proceedings of the 6th International Conference & Exposition on Petroleum Geophysics*, 662-669.

Zhang, R. (2008). Spectral Decomposition of Seismic Data via Wigner-Ville Distribution and Chirplet Transform. *Proceedings of the70th EAGE Conference & Exhibition*. 10.3997/2214-4609.20147874

Zhao, Y., Atlas, L. E., & Marks, R. J. (1990). The Use of Cone-Shape Kernels for Generalized Time-Frequency Representations of Nonstationary Signals. *IEEE Transactions on Acoustics, Speech, and Signal Processing*, *38*(7), 1084–1091. doi:10.1109/29.57537

Chapter 2
Site-Specific Ground Motion Studies for a Deep Soil Site Near Ahmedabad, Gujarat

A. Boominathan
Indian Institute of Technology Madras, India

Krishna Kumar
University of Cambridge, UK

R. Vijaya
Indian Institute of Technology Madras, India

ABSTRACT

Design ground motions are usually developed by one of two approaches: by performing site-specific analyses or from provisions of building codes. Although contemporary codes consider the site effects to an extent, they provide more conservative estimates. Hence, site-specific analysis, which involves both the seismic hazard analysis and ground response analysis, is a preferred approach to obtain design ground motions. This chapter presents a site-specific analysis for a site near Ahmedabad, Gujarat. The seismic hazard analysis was carried out by DSHA approach. The site is predominantly characterized by deep stiff sandy clay deposits. Extensive shear wave velocity measurement is used for site classification and ground response analysis. The ground response analysis was carried out by using two approaches: the equivalent linear approach using SHAKE2000 and the non-linear approach using FLAC2D. The deep-stiff-soil site is found to amplify the ground motion. The response from nonlinear analysis is found to be considerably higher than those obtained from the equivalent linear approach.

DOI: 10.4018/978-1-5225-6948-0.ch002

INTRODUCTION

Earthquakes are one of the most devastating natural hazards that cause great loss of life and livelihood. Seismic waves, generated by an earthquake, propagate through different geological formation before they reach the surface at a specific site. The path of the seismic waves in the upper geological formation is strongly influenced by the local soil condition, the geometry of the site (basin effect) and the effect of surface topography thereby producing a varying effect on the ground surface motion. Thus the ground motion parameters at a particular site are influenced by the source, travel path and the site characteristics. The influence of local soil conditions on the ground response has been recognized for many years and its significance was felt during the 1985 Mexican earthquake. The differences in local site conditions such as variations in geological formations, thickness and properties of soil and rock layers, depth of bedrock and water table, surface and underground topography could have significant effects on the characteristics of the ground motion at the surface, such as the: amplitude, frequency content and its duration (Carlos et al., 2006). The extent of influence depends on the geometry and material properties of the subsurface materials, site topography and on the characteristics of the input motion. If the soil medium at the site exhibits only vertical heterogeneities, one-dimensional (1D) analysis will suffice, whereas if there are lateral homogeneities, it is necessary to do two or three-dimensional (2D/3D) analysis.

The significance of the local site effect on the earthquake-resistant design must be considered during the development of site specific design ground motions i.e. motions that reflect the levels of strong motion amplitude, frequency content and duration of a structure or facility at a given site. Hence the process of estimation of the site-specific design ground motion should include both seismic hazard analysis and ground response analysis. Seismic hazard analysis can be carried out by deterministic or probabilistic based methods, by considering the seismicity and seismotectonic of the region. The two or three-dimensional nonlinear ground response analyses should be carried out depending on the subsurface, to predict the design ground motion parameters. However, the equivalent linear approach is used commonly due to the computational demand. This paper presents an evaluation of site specific design ground motion parameters for a seismically vulnerable site near Ahmedabad, Gujarat by conducting a detailed seismic hazard analysis and ground response analysis. The 1D ground response analysis is carried out using an equivalent linear and a nonlinear approach and the ground responses are compared.

SIGNIFICANCE OF SITE-SPECIFIC ANALYSIS

The characteristics of the design ground motion at a particular site are influenced by the location of the site relative to potential seismic sources, the seismicity of those sources, the nature of rupture at the source, travel path effects, and the importance of the structure or facility for which the ground motion is to be used. Design ground motions are usually developed in one of two approaches: from site-specific analyses or from the provisions of building codes and standards. Although contemporary codes consider site effects, they usually do so by lumping groups of similar soil profiles together so that their provisions apply to broad ranges of soil conditions within which the local conditions of a particular site are expected to fall. Because of this, the design ground motions developed from code provisions are usually more conservative (i.e. correspond to stronger levels of shaking) than those developed from site-specific analyses. The UBC (1997) adopts two basic approaches: a static approach, which considers the effects of ground motions represented by static lateral forces and a dynamic approach in which ground motion is characterized by a design response spectrum. These approaches are based on developed hazard maps and provide zone factors that reflect to an extent the local site conditions. However, these maps do not consider local variations at the site when developing the ground motion parameters. Hence, it is necessary to perform site specific ground response analysis for vulnerable sites and critical structures.

Site specific design ground motions reflect the detailed effects of the subsurface conditions at the sites of interest. The process involves the development of site-specific ground motions by seismic hazard analysis and ground response analysis. The seismic hazard analysis will produce a set of ground motion parameters that may or may not correspond to the conditions at the site of interest. Seismic hazard analyses are based on a set of attenuation equations, which usually correspond to a fairly narrow range of subsurface conditions. If the site of interest is located on a similar profile the parameters developed shall be taken as ground motion parameters. If not, the parameters developed should be suitably modified for the effects of local site conditions. The parameters can be modified by an analytical or empirical approach. The former involves deconvolution and conventional ground response analysis (Kramer, 1996).

A site-specific analysis for the evaluation of design ground motion parameters of a particular site involves the following stages:

1. Site characterization based on geological, geophysical and geotechnical investigation,
2. Site classification using the shear wave velocity profile,
3. Seismic hazard analysis which involves collection of seismicity and seismotectonic of the region, selection of predictive relationship to arrive at controlling earthquake with pga,
4. Ground response analysis, and
5. Evaluation of design ground motion parameters including the development of site specific design response spectra.

The overall process flow of site specific response analysis based on DSHA approach adopted in the present study is presented in Figure 1.

Figure 1. Process flow for site specific response analysis based on DSHA approach

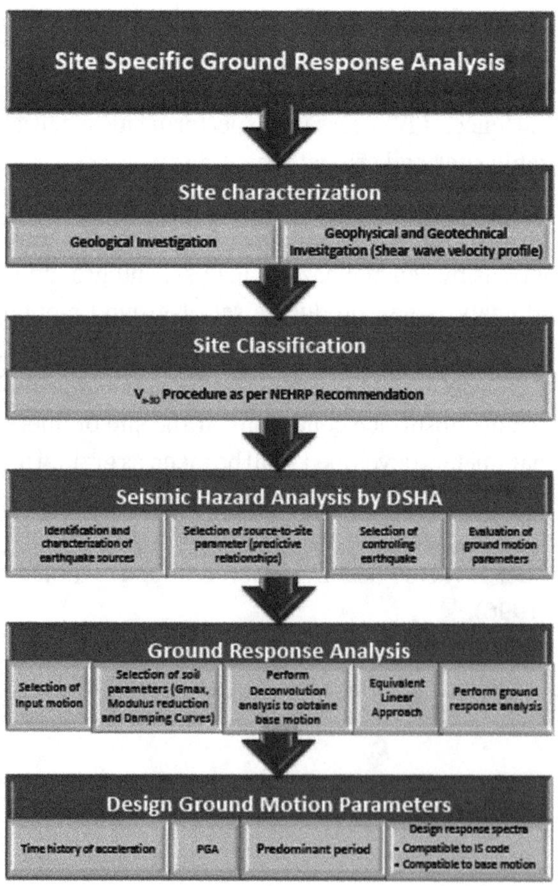

Ground Response Analysis

Majority of the studies on ground response analysis are predominantly one-dimensional as reasonable estimates are obtained at lower cost. Later it was found that such simple models would not adequately reproduce the observations and two-dimensional (2D) models began to be introduced. In the recent days, boundary element method, finite difference method, finite element method and spectral element method (SEM) has proved to be successful tools in modelling and analyzing two/three-dimensional wave propagation problems.

In order to update the seismic code of India, there are several researchers working on site specific analysis across India. However, due to the lack of data on subsurface geometry, studies based on one-dimensional analysis are still widely carried out. For example, the earlier studies in the Delhi region (Sharma et al., 2003; Iyengar & Ghosh, 2004) did not address the local soil effects. Equivalent-linear 1D site specific analyses were carried out by Rao and Ramana (2009) for the Delhi region, and Phanikanth et al., (2011) studied the Mumbai region, by considering the effect of local soil in modifying the ground response.

The incomplete 1D site characterization that lacks information about the deeper geologic formations leads to underestimation of 2D site characteristics such as basin and topography effects (Asimaki & Jeong, 2015). The most recent investigations in this direction carried out by the Network of European Research Infrastructures for Earthquake Risk Assessment and Mitigation, focused on establishing scientifically solid and practically acceptable propositions to incorporate basin and surface topography effects in seismic design codes. In the recent days, spectral element-based tools are found to be efficient to compute the seismic wavefields in soil media with complex geometries (Komatitsch et al., 1998). The effect of geomorphology on the seismic response of the site at Greece has been studied by using 3D Spectral Element Method (Cauzzi et al., 2015). The SEM has been used successfully to simulate seismic wave propagation in the Los Angeles basin (Komatitsch et al., 2004) and in the Taipei basin (Lee et al., 2008). Two-dimensional dynamic finite element analyses considering nonlinearity to study the amplification effect on ground motion due to topography gave consistent results with field observation (Wang et al., 2015). Simulation of 3D elastic wave propagation by finite difference method for the Salt Lake basin predicted some of the overall patterns observed in the amplification of ground motions (Olsen et al., 1996).

SITE DESCRIPTION

The site considered in the present study is located about 12km from Ahmedabad in the state of Gujarat (India). The site lies in the western coastal region of India which has experienced several major earthquakes, including the disastrous Bhuj earthquake in 2001 (Mw = 7.7). Severe damages were observed on multi-story buildings located in close proximity to the Sabarmati river area in Ahmedabad during 2001 Bhuj Earthquake (Sitharam & GovindaRaju, 2004). The site is categorized under Zone III as per Indian seismic code (IS 1893-2002). The site is situated in the Sabarmati River basin where the overburden thickness is reported as high as 300-500 m. Various types of high rise buildings are proposed in the site which spreads over 500 acres of land. The seismic vulnerability of the region and presence of deep soil necessitates performing site specific ground response analysis to arrive design ground motion parameters for the design of various important buildings.

SITE CHARACTERIZATION

Site characterization is the initial phase of the site-specific ground response analysis involves acquiring, processing and interpretation of qualitative and quantitative information of the site such as geological, geotechnical, seismic and seismotectonic details to evaluate the hazard based on the level of site response. The scope of site characterization varies from simple to complex depending upon several major factors including

1. Importance of the structure and the degree of the risk
2. Regional seismicity (the degree of hazard) and adequacy of available data
3. Physiographic conditions (mountains, coastline, plains, etc.)
4. Regional and local geology (hazards, rock type and structure, soil types, and groundwater characteristics)

One of the major problems in geotechnical engineering is the risk of encountering unexpected geological conditions such as sudden variation in the soil profile, rock strata, failure plane and faults in the rock, etc. Failure to anticipate such conditions generally is due to the inadequate geological understanding of the site and may lead to issues concerning design and performance of critical and important structures. Hence it is very important that rigorous geological and geophysical analyses accompany extensive geotechnical investigation to understand the behaviour of the site to seismic loads (Boominathan, 2004). Multichannel Analysis of Surface Waves (MASW) test has been conducted at Chennai city and empirical correlations between Vs and SPT-N value has been obtained for different categories of soil (Maheswari et al[1]., 2010).

Geology

The geological study of the site involves identification of important natural factors such as physiography (mountain ranges, rift, basins, plains, river valleys, coastal regions, etc.) and local geological conditions (geological formations, rock types, faults, fold, lineaments, and terrain variations) that affect the response of the site. Information on physiography is obtained from topographic maps and remote sensing imageries. The physiography of the site has a direct relationship in most cases to the geological hazards, with respect to seismic response and attenuation. The geological formations at the site provide an insight to overall response of site to seismic loading. Numerous methods like radiometric dating are employed to date the geological formations. Terrain analysis and field reconnaissance provide detailed geological data on the surface conditions (Hunt, 2005). Detailed geological studies found abundant in literature provides sufficient data to envisage the likely response of site to seismic excitation.

The study region located on the Sabarmati River basin is characterized by quaternary soil deposit with Mesozoic rocks over thrusting them (Rastogi, 2001). It is located within the Cambay rift flanked by the east and the west Cambay faults. The Cambay rift basin in northwestern India is one of the pre-continental rifts that originated between the early Jurassic and Tertiary, after the breakup of Gondwanaland. The basin is presently covered by thick layers of Quaternary and Tertiary sediments, occurred during the Cenozoic period (Biswas, 1987). Refraction and deep seismic sounding (DSS) studies established the thickness of the Quaternary and Tertiary sediments in most parts of the basin to vary between 3000 and 5500m (Kaila et al, 1981, 1990). The top 300-500 m is characterized by quaternary deposits (Tewari et al., 1995; Rastogi et al., 2001).

Geophysical and Geotechnical Investigation

Geophysical and geotechnical site investigation in seismically active regions should include the gathering of information about the physical nature of the site and its environment that will allow adequate evaluation of seismic hazard. The scope of the investigation depends on the seismicity of the area and nature of the site as well as the proposed or existing construction. In addition to the effects of local soil conditions on the severity of ground motion, the investigation should cover possible earthquake danger from geological or other consequential hazards such as fault displacement, subsidence, liquefaction, landslides, mudflows, etc (Boominathan, 2004).

The Standard Penetration Test (SPT) was conducted to a depth of about 60 to 80 m at 150 locations on the site. The water table is found to be at a depth of 30 m below the ground level. The top 2.5 to 4.0 m layer is clayey sand (SC) with SPT 'N' value varying from 30 to 40. Silty sand (SM) and clays of intermediate plasticity (CI) are also observed in the top layer at certain locations. It is followed by a 25 m thick sandy clay layer with high SPT 'N' value varying from 60 to 100. A very thick clay layer with sand gravel mixtures is encountered 30 m below the ground level which extends to the borehole termination depth of 60 m. The SPT 'N' value in this layer is significantly greater than 100. The laboratory tests conducted on soil samples collected at this layer reveals plasticity index values of above 30. The unconfined compressive strength of this layer varies from 360 to 380 kPa, indicating the hard consistency of the clay. The consolidation test carried out on the above soil samples shows low compression index values less than 0.06 confirming the presence of hard clay stratum. In general the soil strata at the site have a very high strength and stiffness, although rock formations are not encountered within the depth of investigation. A typical bore log of the site is presented in Figure 2.

Field tests which are most commonly employed in geotechnical earthquake engineering can be grouped into those that measure low-strain properties and those that measure properties at intermediate to high strains. Low-strain field tests typically induce seismic waves in the soil and seek to measure the velocities at which these waves propagate. Due to low strain amplitudes, the measured shear wave velocity (V_s) along with soil density (ρ) is used to compute low-strain shear modulus.

$$G_{max} = \rho V_s^2 \tag{1}$$

The shear wave velocity measurement in the field can be carried out either using borings or without borings. Seismic reflection and seismic refraction are staples of conventional geophysical exploration and can provide information on subsurface layer thickness and velocities of propagating waves without the need for soil borings. However, these techniques fall short in detecting soft layers below stiff layers (Kramer and Stewart, 2004). A relatively new technique that can be used to determine subsurface thickness and wave propagation velocities without the necessity of borings is the Multi Channel Analysis of Surface Waves (MASW) test (Nazarian & Stokoe, 1983; Suto, 2007). In this test, series of vertical receivers are placed on the ground in line with an impact source. The output of the receivers is recorded and transformed into the frequency domain. The phase angles between the recorded responses are used to compute an apparent travel time of the surface waves. The surface wave phase velocity is computed as a function of frequency by knowing the distance between the source and the receiver. MASW profiling also

Figure 2. Typical borelog of the site

Depth	Notation	Soil Description	Depth of Sample	Type of Sample	SPT N Value & (Penetration of S.S.S)				Rate of Bit Pressure
					N1	N2	N3	N	0
0.00		**0.00 to 4.80 m**	0.00	DS	-	-	-	-	
1.50		Brownish,fine to very fine	1.50	UDS	-	-	-	-	
3.00		grained, clayey sand with	3.00	SPT	10	16	22	38	
4.50			4.50	UDS	-	-	-	-	
6.00		**4.80 to 15.80 m**	6.00	SPT	17	30	39	69	
7.50		Reddish brown and	7.50	UDS	-	-	-	-	
9.00		greyish, fine to very fine	9.00	SPT	28	78	40-6cm	>100	
10.50		grained, sandy clays of	10.50	UDS	-	-	-	-	
12.00		intermediate plasticity	12.00	SPT	35	81	55-9cm	>100	
13.50		with occational gravels	13.50	UDS	-	-	-	-	
15.00		(CI)	15.00	DS	-	-	-	-	
16.50		**15.80 to 26.50m**	16.50	UDS	-	-	-	-	
18.00			18.00	SPT	14	25	48	73	
19.50		Yellowish brown, fine to	19.50	UDS	-	-	-	-	
21.00		very fine grained, sandy	21.00	SPT	31	48	68	100	
22.50		clays of high plasticity	22.50	UDS	-	-	-	-	
24.00		with little gravels (CH)	24.00	SPT	24	39	45	84	
25.50			25.50	UDS	-	-	-	-	
26.50		**26.50 to 28.90 m**							
27.00		Yellowish brown, fine to	27.00	SPT	35	65		>100	
27.50		medium grained,				10cm			
28.50		water table encountered	28.50	DS	-	-	-	-	
29.00		**28.90 to 31.60 m**							
29.50		Yellowish brown, fine to medium grained, silty sand with little gravels (SM)							
30.00			30.00	UDS	-	-	-	-	
31.50		**31.60 to 34.50 m**	31.50	DS	-	-	-	-	
33.00			33.00	SPT	40	126		>100	
34.50		**34.50 to 40.00 m**	34.50	UDS	-	-	-	-	
36.00		Brownish, greyish, fine to	36.00	SPT	47	50		>100	
36.50		to very fine grained,				5cm			
37.50		clayey sand with some	37.50	UDS	-	-	-	-	
39.00		gravels (SC)	39.00	DS	-	-	-	-	
40.00		**40.00 to 46.50 m**							
40.50		Reddish brown, fine to	40.50	SPT	43	60		>100	
41.00		very fine grained, sandy				6cm			
42.50		clays of intermediate	42.00	SPT	42	88		>100	
43.00		plasticity with little				10cm			
45.00			45.00	UDS	-	-	-	-	
46.50		**46.50 to 50.00 m**							
48.00		Yellowish brown, fine to	48.00	SPT	78	50	-	>100	
48.50		very fine grained, clayey				4cm			
50.00		sand with some gravels	50.00	DS	-	-	-	-	

provides the stiffness of the underlying soil which is the variation of phase velocity with frequency. MASW test is becoming increasingly popular in India with its wide usage in microzonation (Sitharam et al., 2007; Hanumantharao & Ramana, 2008; Uma Maheshwari et.al., 2008; Neelima Satyam & Rao, 2009). However, the MASW test is not effective in deep soil sites.

Low-strain tests that require borings include down-hole test, up-hole test, cross hole test and suspension logger test. Among these methods, the cross-hole test is widely used to measure the shear wave velocity profile, particularly at deep soil sites. The cross-hole test makes use of more than one borehole. A vibratory source is placed in one boring and the receiver is placed at the same depth in each of the other boreholes. An impulsive disturbance is applied at the source and the travel times to each of the receivers is measured. The wave propagation velocity is measured by knowing the distance between the source and the receiver. The cross-hole test is the best method to measure the shear wave velocity of the strata at any depth and it can detect the low velocity layers and it is widely used in earthquake engineering application. Therefore in the present deep soil site, cross hole test was performed to measure the shear wave velocity.

Cross hole test was performed at 28 locations in the site as per ASTM standard 4428 M-84. The test was conducted using three 150 mm size boreholes: a source borehole and two receiver boreholes each spaced 3.0 m apart. The receiver bore holes drilled were cased with a PVC pipe of 150 mm diameter and the gap between the casing and the receiver borehole was filled with fine sand slurry to establish good contact between the surrounding soil medium and the casing. In the present investigation, blows of SPT hammer in the source borehole were used to generate seismic waves. The P and S-waves generated at seismic source are gathered by a borehole-pick which essentially consists of acceleration transducer, pneumatic packer assembly, pneumatic pump and control. The seismic wave signals were acquired and recorded using a HBM make multi-channel Digital carrier frequency amplifier system with data acquisition system and CATMAN Professional software installed in Laptop. The cross hole tests were performed at the site at an interval of 1.5 m up to a depth of 60.0 m. The typical wave traces obtained from the cross hole test is presented in Figure 3. Figure 4 shows a typical variation P-wave and S-wave velocity with depth.

Maximum shear modulus of soil (G_{max}) is the fundamental property of the soil in geotechnical earthquake engineering application. The most reliable methods to determine the maximum shear modulus of soil are those conducted in the field. This is because the laboratory soil testing of undisturbed soil samples is often subjected to errors due to sample disturbance. Even if the disturbance is minor in advanced technique of sampling, time and expense may be substantial. Hence in the present study shear wave velocity obtained from the cross-hole test is utilized to compute the maximum shear modulus of the soil using the formula discussed earlier.

SITE CLASSIFICATION

Site classification can be carried out by different schemes: based on average shear wave velocity in the upper 30 m, V_{s-30}, surface geology and geotechnical data are available to represent the generic response of the site to seismic excitations (Kramer & Stewart, 2004). The V_{s-30} scheme (Borcherdt, 1994) is the most widely used site classification procedure in modern practice. Based on the geophysical and geotechnical data, NEHRP standard based site classification map is developed for Chennai city by Maheswari et al[2]., (2010) by V_{s-30} approach. Hence, in the present study the site classification for the site is also carried out based on V_{s-30} approach.

Figure 3. A typical wave trace obtained from cross hole test

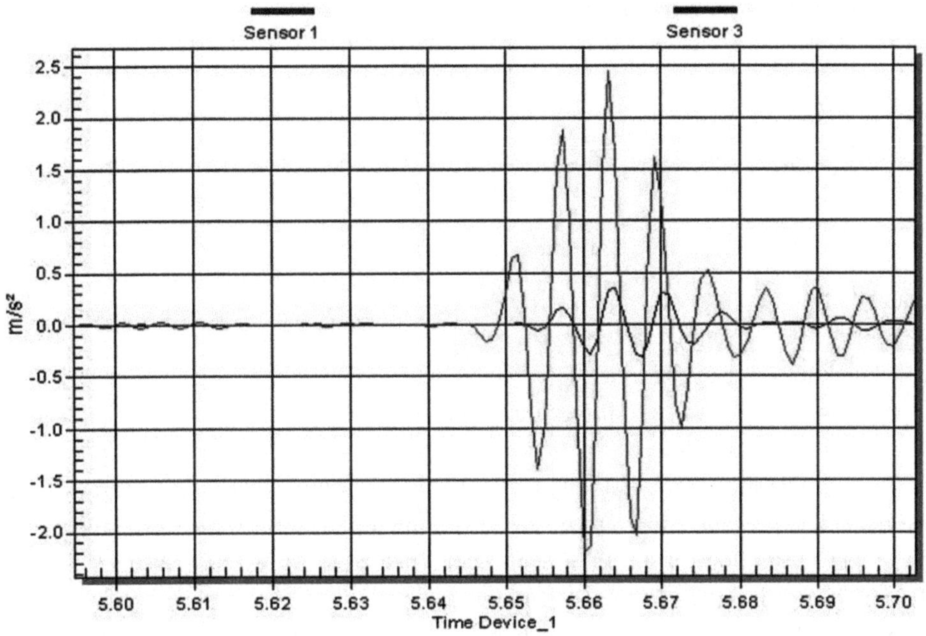

Figure 4. Typical variation P-wave and S-wave velocity in the study area

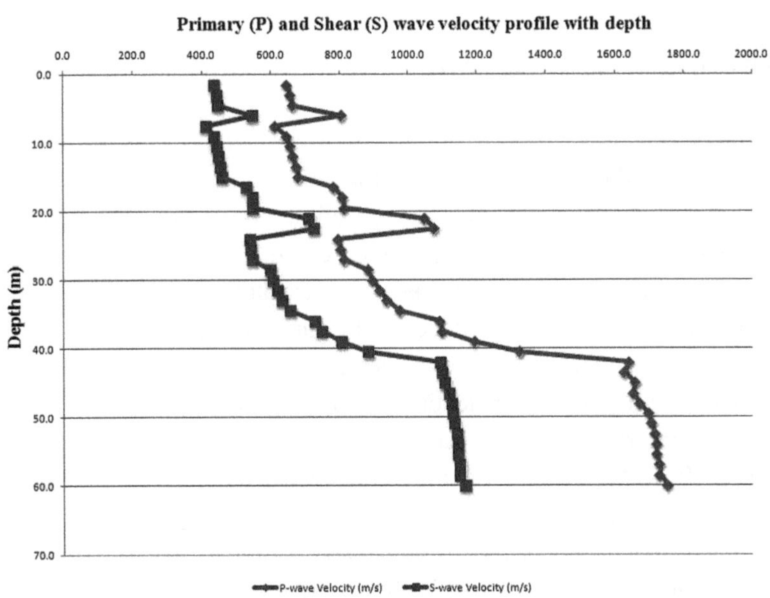

The average shear wave velocity for the upper 30 m is determined by using the following formula (Lew, 2001).

$$V_{s-30} = \frac{\sum_{i=1}^{n} d_i}{\sum_{i=1}^{n} \frac{d_i}{v_{si}}}$$

where:

d_i = Thickness of Layer *i* in m
v_{si} = Shear wave velocity in Layer *i* in (m/s)

The average shear wave velocity for the upper 30m computed using the cross-hole test data. The average shear wave velocity in the top 30m for most of the locations varies between 400m/s to 600m/s and is classified as C class site as per NEHRP (BSSC, 2001) classification. Only very few locations the average shear wave velocity in the top 30m exceeds 750m/s and hence are classified as B class site. The above site classifications are used in the seismic hazard analysis to consider approximately the site effects.

SEISMIC HAZARD ANALYSIS

Seismic hazard analyses involve the quantitative estimation of ground shaking hazards at a particular site. To evaluate the seismic hazard for a particular site or region all possible sources of seismic activity must be identified and their potential for generating future strong motion evaluated. Seismic hazards may be analyzed deterministically, as when a particular earthquake scenario is assumed or probabilistically, in which certain uncertainties are considered explicitly (Kramer, 1996). The earthquake ground motions that ultimately are selected for engineering design depend mainly on the criticality of a site or structure and the engineering analyses that are to be performed (Krinitzsky, 2002).

In the past 20 to 30 years the use of probabilistic concepts has allowed uncertainties in the size, location and rate of recurrence of earthquakes and in the variation of ground motion characteristics with earthquake size and location to be explicitly considered in the evaluation of seismic hazards. Probabilistic seismic hazard analysis (PSHA) provides a framework in which these uncertainties can be identified, quantified, and combined in a rational manner to provide a complete picture of seismic hazard. The proper performance of a PSHA requires careful attention to the problems of source characterization and ground motion parameter prediction and to the mechanics of the probability computations.

Deterministic Seismic Hazard Analysis (DSHA) analysis is simple and more prevalent methods of hazard analysis in geotechnical earthquake engineering. DSHA involves the development of a particular seismic scenario upon which a ground motion hazard evaluation is based. The scenario consists of the postulated occurrence of an earthquake of a specified size occurring at a specified location. DSHA is more logical, more transparent, and more appropriate for requirements in engineering design. DSHA should be considered as the basis for the design of critical construction. DSHA identifies individual faults with their estimated MCEs affecting a site and gives site-specific hazard estimates from the earthquake potential of each fault (Krinitzsky, 2003). Hence in the present study seismic hazard analysis is carried out by the deterministic approach. The deterministic based seismic hazard analyses were also performed for microzonation of various cities in India (Sitharam et al., 2007; Boominathan et al., 2008)

Deterministic seismic hazard analysis involves the following steps (Reiter, 1990) which are discussed in detail in the succeeding sections.

- Identification and characterization of all earthquake sources capable of producing significant ground motion at the site.
- Selection of source to site distance parameter for each source zone.
- Estimation of PGA for all earthquake sources considered using appropriate attenuation equation
- Selection of the controlling earthquake, which is expected to produce the maximum PGA.

Identification and Characterization of Earthquake Sources

Seismicity and Seismotectonics

Historical seismicity of the study region is pivotal in site specific ground response analysis. Details of historical seismic activities are made available through catalogues of earthquake provided by USGS Earthquake Hazards Program, IMD, GSI and other reliable agencies. Seismic susceptibility of the area can be assessed with the help of the seismicity data on the occurrence of past earthquakes and seismotectonic details that describes the tectonic features of the site. Seismological information and seismotectonic features are provided by Seismotectonic Atlas or shall be obtained from terrain analysis employing remote sensing imageries to locate fault structures, length and to identify capable faults. Seismicity data are alone not sufficient for the purpose of identification and characterization of earthquake sources and are to be supplemented by historical and instrumental seismicity data.

In the present study, the details of the faults and lineaments that lie within 250 km radius of the study area were collected from the seismotectonic atlas of India (GSI 2000). The mere presence of a fault, however, does not indicate the likelihood of future earthquakes. Hence it is required to identify the *capable faults* which are capable of generating strong motion. Thirteen earthquake sources were identified within 250km from the study region. The identified sources are shown in the base map which is developed using the Geographic Information System (GIS) tool ArcGIS 9.2 (Figure 4).

The earthquake data from 1668 to 2008 were collected from various sources such as IMD, USGS, GSI, ISR, ISC, and GERI which includes details on time of occurrence, location, depth, magnitude and intensity. These earthquake details were mapped onto the faults based on the location and depth of the earthquake and the length of the fault as shown in Figure 5. This information is utilized while assigning the maximum magnitude for each fault source by considering the seismicity around the particular fault source. The maximum magnitude of a particular seismic source was taken as the largest observed past magnitude plus 0.5 (Kijko and Graham, 1998; Sokolov et al., 2001). In the present study the source to site distance is selected as the shortest epicentral distance from the sources to the site.

Controlling Earthquake

The earthquake that is expected to produce the strongest level of shaking is the controlling earthquake, generally expressed in terms of some ground motion parameter, at the site. The selection is made by comparing the levels of shaking produced by different earthquake sources identified in the earthquake source characterization phase. The controlling earthquake is described in terms of the size (magnitude) and distance from the site.

The peak ground acceleration is the most commonly used parameter to evaluate the effect of ground shaking due to different earthquakes identified. The peak ground acceleration at the ground surface is obtained using attenuation equations or predictive relationships. Predictive relationships, which express a particular ground motion parameter in terms of the quantities that affect it most strongly, are used to estimate ground motion parameters. A large number of useful attenuation relationships for different geographic and tectonic environments are available to predict the ground motion parameters (Campbell, 1981; Boore et al., 1983; Frankel et al., 1996). The predictive relationship that characterizes the source, travel path and that idealizes the site condition should be chosen from the available relationships. The predictive equations shall be checked with the recorded PGA value at or near the site.

Figure 5. Base map of the study region

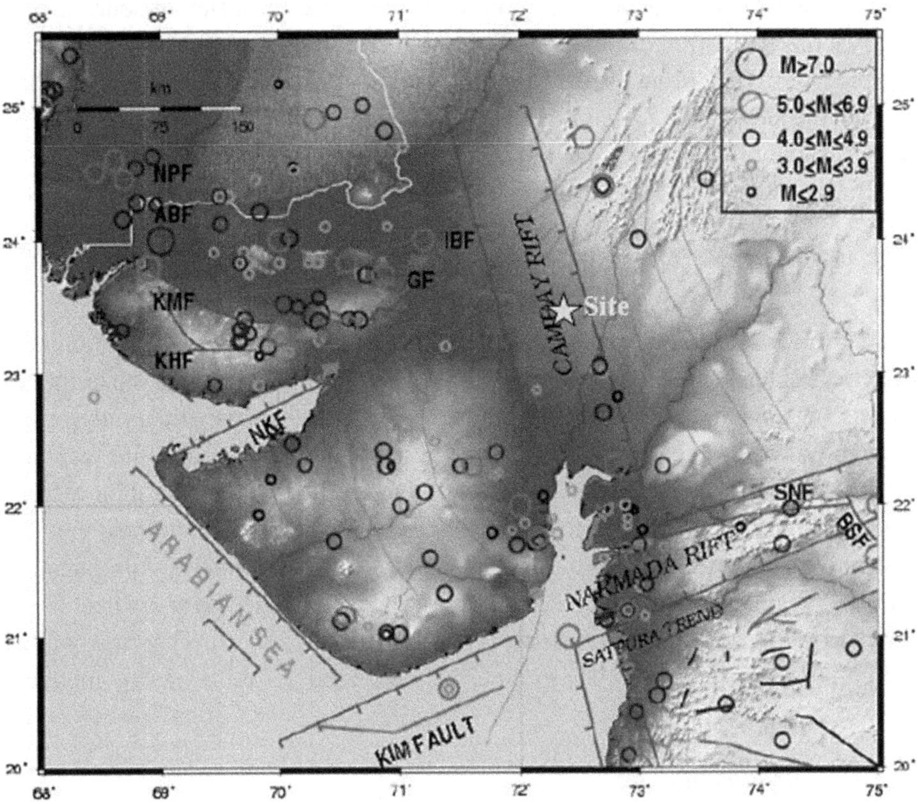

It is observed that the crustal intraplate relation of Frankel et al. (1996) yields ground motions similar to the strong ground motion data recorded from the 2001 earthquake at large distances (Cramer & Wheeler, 2001). Hence in the present study, the PGA value is estimated based on the predictive relationship proposed by Frankel et al (1996). The crustal intraplate relation developed is for the site condition specified as the boundary between NEHRP classes B and C, having an average shear-wave velocity of 760m/s in the top 30 m. The study area considered also have an average shear wave velocity of about 800m/s in the top 30m, justifying the use of Frankel attenuation relation. It is also observed that the PGA of 0.1g at Ahmedabad recorded by seismogram located at the ground floor of regional passport office during Bhuj 2001 earthquake (Hazarika & Boominathan, 2009) compares well with the predictive relation proposed by Frankel having a PGA of 0.14g for the magnitude of 7.7. As the Cambay rift region is more susceptible to shallow focus earthquakes, a focal depth of 15km was adopted. The Frankel predictive relationship was adopted to obtain the PGA at the surface. The earthquake that is expected to produce the strongest level of shaking is referred as the controlling earthquake and is obtained by plotting the variation of peak ground acceleration with distance for different sources. The details of the estimation of PGA for all the sources are presented in Table 1. It can be found from Table 1 that the East Cambay fault located at a distance of about 20.5 km from the site for the magnitude of 6.2 causes the maximum PGA of 0.46g and hence it is identified as the controlling earthquake source.

SITE SPECIFIC GROUND RESPONSE ANALYSIS

Local site effects profoundly influence all of the important characteristics: amplitude, frequency content, and duration of strong ground motion. The extent of the influence depends on the geometry and material properties of the subsurface materials, onsite topography and on the characteristics of the input ground motion. Site response analysis includes 1-D ground response effects, basin effects and topographic effects. Ground response analysis is performed either using 1D models or nonlinear models. Most ground response analysis models solve for one dimensional wave propagation problems. Nonlinear behavior of the soil can be approximated by an equivalent linear characterization of dynamic properties (Seed and Idriss, 1970). The most widely used computer program for 1-D ground response analysis utilizing this model is

Table 1. Estimation of PGA

S. No	Name of the Fault	Moment Magnitude (M_w)	Distance From Site (km)	PGA (g)
1	East Cambay Fault	6.2	20.51	0.460
2	West Cambay Fault	5.1	15.72	0.110
3	West Coast Lineament	5.5	100.30	0.030
4	Island Belt Fault	6.1	130.00	0.024
5	Katrot Bhuj Fault	6.5	236.50	0.016
6	Kutch Mainland Fault	8.3	169.50	0.140
7	North Kaitwar Fault	4.9	95.00	0.007
8	Chambal Jamnagar Lineament	6.2	100.00	0.027
9	Kim Fault	4.3	292.00	---
10	Son Narmada Fault	5.1	207.50	0.003
11	Tapti North Fault	6.2	230.75	0.006
12	Paldi Fault	5.9	141.50	0.016
13	Allah Bund Fault	5.6	255.00	0.004

currently SHAKE91 (Idriss and Sun, 1992), which is a modified version of the program SHAKE (Schnabel et al., 1972). The program uses an equivalent linear, total stress analysis procedure to compute the response of a 1-D, horizontally layered viscoelastic system subjected to vertically propagating shear waves. The program uses the exact continuum solution to the wave equation adapted for use with transient motions through the Fast Fourier Transform algorithm. The equivalent linear method models the nonlinear variation of soil shear moduli and damping as a function of shear strain. The hysteretic stress-strain behavior of soils under symmetrical cyclic loading is represented by an equivalent modulus, G, corresponding to the secant modulus through the endpoints of the hysteresis loop and equivalent linear damping ratio, β, which is proportional to the energy loss from a single cycle of shear deformation. For a given soil layer, G and β are assumed to be constant with time during the earthquake shaking. An iterative procedure, based on linear dynamic analysis, is performed to find the shear moduli and damping ratios corresponding to the computed shear strains.

Nonlinear models solve the one-dimensional wave equation by direct numerical integration in the time domain. A variety of material models are used, which range from relatively simple cyclic stress-strain relationships to advanced constitutive models incorporating yield surfaces, hardening laws and flow rules (Wang, 1990). Nonlinear methods can be formulated in terms of effective stresses to allow modelling of the generation, redistribution and eventual dissipation of excess pore pressure during and after earthquake shaking, whereas equivalent linear methods can only perform total stress analysis. Cyclic nonlinear models generally consist of a backbone curve and rules that describe unload-reload behaviour, pore-pressure generation and cyclic modulus degradation. DYNA1D, SUMDES, SPECTRA and AMPLE are some of the nonlinear analysis programs, which includes advanced constitutive models. Nonlinear multi-dimensional ground response analyses are sometimes preferred for critical structures.

In the present study, the equivalent linear (using SHAKE2000) and nonlinear approaches (using FLAC2D) are used to perform one dimensional ground response analysis.

Input Motion

Earthquake response of structures and their foundations is an outcome of the complex interaction between the random input ground motion and the continuously changing dynamic characteristics of the system subjected to the ground motion. The site specific ground response analysis involves the determination of input acceleration time history characterized by amplitude, frequency and duration content. The time history of the ground motions can be obtained by two approaches: use of natural strong motion earthquake data and artificial ground motion generation. The ideal procedure for the selection of strong motions for use in analysis is to obtain records generated under conditions that are identical to those of the seismic design scenario. Bolt (1978) showed that if all the characteristics of the design earthquake could be matched to those of a previous earthquake, the probability of the characteristics of the record matching would be unity. Hence in the present study, the input acceleration time history is obtained by scaling the acceleration time history of a previously recorded natural earthquake which has magnitude, distance and local site characteristics that are similar to those of seismic scenario at the site. The bracketed duration was estimated based on the magnitude and the epicentral distance from the causative fault as 18s (Chang and Krinitzsky, 1977). The accelerogram of 1999 Chi-Chi earthquake with a similar site condition was scaled to a required PGA of 0.46g as estimated from the DSHA and it represents the required surface acceleration time history (Figure 6). The Fourier spectrum for the adopted input motion is plotted in Figure 7 and it shows the predominant frequency of 2Hz.

Deconvolution Analysis

Deconvolution is the process of obtaining the bedrock motion from a known free surface motion. The average shear wave velocity variation at the site was used to characterize the strata for deconvolution analysis. The deconvolution analysis was carried out for the surface acceleration time history using SHAKE2000 to obtain the base motion at a depth of 60m. The obtained time history at the base level is shown in Figure 8 and its Fourier spectra are shown in Figure 9. It is evident from Figures 8 and 9 that the deconvolution resulted in a decrease in the PGA value to 0.3 g as well as a shift in the predominant frequency from 2 to 0.6 Hz. The deconvoluted accelerogram is used as input at the base level for further site specific ground response analysis.

Input Soil Data

The overall soil profile of 60 m thickness is divided into layers of 1.5 m thick is considered. Each layer is characterized by the unit weight of soil obtained from SPT data and shear wave velocity obtained from the seismic cross hole test data. A typical input soil data used in SHAKE analysis is presented in Figure 10 and 11. The water table is taken to be at a depth of 30 m below the ground level.

The nonlinear behaviour of soil, reduction in stiffness and increase in the damping with an increase in the shear strain, is accounted by selecting the standard modulus reduction and damping curves. In the present study, the standard modulus reduction

Figure 6. Input acceleration time history at the surface from DSHA

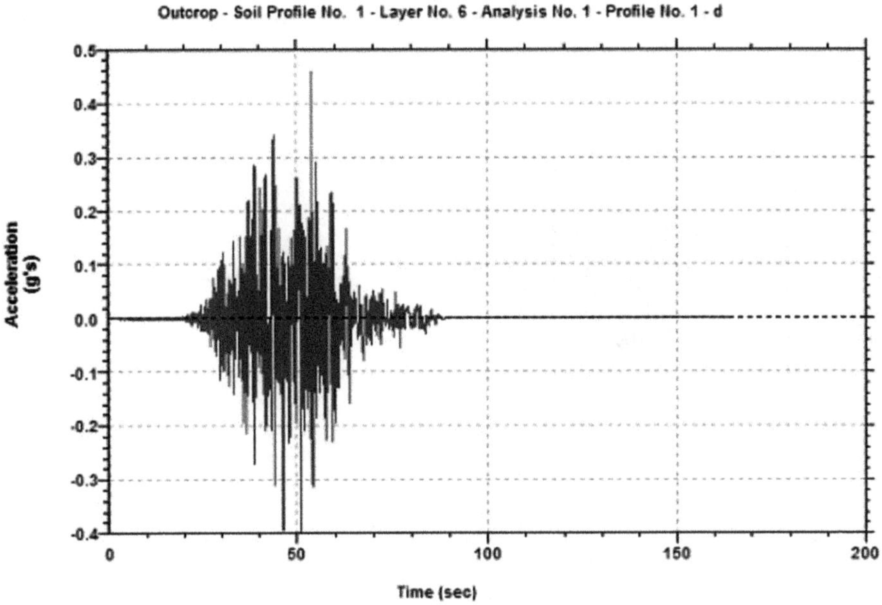

Figure 7. Fourier response spectrum for input acceleration time history

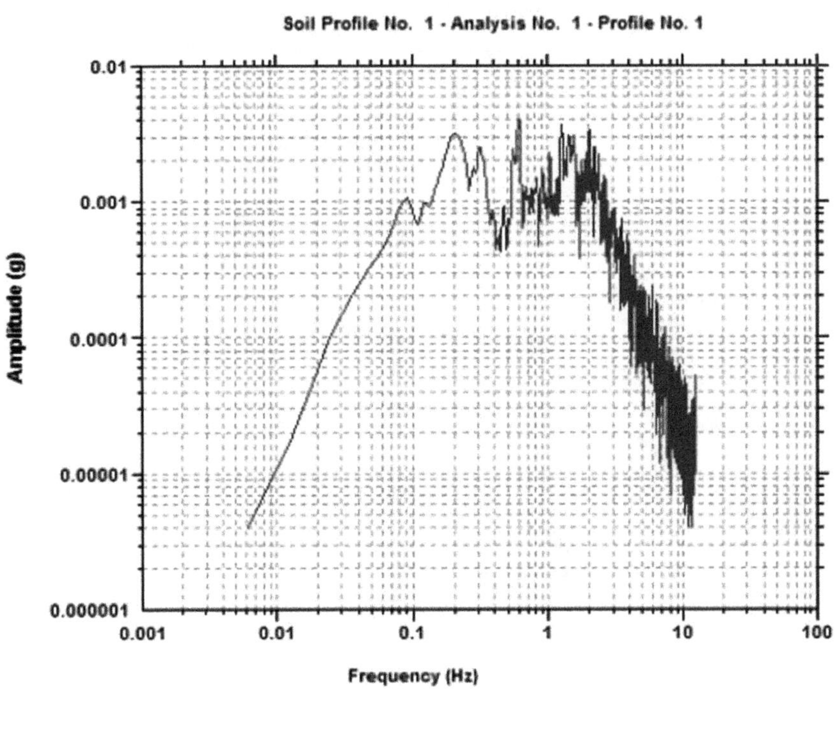

Figure 8. Acceleration time history at the base obtained from deconvolution analysis

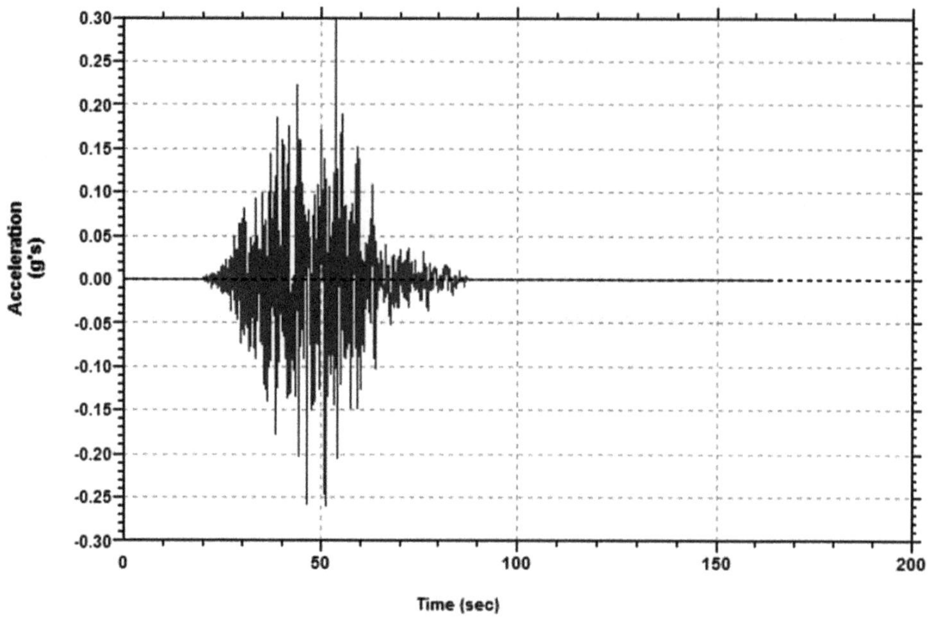

Figure 9. Fourier spectrum for base motion obtained from deconvolution analysis

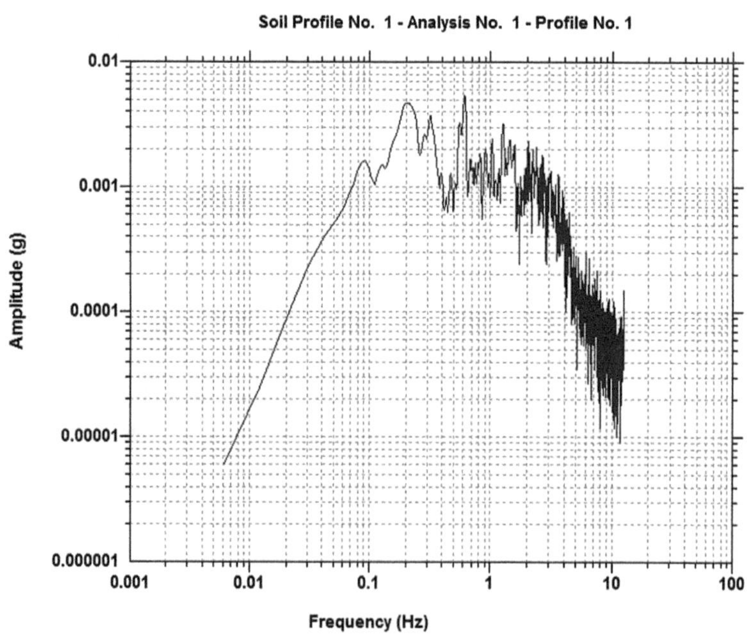

Figure 10. Unit weight, Shear wave velocity and shear modulus variation with depth soil profile for C class site

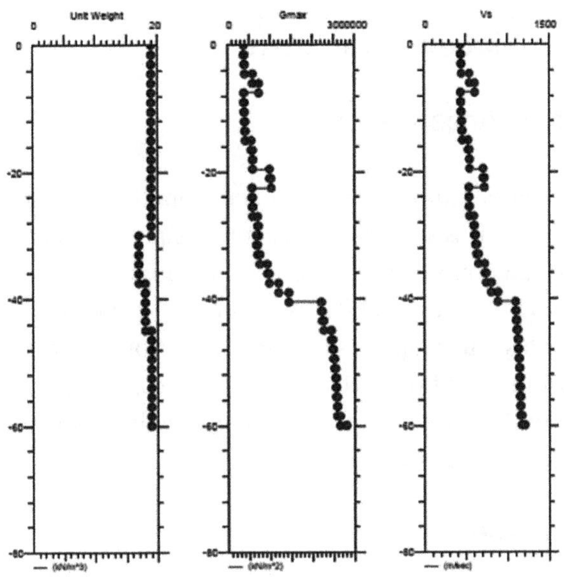

Figure 11. Unit weight, Shear wave velocity and shear modulus with depth for typical soil profile for B class site

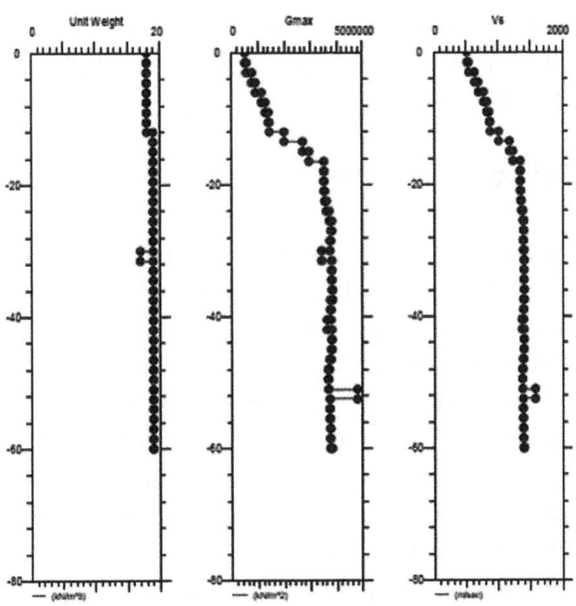

and damping curves proposed by Sun et al (1988) are selected based on the soil characteristics and confining pressure which is depicted in Figure 12 and Figure 13.

Design Ground Motion Parameters

History of Surface Acceleration

The surface acceleration time history obtained from the ground response analysis for two typical sites C class site and B class site are presented in Figures 14 and 15 respectively. It can be noticed from Figure 14 and Figure 15 that the surface peak ground acceleration for C class and B class site is 0.53 g and 0.31g respectively. It is evident that the C class site amplifies the ground motion in comparison to the input ground motion with PGA of 0.3 g. However, the B class site does not amplify the input motion due to the presence of a thick hard sandy clay deposit characterized by very high shear wave velocity.

Predominant Frequency

The Fourier spectra for the surface motion obtained from the ground response analysis for site class C and B are presented in Figures 16 and 17 respectively. It is evident from the figures that the frequency content of surface motion is found to be distributed between 0.6 to 2.0 Hz for both class sites. However, the C class site exhibits a shift in the predominant frequency to a lower frequency of 1.7 Hz in comparison to the base motion frequency of 2.0 Hz, but the predominant frequency remains unaltered in B class site category.

Site Specific Response Spectra

The design response spectra i.e. the spectral acceleration versus the period of the structure obtained from the surface time history of acceleration for 5, 10 and 20% damping for both class sites are presented in Figure 18. It can be observed from the figure that the peak spectral acceleration for 5% damping in C and B class sites are 2.6 g and 1.3 g respectively. However, the peak spectral acceleration occurs practically at the same period of about 0.6 s for both site classes.

The design response spectrum is also obtained by the Ratio of Response Spectral (RRS) analysis method. The RRS analysis involves obtaining the spectral acceleration for the surface and the base motion, then dividing the surface spectrum by the base spectrum for each period. In the present case, the surface spectra obtained from the ground response analysis covering the entire site is utilized to obtain the Ratio of the Response Spectra (RRS) curves. The RRS curves obtained from the ground

Figure 12. Modulus reduction curve for ground response analysis

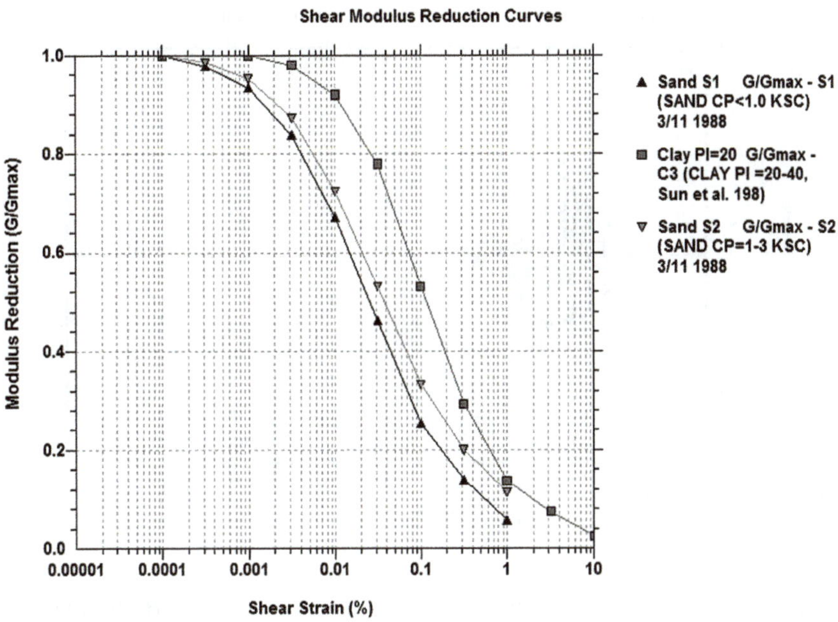

Figure 13. Damping curves used for ground response analysis

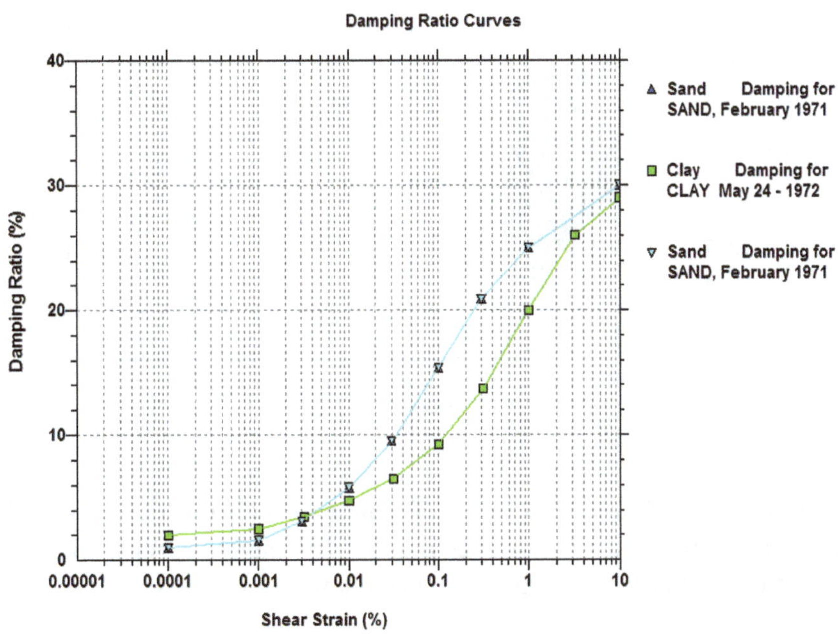

Figure 14. Acceleration time history obtained at the surface for C class site

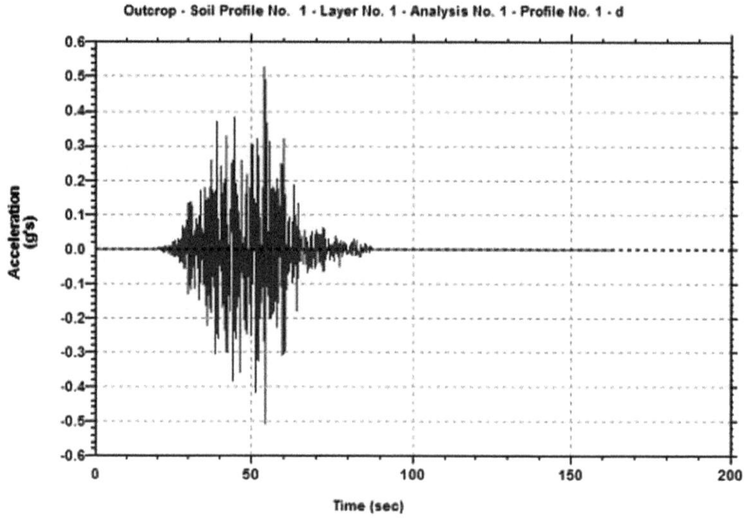

Figure 15. Acceleration time history obtained at the surface for B class site

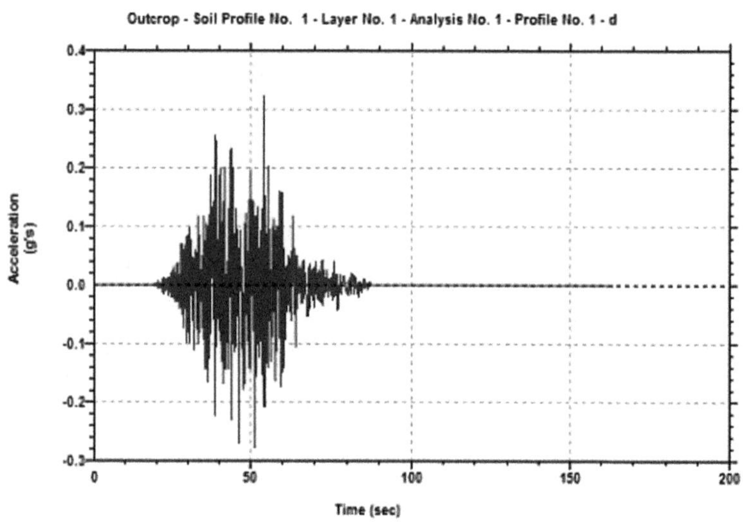

response analysis is shown in Figure 20. The mean and the median values of RRS curves are multiplied with the base spectrum to arrive the site-specific spectra. The base spectrum in this case considered by the following two types:

1. The response spectra of the base motion obtained from the present study.

Figure 16. Fourier spectrum for obtained surface motion for C class site

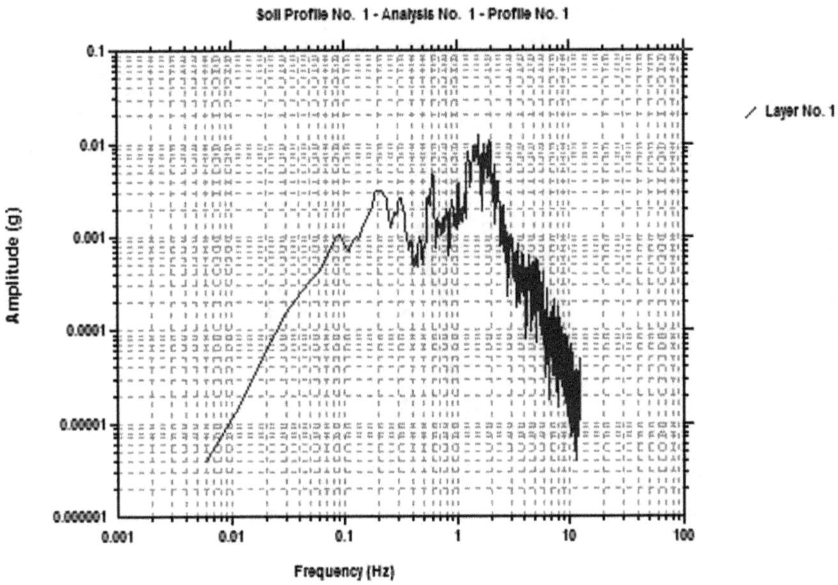

Figure 17. Fourier spectrum obtained surface motion for B class site

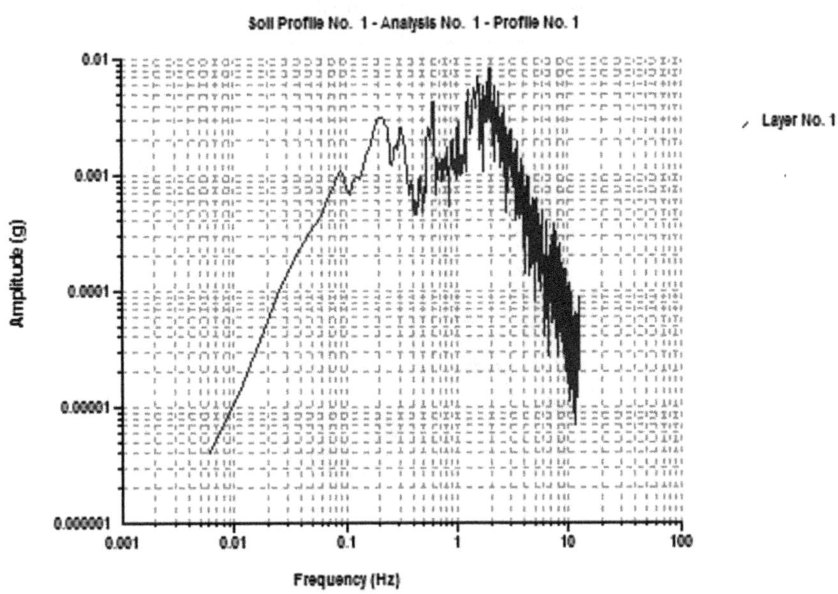

Figure 18. Response spectrum at surface for C class site

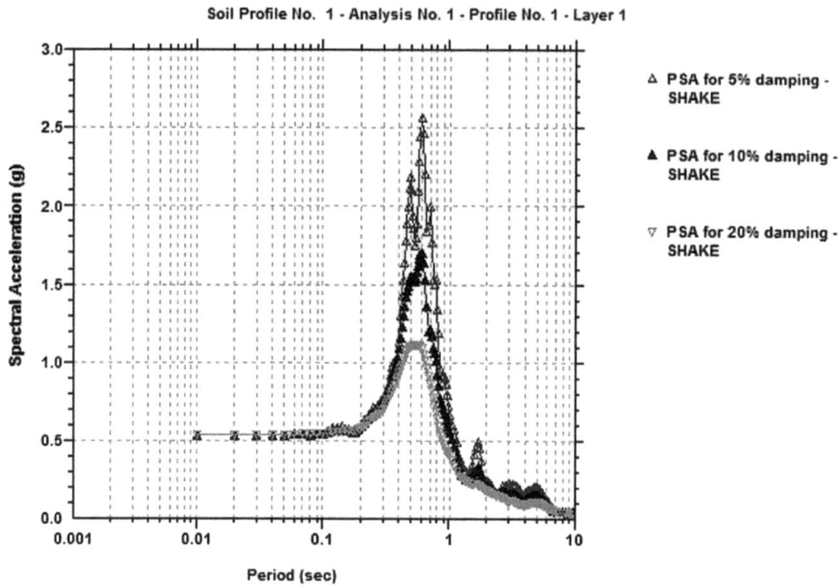

Figure 19. Response spectrum at surface for B class site

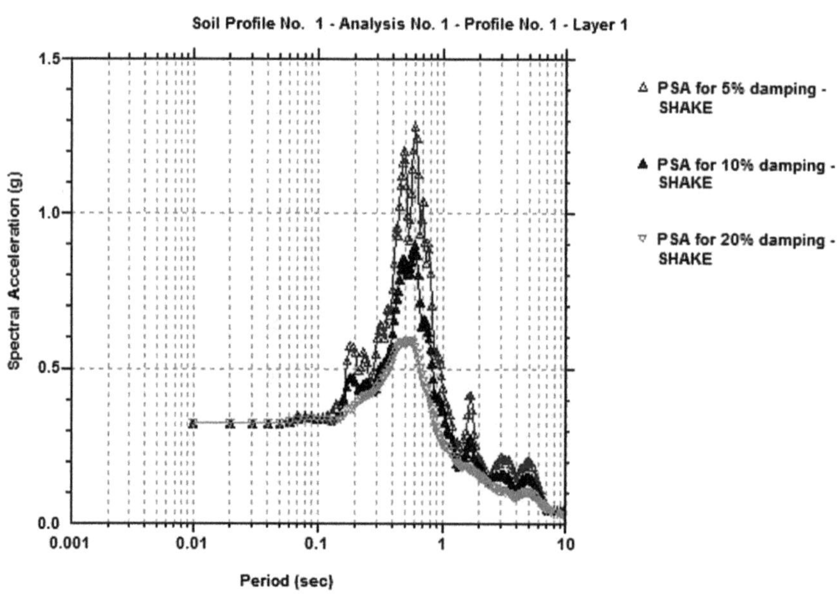

2. IS 1893-2002 design response spectra.

The site-specific spectra arrived from the RRS analysis considering the response spectra of the base motion obtained from the present study is shown in Figure 21. It can be found from the figure that the maximum spectral acceleration is about 1.8 g at the range of period of 0.4 to 0.6 s.

The site-specific spectra arrived from the RRS analysis considering the response spectra specified in IS 1893-2002, i.e., site specific spectra compatible to IS design spectra is shown in Figure 16. The IS design spectra for Type I soil is also shown in Figure 22. It can be noticed from the figure that the spectral ratio (Sa/g) at the short and mid period ranges up to about 0.6s is 20% higher than the IS spectral ratio values.

Nonlinear Analysis of 1D Soil Column

The soil is discretized into a finite difference mesh composed of quadrilateral elements of 0.75m element size to represent the one-dimensional soil media. The bottom of the soil column is fixed in all the directions. The sides are allowed to move freely along the direction of application of loading. The nonlinear elastic model is used for the ground response analysis. The strain dependent modulus and damping has been incorporated into a hysteretic damping model. The surface acceleration time

Figure 20. Ratio and response spectra (RRS) mean curves considering response spectra of all sites

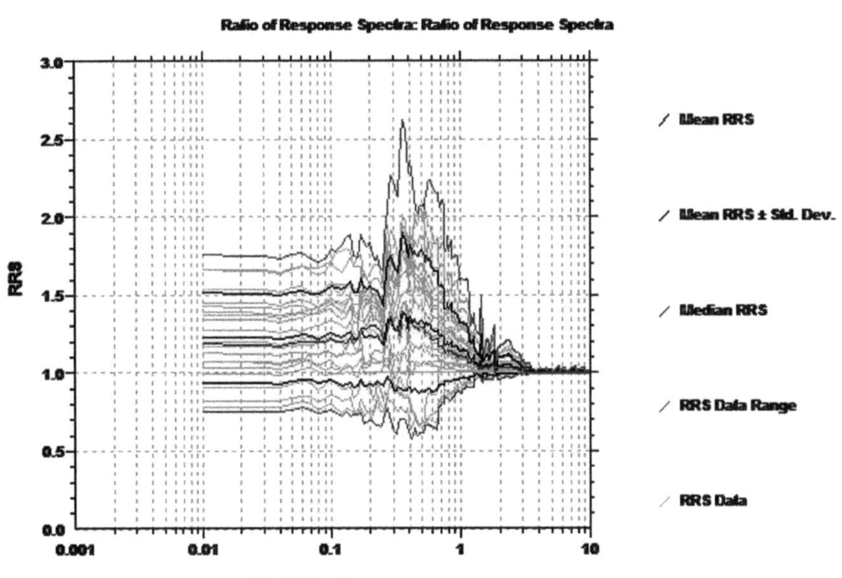

Figure 21. Response spectrum compatible to the base ground motion obtained from the present study

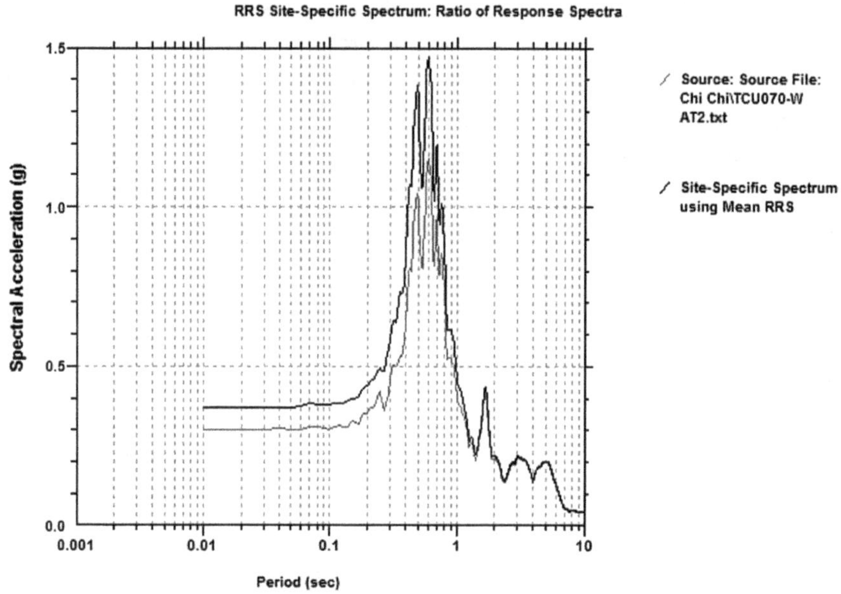

Figure 22. Normalized site-specific response spectra compatible to IS 1893-2002 design spectra

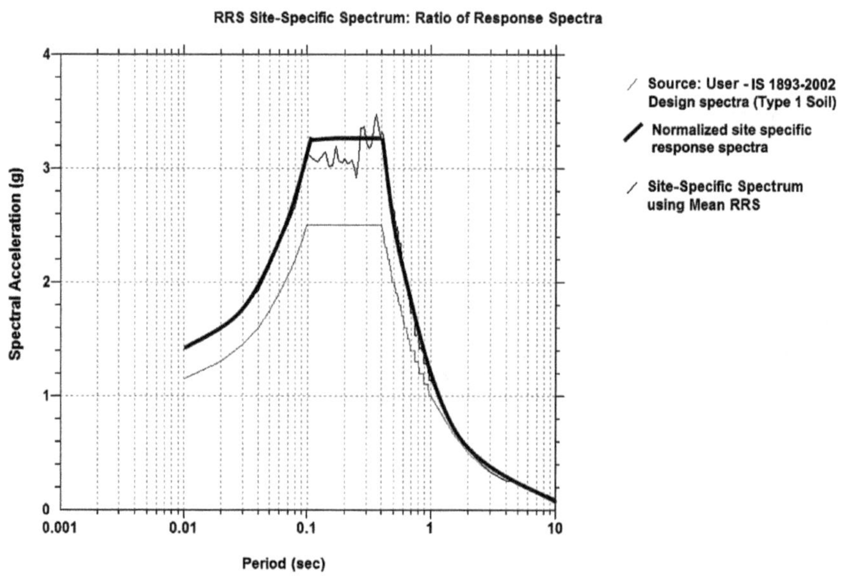

history obtained from the ground response analysis for two typical sites C class site and B class site are presented in Figures 23 and 24, respectively. It can be noticed from Figure 23 that the surface peak ground acceleration for C class and B class site is 0.71g and 0.58g, respectively. It is evident that both the B and C class site amplifies the ground motion in comparison to the input ground motion with PGA of 0.3 g. The amplification of input motion observed in the equivalent linear analysis is comparatively lesser than that of the nonlinear analysis.

SUMMARY AND CONCLUSION

A site-specific evaluation of design ground motion parameters for a seismically vulnerable site located 12 km from Ahmedabad (Gujarat) was carried out involving both seismic hazard analysis and ground response analysis. The seismic hazard analysis was performed by DSHA approach considering the seismicity and seismotectonic details for 250 km radius from the site using Frankel et al., (1996) predictive relationship. It is found that the East Cambay fault with a moment magnitude of 6.2 is the controlling source that causes surface PGA of 0.46 g at the site.

The acceleration time history at the surface for the site considered is obtained from the recorded ground motion data considering similar magnitude, distance and frequency content of the controlling earthquake. The 1999 Chi-Chi earthquake accelerogram is adopted and is scaled to the required PGA. Deconvolution analysis was performed to obtain the acceleration time history at the base (60m depth).

Based on the extensive shear wave velocity measurement carried out at the site by cross hole test, most locations at the site are classified as C class except a few locations as B class as per NEHRP classification.

The site-specific ground response analysis performed by equivalent linear approach shows amplification for C class site with PGA of 0.53 g, whereas B class site does not amplify the ground motion. The predominant frequency for C class site shifted from 2Hz to 1.7Hz whereas B class site did not alter the frequency content.

The site specific elastic design spectra were developed compatible to Indian standard IS 1893-2002 design spectra (Type 1 soil) by RRS analysis and it indicates that the spectral ratio (Sa/g) at the short and mid period ranges up to about 0.6s is 20% higher than the IS spectral ratio values.

The PGA of 0.71g and 0.58g for site class C and B respectively, obtained from the nonlinear analysis shows higher amplification than that from the equivalent linear approach.

It is evident from the present site-specific study that the deep very stiff soil sites also can significantly amplify the ground motion and hence it is important to conduct site specific study even for stiff soil sites.

Figure 23. Acceleration time history obtained at the surface for B class site

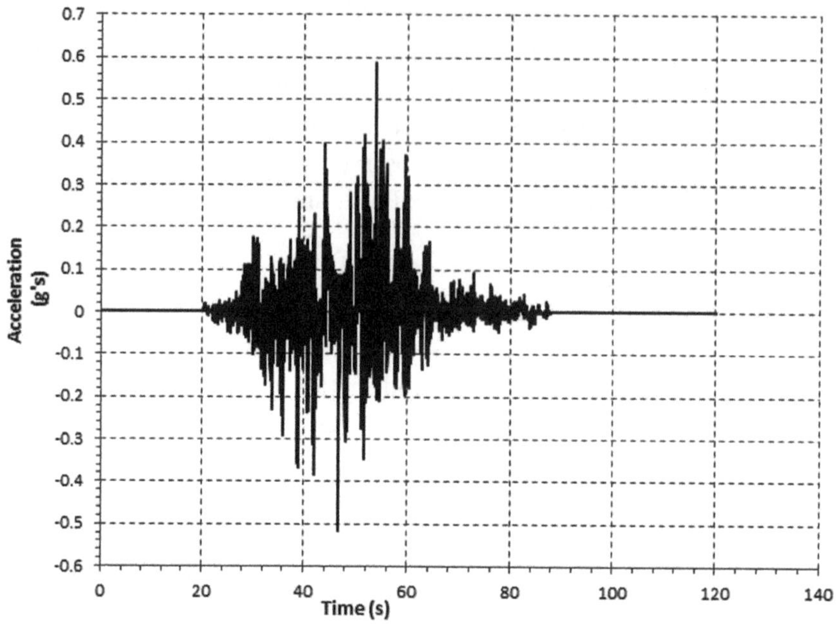

Figure 24. Acceleration time history obtained at the surface for C class site

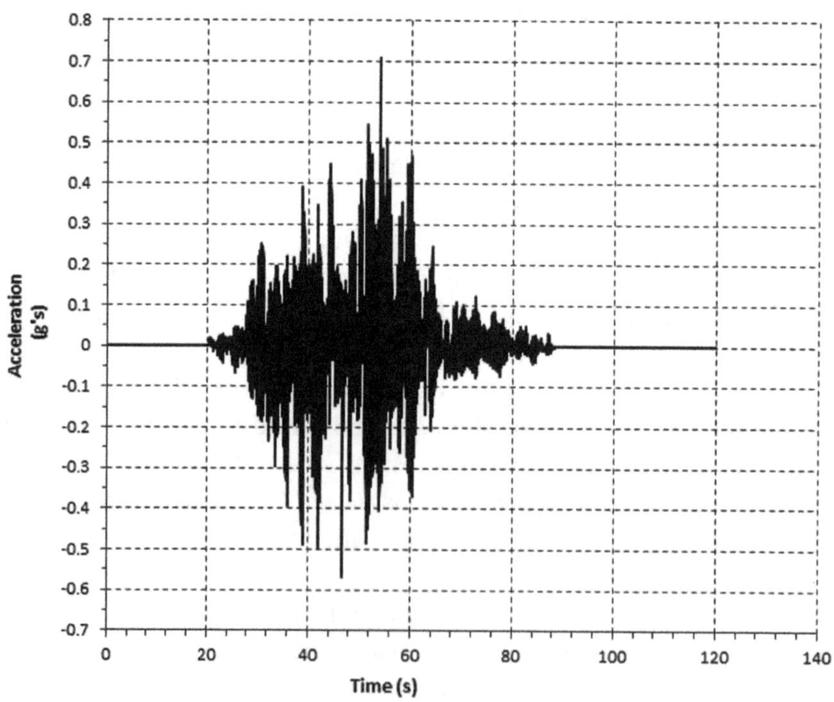

REFERENCES

Asimaki, D., & Jeong, S. (2015). Two-Dimensional Site Effects for Dry Granular Soils. *International Conference on Earthquake Geotechnical Engineering*.

ASTM Standard D 4428-M. (2007). *Standard test methods for seismic cross hole testing*. ASTM International. DOI: 10.1520/D4428_D4428M-07

Biswas, S. K. (1987). Regional tectonic framework, structure and evolution of western marginal basins of India. *Tectonophysics*, *135*(4), 305–327. doi:10.1016/0040-1951(87)90115-6

Boominathan, A. (2004). Seismic site characterization for nuclear structures and power plants. *Current Science*, *87*(10), 1388–1397.

Boominathan, A., Dodagoudar, G. R., Suganthi, A., & Uma Maheshwari, R. (2008). Seismic hazard assessment of Chennai city considering local site effects. *Journal of Earth System Science*, *117*(S2), 853–863. doi:10.100712040-008-0072-4

Boore, D. M. (1983). Stochastic simulation of high-frequency ground motions based on seismological models of the radiated spectra. *Bulletin of the Seismological Society of America*, *237*(6), 68–78.

Borcherdt, R. D. (1994). Estimates of site-dependent response spectra for design (methodology and justification). *Earthquake Spectra*, *10*(4), 617–653. doi:10.1193/1.1585791

BSSC. (2001). NEHRP recommended provisions for seismic regulations for new buildings and other structures 2000 edition, part 1: Provisions. Report no. FEMA 368, Building seismic safety council for the federal emergency management agency.

Campbell, K. W. (1981). Near source attenuation of peak horizontal acceleration. *Bulletin of the Seismological Society of America*, *71*, 2039–2070.

Carlos, S. O., Antoni, R., & Xavier, G. (2006). *Assessing and managing earthquake risk*. Springer.

Cauzzi, C., Kalogeras, I., Melis, N., Stupazzini, M., Mazzieri, I., & Clinton, J. (2015). Seismic Response of the Acropolis of Athens (Greece) through Recorded Earthquake Data and Numerical Simulations. *International Conference on Earthquake Geotechnical Engineering*.

Chang, F.K., & Krinitzsky, E.L. (1977). *Duration, spectral content and predominant period of strong motion earthquake records from Western United States*. Miscellaneous paper 5-73-1, U.S. Army Corps Engineers Waterways Experiment Station.

Cramer, C.H., & Wheeler, R.L. (2001). The 2001 Gujarat, India earthquake and seismic hazard in central and Eastern North America. *Abstract in Seismological Research Letters, 72*, 396.

Frankel, A., Mueller, C., Barnhard, T., Perkins, D., Leyendecker, E.V., Dickman, N., Hanson, S., & Hopper, M. (1996). *National seismic hazard maps: documentation June 1996*. Open-File Report (United States Geological Survey) 96-532.

GSI. (2000). *Seismotechnic Atlas of India*. New Delhi: Geology Society of India.

Hanumantharao, C., & Ramana, G. V. (2008). Dynamic soil properties for microzonation of Delhi India. *Journal of Earth System Science, 117*(S2), 719–730. doi:10.100712040-008-0066-2

Hanumantharao, C., & Ramana, G. V. (2009). Site Specific Ground Response Analyses at Delhi, India. *The Electronic Journal of Geotechnical Engineering*, 14.

Hazarika, H., & Boominathan, A. (2009). Liquefaction and ground failures during the 2001 Bhuj Earthquake India. *Proceedings of International Conference on Performance-Based Design in Earthquake Geotechnical Engineering from case history to practice*. 10.1201/NOE0415804844.ch13

Hunt, R. E. (2005). *Geotechnical Investigation Handbook* (2nd ed.). Boca Raton, FL: Taylor & Francis. doi:10.1201/9781420039153

Idriss, I. M., & Sun, J. I. (1992). *SHAKE91: A computer program for conducting equivalent linear seismic response analyses of horizontally layered soil deposits*. Davis, CA: Center for Geotechnical Modeling, Department of Civil and Environmental Engineering, University of California.

IS 1893 (Part 1). 2002 *Indian Standard Criteria for Earthquake Resistant Design of structures: Part 1 General Provisions and Buildings* (5th Revision).

Iyengar, R. N., & Ghosh, S. (2004). Microzonation of earthquake hazard in Greater Delhi area. *Current Science, 87*(9), 1193–1202.

Kaila, K. L., Krishna, V. G., & Mall, D. M. (1981). Crustal structure along Mehmedabad-Billimora profile in the Cambay basin, India, from deep seismic sounding. *Tectonophysics, 76*(1-2), 99–130. doi:10.1016/0040-1951(81)90255-9

Kaila, K. L., Tewari, H. C., Krishna, V. G., Dixit, M. M., Sarkar, D., & Reddy, M. S. (1990). Deep seismic sounding studies in the north Cambay and Sanchor basins, India. *Geophysical Journal International, 103*(3), 621–637. doi:10.1111/j.1365-246X.1990.tb05676.x

Kijko, A., & Graham, G. (1998). Parametric-historic procedure for probabilistic seismic hazard analysis, Part I: Estimation of maximum regional magnitude Mmax. *Pure and Applied Geophysics*, *152*(3), 413–442. doi:10.1007000240050161

Komatitsch, D., Liu, Q., Tromp, J., Suss, P., Stidham, C., & Shaw, J. H. (2004). Simulations of ground motion in the Los Angeles basin based upon the spectral-element method. *Bulletin of the Seismological Society of America*, *94*(1), 187–206. doi:10.1785/0120030077

Komatitsch, D., & Vilotte, J. P. (1998). The spectral-element method: An efficient tool to simulate the seismic response of 2D and 3D geological structures. *Bulletin of the Seismological Society of America*, *88*, 368–392.

Kramer, S. L. (1996). *Geotechnical Earthquake Engineering*. Prentice-Hall, Inc.

Kramer, S. L., & Stewart, J. P. (2004). Geotechnical aspects of seismic hazards. In *Earthquake Engineering from Engineering Seismology to Performance Based Engineering*. CRC Press.

Krinitzsky, E. L. (2002). How to obtain earthquake ground motions for engineering design. *Engineering Geology*, *65*(1), 1–16. doi:10.1016/S0013-7952(01)00098-9

Krinitzsky, E. L. (2003). How to combine deterministic and probabilistic methods for assessing earthquake hazards. *Engineering Geology*, *70*(1-2), 157–163. doi:10.1016/S0013-7952(02)00269-7

Lee, S. J., Chen, H. W., Liu, Q., Komatitsch, D., Huang, B. S., & Tromp, J. (2008). Three Dimensional Simulations of Seismic-Wave Propagation in the Taipei Basin with Realistic Topography Based upon the Spectral-Element Method. *Bulletin of the Seismological Society of America*, *98*(1), 253–264. doi:10.1785/0120070033

Lew. (2001). *The Seismic Design Handbook*. Kluwer Academic Publishers.

Nazarian, S., & Stokoe, K.H. (1983). *Use of spectral analysis of surface waves for determination of moduli and thicknesses of pavement systems*. Transportation Research Record No. 954.

Neelima Satyam, D., & Rao, K. S. (2009). Dynamic site characterization in Delhi region using MASW testing. *International Journal of Earth Sciences and Engineering*, *2*(1), 32–42.

Olsen, K. B., Pechmann, J. C., & Schuster, G. T. (1996). An Analysis of Simulated and Observed Blast Records in the Salt Lake Basin. *Bulletin of the Seismological Society of America*, *86*(4), 1061–1076.

Petersen, M. D., Rastogi, B. K., Schweig, E. S., Harmsen, S. C., & Gomberg, J. S. (2004). Sensitivity analysis of seismic hazard for the northwestern portion of the state of Gujarat. Academic Press.

Phanikanth, V. S., Choudhury, D., & Rami Reddy, G. (2011). Equivalent-Linear Seismic Ground Response Analysis of Some Typical Sites in Mumbai. *Geotechnical and Geological Engineering*, *29*(6), 1109–1126. doi:10.100710706-011-9443-8

Rastogi, B. K. (2001). Ground deformation study of Mw 7.7 Bhuj earthquake of 2001. *Episodes*, *24*, 160–165.

Reiter, L. (1990). *Earthquake hazard analysis – Issues and Insights*. New York: Columbia University Press.

Schnabel, P. M., Lysmer, J., & Seed, H. B. (1972). *SHAKE: A computer program for earthquake response analysis of horizontally layered sites. Report No. EERC 72/12*. Berkeley, CA: Earthquake Engineering Research Centre, University of California.

Seed, H. B., & Idriss, I. M. (1970). *Soil modules and damping factors for dynamic response analyses. Report EERC 70–10*. Berkeley, CA: Earthquake Engineering Research Center, University of California.

SHAKE. (2000). *A computer program for conducting equivalent-linear seismic response analyses for horizontally layered soil deposits. A modified PC version of the original SHAKE program published in 1972 by Schnabel, Lysmer and Seed (modifications made by Idriss IM, Sum JI). EERI*. Berkeley, CA: University of California.

Sharma, M. L., Wason, H. R., & Dimri, R. (2003). Seismic zonation of the Delhi region for bedrock ground motion. *Pure and Applied Geophysics*, *160*(12), 2381–2398. doi:10.100700024-003-2400-6

Sitharam, T.G., & Anbazhagan, P. (2007). Seismic hazard analysis for Bangalore region. *Journal of Natural Hazards, 40*, 261-278.

Sitharam, T. G., & Govindaraju, L. (2004). Geotechnical aspects and ground response studies in Bhuj earthquake, India. *Geotechnical and Geological Engineering*, *22*(3), 439–455. doi:10.1023/B:GEGE.0000025045.90576.d3

Sokolov, Y. V., Loh, C. H., & Wen, K. L. (2001). Empirical models for site- and region-dependent ground motion parameters in the Taepei area: A unified approach. *Earthquake Spectra*, *17*(2), 313–332. doi:10.1193/1.1586177

Sun, J. I., Golesorkhi, R., & Seed, H. B. (1988). *Dynamic moduli and damping ratios for cohesive soils. EERC 88-15*. Berkeley, CA: University of California.

Suto, K. (2007). Multichannel analysis of surface waves (MASW) for investigation of ground competence: an introduction in Engineering Advances in Earthworks. In *Proceedings of the Sydney Chapter 2007 Symposium* (pp. 71-81). Australian Geomechanics Society.

Tewari, H. C., Dixit, M. M., Sarkar, D., & Kaila, K. L. (1991). A crustal density model across Cambay basin, India and its relationship with the Aravallis. *Tectonophysics*, *194*(1-2), 123–130. doi:10.1016/0040-1951(91)90276-X

UBC. (1997). Uniform Building Code. *Int. Conf. Building Officials*, 2.

Uma Maheshwari, R., Boominathan, A., & Dodagoudar, G. R. (2008). Development of empirical correltation between shear wave velocity and standard penetration resistance in soils of Chennai. *Proceedings of 14th World Conference on Earthquake Engineering*.

Uma Maheswari, R., Boominathan, A., & Dodagoudar, G. R. (2010a). Use of Surface Waves in Statistical Correlations of Shear Wave Velocity and Penetration Resistance of Chennai Soils. *Geotechnical and Geological Engineering, 28,* 119-137.

Uma Maheswari, R., Boominathan, A., & Dodagoudar, G. R. (2010b). Seismic site classification and site period mapping of Chennai City using geophysical and geotechnical data. *Journal of Applied Geophysics, 72*, 152–168.

Wang, L. M., Wu, Z. J., & Xia, K. (2015). Study on the Effect of Loess Sites on Seismic Ground Motion and Its Application in Seismic Design. *International Conference on Earthquake Geotechnical Engineering*.

Wang, W. (1990). *Bounding surface hypoplasticity model for granular soils and its applications*. Davis, CA: University of California.

Chapter 3
A Novel Meta-Integrative Platform for Effective Disaster Management

Srikanth Venkatesan
Victoria University, Australia

Abbas Rajabifard
University of Melbourne, Australia

Nelson Lam
University of Melbourne, Australia

Emad Gad
Swinburne University of Technology, Australia

Helen Goldsworthy
University of Melbourne, Australia

Ged Griffin
University of Melbourne, Australia

ABSTRACT

Disaster management is recognized as a complex task. Despite current facilities and advancements, each disaster continues to frustrate the government and the community. Well-known issues include lack of awareness, direction, preparation and planning, response, recovery, and policy vacuum, besides many other factors. Other issues include emerging drivers of change such as globalization, climate change, technology, social construct, global finance, and education. Despite the recognition of such complex issues, the pace of disaster management strategies seems to be inadequate. This is mainly due to the lack of an approach that can integrate the myriad of issues with the roles and responsibilities of stakeholders such as governments, practitioners, and the community. This chapter identifies a spatially enabled platform as a tool to overcome the aforementioned issues. A preliminary roadmap with "stakeholder position" as the central point of integration is presented. It is anticipated that the roadmap will provide governments and decision makers with the direction needed for future planning.

DOI: 10.4018/978-1-5225-6948-0.ch003

INTRODUCTION

With the recent developments in the field of information and technology any disaster related news (such as disaster alerts) is available almost instantly in the palms of an individual. There are also significant technological developments such as GPS and GIS that has been used to improve the decision making processes in disaster management. Despite such advancements in knowledge and practice, each major disaster continues to frustrate governments and communities because of the imposed economic losses and emotional pain. Difficulties in the conceptualisation and operationalization of disaster management are widely known. In this paper, we analyse some of these difficulties and investigate whether new knowledge can be used to improve current practices. An effective way of integrating the needs and requirements to effectively address the challenges in disaster management are presented.

BACKGROUND

Figure 1 presents the increase in number of disasters worldwide since the 1900's (Source: EM-DAT http://www.emdat.be/country-profile). The increasing trend since the 1900's and the more significant upward trend around the 1980's are apparent. Some critics argue that since the 1980's disaster information has been well documented leading to the upward trend. Despite this criticism, many events require no reference to the readers: e.g. the devastating tsunami in 2004 that damaged most Asian coastlines, the Fukushima disaster in 2011, bushfires in Australia, hurricane Katrina and the 9/11 terrorist attacks. Apart from the financial damage and loss of lives, disasters leave a huge emotional imprint on the survivors. Although most governments have developed new frameworks, action plans and forming apex bodies of national disaster or emergency management, most of the challenges and issues remain unresolved.

CHALLENGES AND ISSUES IN DISASTER MANAGEMENT

One of the foremost issues to be highlighted is the scale of disasters in relation to management aspects. Mener (2007) mentions that during the 9/11 attacks, the emergency responses were effective even though there was significant confusion and chaos. On the contrary events such as hurricane Katrina and Andrew left the system paralysed and ineffective. A much higher scale of devastation can be attributed to the 2004 tsunami and the Fukushima event. A particular coastal town in the Southern part of India was almost wiped out due to the 2004 tsunami. Local

Figure 1. Trends in natural disasters between 1900 and 2011

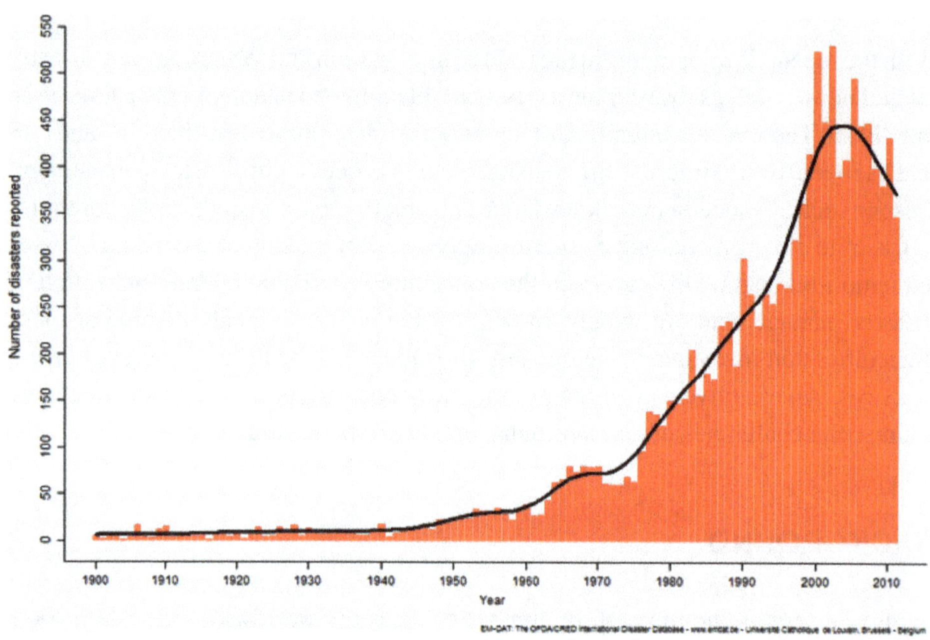

authorities had to seek help from state and central government services. The situation was so unbearable and overwhelming that even some well-trained relief workers and some volunteers needed counselling, as many have not experienced such situations in the past (ST060088, 2013). Nine years on the town has recovered and normal life seems to be in order. However plans or preparations to face and overcome a similar disaster (if exposed) are not widely publicised. Similar situations exist in the neighbouring country of Sri Lanka as well. The enduring image of a house swept to sea (flashed widely in the media) post the Fukushima disaster clearly captured our inability to cope with mega scale disasters. It is not our intention to criticise a country or a government but the scale of a disaster can sometimes be so large and overwhelming that even technologically and operationally advanced countries have problems coping with the challenges.

Apart from the scale of disasters which is beyond our control, we analyse the challenges in disaster management from the perspectives of three key stakeholders: governments (includes all levels such as central, state and local), practitioners (includes academics, researchers and responders) and the community (includes individuals, families, social groups and business groups). Although there is an overlap between the three key stakeholder positions or multiple representations; we define the stakeholder position in relation to the role, responsibility and expectation of an individual.

Governments

Out of all the stakeholders, governments have the highest level of responsibility. They are required to develop policies, implement and control all operations related to disaster or emergency management. The most preferred arrangement by the governments is the establishment of an apex body such as Federal Emergency Management Agency (FEMA) or the National Disaster Management Authority (NDMA). Whilst most governments cannot be faulted on their initiatives, it is the speed of implementation and a lack of coordination and the lack of control at strategic points that breaks the chain and the system. A high level group of officials in Australia identified the following national challenges in 2002:

- The need to develop a national disaster assessment capability which can speedily provide critical information on the scope of a disaster, the needs of affected communities, and problems associated with mounting a response;
- The decline in the availability of volunteers in the emergency management sector;
- Difficulties of providing counter-disaster preparedness and response capabilities in remote indigenous communities;
- The need to improve inter-operability of emergency management agencies across the nation;
- More systematic use of gis and remote sensing products for more effective mitigation, response and recovery functions and decision making;
- Resolution of the issue of mandatory reporting by radio and television broadcasters of emergency warnings issued by authorised emergency management authorities;
- The need to develop improved capacity for real-time, widely understood disaster warnings to communities on an 'all hazards' basis;
- The need for a more sophisticated and effective approach to community disaster awareness and education;
- The enhancement of local government preparedness and response capabilities;
- The need for a national approach for dealing with catastrophic disasters; and
- The need to establish a national coordination framework, and ministerial and high level official oversight, for disasters of national significance.

The above challenges can be broadly grouped into issues related to policy development, implementation and control in general. In addition to the above points, FEMA (www.fema.gov) identifies emerging drivers of change such as globalisation, technological development, and social construct of communities, finance, education, ageing population, and infrastructure and data interoperability to be considered by governments in the planning and preparatory stages of disaster management. Despite high levels of responsibility on the parts of the government, it should be noted that disaster management essentially requires coordinated efforts from all other stakeholders as well.

Practitioners

Based on the volume of literature available, it appears that practitioners, as defined in this paper, analyse issues and challenges related to natural, biological, technological and human induced disasters to provide findings and recommendations to other stakeholders. The analysis undertaken by practitioners may include (but not limited to) the development of frameworks, hazard modelling, risk assessments and normal operational issues. For example, Price-Robertson and Knight (2012) present a framework for analysing community resilience. Such analysis has been based on a number of factors influencing the construct of community resilience. Mener (2007) in his analysis points to a lack of coordination between FEMA and state services, lack of leadership, coordination, communication, resources, facilities and infrastructure during some of the emergency operations in the aftermath of hurricanes in the United States. Similarly, there have been reports of conflicting information being presented to the public during the 2009 floods in Queensland. Miscommunication and lack of coordination has been well documented during the bushfires in Perth, Australia by practitioners. It is plausible that such criticisms have been levelled at governments worldwide given the democratic views and in resentment to bureaucratic influences.

Some practitioners identify the availability of information especially at the initial phases as central to the decision making process. The need to engage community and measures to improve organisational resilience have also been well stated. Another important dimension that is usually ignored and left to experts in the field is hazard modelling. The wide promotion of disaster risk assessment, reduction and mitigation requires a better understanding of hazards. It is noted that knowledge on a particular hazard has direct implications in the drafting of building codes of practice. For example, earthquake design codes of practice are fundamentally derived from an understanding of earthquake hazard at a given site. The inherent difficulty lies in the choice of appropriate methods as each of these methods may have implicit assumptions. Although the methods may be rigorous, resulting uncertainties are usually not known to decision makers and responders. Similarly, reliable vulnerability modelling (e.g. infrastructure) and consequence modelling are essential in risk assessments. It has to be mentioned that all modelling approaches in this context have not yet been fully developed. Significant research efforts are warranted in the immediate future to address the deficiencies. Therefore an understanding of the limitations of such models is paramount. These statements noted above call for the need to include expert opinions at every stage of decision making. It is worth noting that a high level delegation in Australia reviewed current disaster management capacity and identified that the major weakness in disaster management lies in the lack of independent and comprehensive analysis of natural disaster risk assessments, lack of natural disaster data collection and analysis (Australian Emergency Management, 2002).

The above mentioned high level delegation recommended 66 measures to improve current practice. The first of these recommendations calls for the use of common terminologies to be adopted consistently amongst all stakeholders. It also recommends a National Framework Objective underpinned by a unified national approach to disasters under which all stakeholders can play a role. Clarity of roles and responsibilities for all government levels has been suggested by the delegation. Measures to advance cost-effective mitigation through effective use of land planning have also been suggested. Improving public awareness, engaging rural communities, distribution of funds, building resilient infrastructure are some of the other recommendations made. The 66 recommendations can be grouped under policy development, implementation and control as noted in the previous section. Research, education, community engagement have been identified as top priorities by these stakeholders.

From the above discussions it can be clearly seen that there are wide ranging agreements on issues related to policy, implementation and control measures between these two stakeholders (i.e governments and practitioners) including the acknowledgement of myriad of challenges and issues.

Community

One of the significant challenges is the intention of communities to return to pre-disaster life position (or normal life) whilst governments intend to build back better. Most communities may not be satisfied with response and recovery measures. Whilst governments would like the communities to build adaptive capabilities and increase resilience, communities may not be aware of local emergency plans. Thus conflicts of interests may exist between the governments and the communities that may pose huge challenges in relation to disaster management. As noted earlier the social construct of the families and ageing population are significant drivers of future directions. Most governments have shifted to an inclusive approach of engaging community at all phases of disaster management. This suggests that communities should be aware that they are not necessarily the beneficiaries but have a role to play in combatting the negative impact of disasters. For example the National Strategy for Disaster Resilience (NSDR, 2011) spells out the characteristics of a disaster resilient society and points out the necessary measures to be undertaken by the community. Although communities do not play a role in research directly, education, awareness and training are emphasised by government at all levels. Thus it is clear that policy, implementation and control measures are directly related to the communities as well.

The above discussions which is based on the views of a stakeholder position presents a convenient way of analysing the challenges and issues related to disaster management despite the complexities. This suggests that the stakeholder position should be maintained as the focal point of all discussions. It might rightly be argued that this paper has not discussed all of the challenges and issues associated with disaster or emergency management. We believe that the above conclusion may still apply even if other issues have been taken into account.

TECHNOLOGICAL DEVELOPMENTS AND OTHER ADVANCEMENTS INFLUENCING BEST PRACTICES

In this section we review some of the recent advancements in the field of information technology including the use of smart phones, sensors, spatially enabled platforms and simulation technologies, hazard and vulnerability modelling to name a few. However the open ended nature of such developments cannot be discussed within the confines of a single paper. Thus salient examples are presented herein.

Smart Phones, Social Media, and Network Centric Models

Even before the advent of smart phones, warning messages have been transmitted via handheld phones. With the advent of smart phones there are number of apps (e.g. Disaster alert, Quakefeed, Disaster readiness guide, Disaster survival plan) that can be used to send disaster related alerts and information. Some organisations like the Victorian Country Fire Authority (CFA) in Australia provide space for uploading photos of incidence thus ensuring direct engagement and connectivity. There are a few apps developed in isolation and there is a need to have them linked to emergency response systems. There are also specific apps designed and developed for emergency responders such police personnel which are not available in the public domain. For maximising the potential of these apps, there is a need to integrate as well as segregate these under a strategic framework.

The term social media is usually reserved for internet based applications that enable people to communicate and share resources and information (Lindsay 2011). Facebook, Youtube, LinkedIn, Flikr, Instagram, Tumblr are some common examples. In regards to disaster management, the communication that relates to early observation and reports of events becomes vital and are captured through crowd sourcing applications. For example, during the international response to the 2010 Haiti earthquake the United States Military's Southern Command (SOUTHCOM) developed the SOUTHCOM 3D User-Defined Operational Picture (UDOP). This system allowed the diverse relief community to collect a wide range of data and information and then fuse the information into an interactive common operating picture. A key feature of this tool was that it allowed independent users to conduct rapid damage assessments of key facilities and have the results and imagery of those assessments shared within the entire relief community in real-time. This was instrumental in directing the strategic planning and coordination of the relief effort. There are also other applications such as Sahana Eden software. Some common problems identified in using this Volunteered Geographic Information (VGI) are mischief, malice and criminal intents (Coleman, 2010). One significant advantage of this information is the speed and volume of information that arrives in the initial stages of an event or disaster. As stated in the earlier section of the paper, this information could be vital in the early stages of decision making. The above discussions clearly highlight the advancements in emergency management underpinned by the interaction/communication between the communities and the responders. Despite the advancements, significant research and analysis is required to authenticate the information in real time.

Network-Centric Warfare (NCW) is a term that has been developed to describe how military forces organise and coordinate to fight in the Information Age. Network-Centric Warfare develops an information superiority-enabled concept of operations to achieve shared awareness, increased speed of command, higher tempo of operations, greater lethality, increased survivability and a degree of self-synchronisation. Note that the application of this concept is directly related to the need of governments to increase the tempo and capability of the response team during large scale disasters. Thus a solution is already being attempted.

Both the military and emergency managers use a command, control and coordination approach. Therefore a comparison between these two types of organisations is insightful and will inform future practice. The NCW approach has transformed our understanding of how a modern military force engages in combat and it represents a significant paradigm shift. Military commanders can now achieve far greater effects with fewer personnel. Being asked to do more with less is a common situation that confronts public safety officials. Whilst NCW appears to be providing significant enhancements to our military operations, the development of Network-Centric Public Safety (NCPS) is still an unrealised opportunity.

The successful adoption of a network-centric approach will require strong leadership, cultural change and technological investment. It cannot be achieved without widespread discussion, debate, experimentation and ultimately broad acceptance. At this stage the potential of network-centric operations within emergency management and the public safety arena has not been realised. Public safety officials need to consider how military forces have embraced technology and have it transformed into information age organisations. The military's experience and foundation documentation will help to rapidly develop a strategy that can be used to transform policing and public safety organisations into information age enterprises.

Essentially there four critical elements or nodes within a NCW concept, namely the:

- Command and control systems or command and control grid where command decisions are made;
- Sensor systems or sensor grid which acquires information necessary to inform decisions and raise situational awareness;
- Engagement systems or engagement grid where resources are tasked to perform roles and execute the commanders intent; and
- Information grid and supporting network which supports the receipt, presentation, transmission and retention of information. (Alberts *et al* 2002; Department of Defence, DoD 2009)

These elements are depicted graphically in Figure 2. Within these four critical elements some of the entities will have complex functionality and perform dual roles across these elements or nodes. For example combat forces may have to perform the role of sensing and engaging at the same time.

Figure 2. Elements of network centric warfare
Source: Alberts et al 2002; DoD 2009.

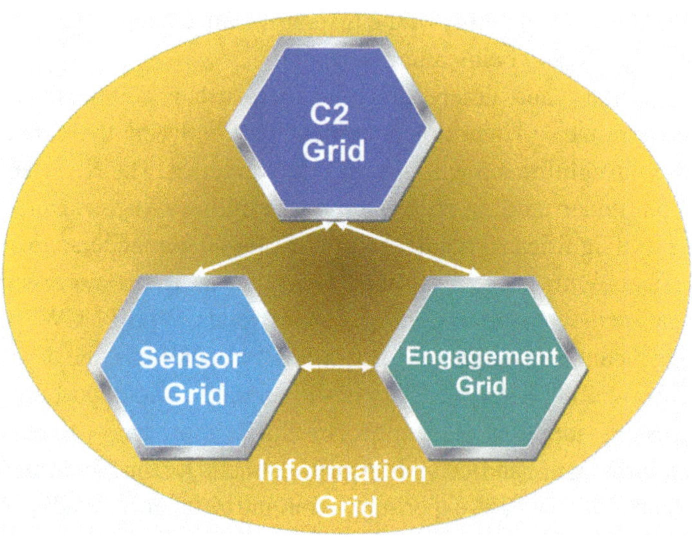

Sensors include all entities that contribute to battle space awareness and the Sensor Grid involves both a range of technical equipment, for example radar and airborne detection systems as well as human sources for example combat soldiers. Collectively, the sensor grid captures information and provides situational awareness across the network. Sensors do not have to be individually linked to each other and depending on the circumstances there may be multiple sensing networks (Alberts *et al* 2002; DoD 2009).

The Engagement Grid consists of major combat units and weapons systems that can be used to engage a target(s) in accordance with the commander's intent. This grid can also involve actors, such as logistics and support personnel, which are responsible for supporting combat units and weapons systems (Alberts *et al* 2002; DoD 2009).

The Information Grid, sometimes referred to as the Global Information Grid (GIG) enables the access to, exchange, and use of information and services throughout the network. The GIG consists of information, equipment, infrastructure, information technology and associated people and processes that support networked personnel and organisations in accomplishing their tasks and missions. Ideally the GIG enables the visibility, accessibility, sharing and understanding of information and services among all networked defence personnel and mission partners through well-defined interfaces. This information can be available on either a push or pull basis, that is, supplied automatically or supplied on request. For example, a push system may involve automated alerts where the sensor grid detects a weapon launch, whereas a pull system can involve a 'reach back' service where analysis and intelligence is supplied in accordance with a mission commander's request. The GIG helps to enhance battle space awareness and creates a common operational picture (COP). The information contained on the GIG should be viewed as the collective property of the networked forces. The way in which information is accessed and presented will vary depending upon the level or role of the user (Alberts *et al* 2002; DoD 2009).

Ultimately the commander must be able to translate their decisions into effective actions in the battle space; however, command and control is an inherently iterative decision making process whereby commanders attempt to achieve a given task(s) or solve a problem(s). These issues can be either tactical or strategic problems that need to be resolved. During their deliberations commanders often consider their objectives, options, enemy capabilities and intent, and the potential impact on non-combatants. The Command and Control Grid allows commanders to consider a wider range of options and employ force with greater precision and granularity. The increased level of situational awareness allows commanders to constantly reassess the impact of their decisions and, if necessary, modify their plans on how best to apply their combat power (Alberts *et al* 2002; DoD 2009).

The benefit of the NCW model is that the integrated network facilitates enhanced situational awareness, collaboration and offensive potential. 'Personnel within the networked force rely on secure and responsive linkages (network connectivity) that allow the correct information to be accessed at the right time by the right force elements' (DoD 2005). This type of information needs to be protected by a wide range of security mechanisms such as encryption, anti-virus programs and firewalls. 'Increases in combat power from being a networked force are derived from the quality and timeliness of shared information and through the exploitation of new system and command relationships' (DoD 2005).

It is clear that adopting these kinds of models supported by technological advancements is certain to improve current disaster management practice.

Spatial Platforms, Simulation Technologies, and Advancements in Hazard Modelling

According to Rajabifard et al. (2010a) the term spatial refers to the use of location-based information and there are number of applications in practice. This availability of location based information has improved the decision making capability (Figure 3) and by the examples noted herein. Geospatial technology has been used in restoration of ecosystems and in the mapping of current river boundaries and coastal boundaries. By comparing old river boundaries encroachments have been identified. The significance of the analysis led to the finding that the mouth of river discharge had been shortened which had led to flash flooding and increased flood heights. Efforts were then undertaken to provide additional culverts at strategic points to facilitate storm water flow. In terms of comparisons based on coastal boundaries, effects of climate change have been analysed and the significance of erosion phenomena had been investigated. Using such analyses coupled with marine satellite data, climate disaster resilience indices had been developed for a region in India as shown in Figure 4.

Apart from such useful analysis presented above, the emergence of simulation technologies has been very attractive to emergency responders. Examples include modelling of flash flooding for councils in Australia (Source: Prof. Lin Paddham and Dr. Dhirendra Singh, RMIT University, Australia) and using real-time simulated response of residents. For example, the simulations can be carried out for people responding to flash flooding by sand bagging the periphery of their houses and how effective it could be in reducing damage. There are also simulation models for evacuation from bush fires, bomb blasts and power blackouts. (Source: Dr. Tuan Ngo, University of Melbourne, Australia). There are also tsunami related simulations that helps viewers to make preparations to act in response to visible signs observed in the sea. It is important to note that visual presentations of disaster related information

Figure 3. Location as the fourth element of decision making
Source: Scott & Pigram, 2011.

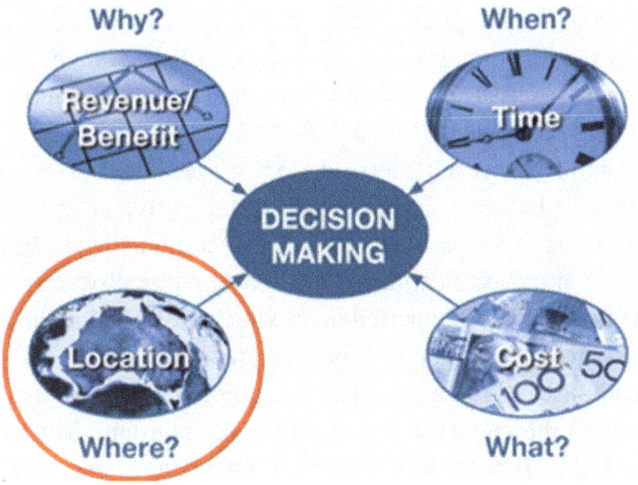

Figure 4. Example of analysing social resilience due to climate change
Adapted from ST060088, 2013.

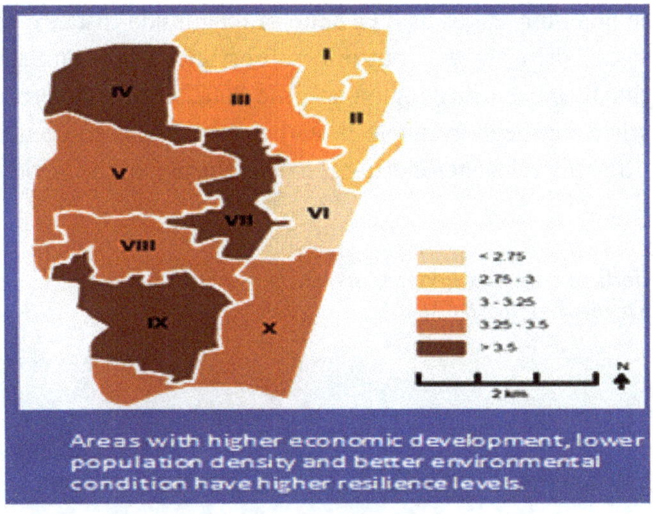

is very effective in terms of communication and time and is also very appealing and attractive to all stakeholders.

The power of spatial and simulation technologies have been used in hazard modelling as well. ARUP, a leading consultancy organisation had modelled the complex seismic source zones around Japan. It is well known that Japan is at the intersection of major tectonic plates with the Pacific plate and the Philippine sea plate subducting under the Amurian and Okhotsk plate along the trench. Traditionally seismologists would employ a 2D approach to ascertain source zones. With the advent of spatial platforms a clear 3D view of earthquake sources could be easily visualised. This approach had helped in redefining the source zones and the calculation of recurrence intervals. (Refer Figure 5).

There are examples of results derived from geotechnical investigations being presented in spatial platforms. For example, surface wave analysis conducted to model shear wave velocity of particular sites can be combined with digital elevation maps to produce a site classification system for an entire city. Such analyses can then be extended to produce site period maps which are useful in the estimation of displacement demands imposed by earthquake loading.

Even in a complex field like earthquake engineering, it is imperative to relate potential hazards to the end user. For example, the building drift (which can be measured from spatial data) can be used as a parameter to assess the damage of buildings in real time. Such fast track methods of estimating structural damage induced by earthquake shaking can be estimated using the ratios of peak displacement demand (as mentioned before and derived from hard core modelling) and the drift angle of the building obtained from real time data. Pseudo-assessments can also be made in the planning stages using a suite of magnitude-distance combinations and estimations of drift angles from prior observations. This modelling can be extended to global scales using region specific data. Such an approach of global scale analysis had not been available in earlier times. It is to be noted that this information is directly relevant for disaster management experts to identify areas of

Figure 5. Modelling complex geometries using spatial platforms
Reproduced with permission from So, M.M.L.

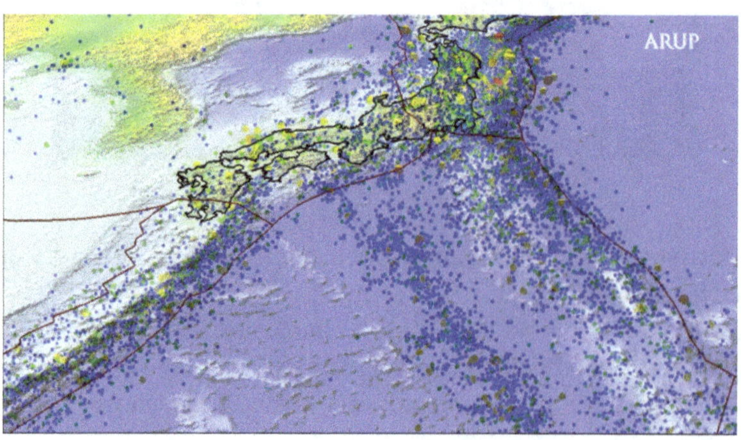

building damage. Reliability of existing models of risk prediction (Refer Figure 6) can be improved with the availability of fast tracked information in real time thus reducing uncertainties and probabilities of failure. When such data and analysis are integrated with probabilistic models on a global scale, useful information can be derived for the insurance industry. Many readers might well be aware of shake maps presented in spatial platforms. In recent years structural design engineers refer frequently to such information.

The above section presented some useful examples in which advancements in technology and knowledge can be used to improve current disaster management practices. However to address the challenges and issues directly for all stakeholders such information should be available in a single platform.

NEED FOR AN INTEGRATED PLATFORM AND A SUGGESTIVE ROADMAP

As noted in the technological developments section, there is an absolute need to integrate the available technologies for the ultimate benefit of effective disaster management. In the challenges and issues section, it was identified that the myriad of challenges and issues would stand alone if a unified approach is not undertaken. Literature identifies that a spatially enabled platform could be the viable option for the meta-integration of suggested technologies to fit with the requirements of stakeholders. Fundamentally, hazards and disasters are location specific; hence it is logical to adopt a location based platform.

Figure 6. Enhancing risk prediction using real time data with hard core modelling techniques

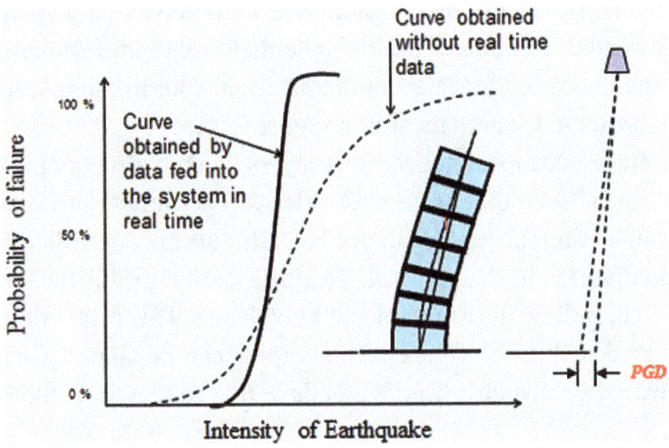

As noted earlier popular uses of spatial and IT technologies lies in the transformational use of such technologies to improve governance and society. Spatial enablement provides a convenient platform for integrating data and information across a multitude of sources. It also presents information in a visual form to various stakeholders. In its present form, spatial enablement is being utilised in managing data and information on population, health, land use, coastal management, hydrology, agriculture, aerial photography and climate. Spatial enablement can also be used in disaster risk reduction at a local governmental level as demonstrated by Sutanta et al. (2010).

As previously stated hazard is location specific and can be directly integrated from hazard maps. In addition, exposure, vulnerability, elements at risk can also be mapped and integrated spatially. It is also possible to ascertain the cumulative hazard at a specific location due to sequential and multiple hazards. The integrated risk assessment can then be obtained by apportioning different weights for each hazard followed by appropriate exposure and vulnerability as outlined by Greiving, Fleischhauer and Wanczura (2006). The above (for example) pertains to specific expertise, the knowledge component. Risk reduction can be further enhanced by appropriate land use such as protecting vulnerable areas of flooding, preventing construction on lands vulnerable to landslides, delineating catchment areas and retarding basins, delineating retreat areas for emergency evacuation and reducing building density in earthquake prone / bush fire areas, Fleischhauer, Greiving and Wanczura (2005). In all cases of risk reduction, the acceptable level of risk must be agreed upon during planning stage and if unacceptable alternate strategies identified. Well-coordinated data can also be used to calculate insurance losses over a specific region. The above pertains to the planning process. Thus integrating specific knowledge with the planning process is feasible. The above mentioned approaches may provide policy makers a greater degree of confidence in the management of hazards.

Based on the above-mentioned capabilities, we believe that a spatially enabled platform as presented in Figure 7 has the potential to integrate technology, policy, processes, people, knowledge, data, communication, coordination, implementation and control as presented schematically in Figure 7.

The above figure encapsulates various layers of embedments in the platform. For a start, it can be seen that any stakeholder can get inputs or outputs from the platform relevant to their access/clearance level. Inputs are controlled by data inter-operability and diverted to the network. The information grid is the source of data, analysis, research, policy, information and knowledge. These are controlled by the commander's grid (C2 in the figure). The outputs can be shared via either of the networks in textual or visual forms. We believe this roadmap would be useful for governments.

Figure 7. Suggested roadmap to a spatially enabled meta-integrative platform.

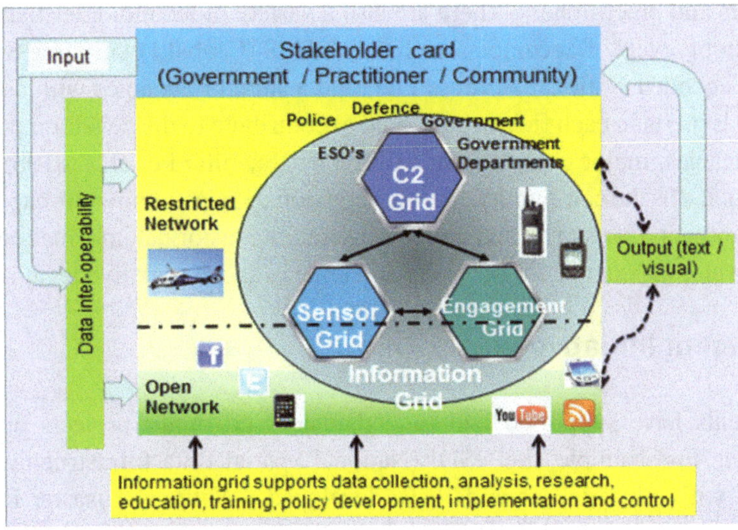

Again for example, the data could be hazard maps, vulnerability data, exposure data which could define risk levels (information). With spatial planning this could then be transformed into knowledge about a particular region to improve planning and preparedness. Similarly with the use of a spatial platform the increase in water levels in a dam due to a storm could be dynamically monitored. Coupled with the information on maximum water levels, weather patterns, areas susceptible to flooding, a decision on evacuation could be discussed between authorities, agencies and governments and a consensus quick decision arrived before informing public through the media. Thus linearising communication to a single voice leading to improved emergency response can be achieved. Recovery is enhanced by prompt support and efficient recovery of infrastructure, assets and facilities. It can be noted that the platform has the flexibility to allow the flow of information or action to be non-linear while maintaining a single voice. Such a facility would certainly help future disaster management practice in a most effective manner.

Challenges in Implementing Spatial Enablement as an Integrative Platform

Although the concept presented in Figure 7 appears viable, a complex task such as disaster management requires major efforts from all the stakeholders and requires an inter-disciplinary framework. Commonly noted challenges start with the use of place as a value, positioning accuracies and address coding. Definitions of data, metadata, organising and managing data, reliability and training have also been noted by researchers and practitioners. There are also a variety of technologies that operate with different levels of accuracies such as the GPS (Global Positioning Systems), GNAF (Geocoded National Address File) and a suite of agencies with associated expertise. Experts in each field would need to contribute to the development. Apart from intricacies, major challenges identified by Rajabifard et al. (2010b) include broader uptake by the community, inter-jurisdictional issues, bandwidth capabilities, internet protocols, e-security and policies in handling historical information, access issues, developing standards, licensing agreements and associative legal issues.

Government Initiatives

Governments have supported and funded major and minor projects on spatial enablement. For example, the NSDI (National Spatial Data Infrastructure, India) combined with other departments have developed a National Disaster Response Force map across the country. It also initiated mapping procedures with high-end technology and expertise, developed tsunami warnings and aerial photography to name a few. Similarly Geo Science Australia (GA) has undertaken significant works in developing vulnerability assessment models and geo spatial models for major cities in anticipation of damages due to earthquakes, strong wind, tsunamis and landslides. The state of Victoria has initiated a spatial information management framework; Western Australia has implemented a `Shared Land Information Platform (SLIP); New South Wales has supported a number of initiatives reflecting international approaches. Finally the Federal Government has supported the development of Commonwealth Spatial Data Initiative (CSDI).

Benefits of Spatial Enablement Approach

The 2007 International Workshop on Spatial Enablement in Seoul, Korea, proposed that a spatially enabled approach will lead to:

- Improved decision making;
- Reduction in administrative costs;
- Whole of government approach;
- Enhanced industry development opportunities.

Notwithstanding the above spatial enablement promotes e-governance, policy development and better stakeholder relationships with the society. The creation of social stability, environmental protection, economic wealth can be facilitated through the development of products and services based on spatial information collected and shared by all levels of societies including governments, private sectors, non-profit organisations and citizens. Thus the shared responsibility agenda of the disaster management approach can be easily achieved. In current practice, the four concepts of disaster management namely the all hazards approach, the comprehensive approach, the all agencies approach and an all society approach are treated individually. This has led to confusion, chaos and dissatisfaction across stakeholders. As spatial enablement facilitates a whole of government approach and a whole of society approach (including agencies as shown in Figure 3) a meta-integrative platform of the above four concepts is feasible. This would certainly lead to improved disaster management practices.

CONCLUSION

In this paper we have presented a roadmap that can be incorporated by government representatives at higher levels to overcome common challenges and issues identified in disaster management. In particular the roadmap suggests the use of a spatial platform that can easily take into account of stakeholder requirements, underpinned by a shared concept of information and control. The information grid supports data, research, education, training, policy development and other sources. The research also shows that the role of a stakeholder is vital. By harnessing recent advancements in technology and knowledge, a way of moving forward to improve disaster management practices and to achieve risk reduction appears feasible. However such an approach will involve a shared understanding between stakeholders and a major paradigm shift.

ACKNOWLEDGMENT

The authors of this paper acknowledge all participants of the workshop entitled "Spatially enabled nations: A meta-integrative framework for improving disaster management practices in India and Australia" held in Chennai, India in January 2013. Some of the statistics, information and facts presented in the paper are derived from the workshop presentations in addition to the literature referred herein. It is noted that all presentations were delivered with sufficient evidence to the claims. Further details can be found in the referenced document ST060088.The funding support provided by the Commonwealth Government of Australia and the Government of India towards conducting the workshop is duly acknowledged herein.

REFERENCES

Alberts, D., Garstka, J., & Stein, F. (2002). *Network Centric Warfare, Developing, and Leveraging Information Superiority*. Washington, DC: Department of Defence Command and Control Research Program.

Australian Emergency Management. (2002). *Natural disasters in Australia – Reforming relief, recovery and mitigation arrangements, A report to the Council of Australian Governments by a high level officials group, 2002*. Retrieved from http://www.em.gov.au/Documents/Natural%20Disasters%20in%20Australia%20-%20Review.pdf

Coleman, D. J. (2010). Volunteered Geographic Information in Spatial Data Infrastructure: an Early Look at Opportunities and Constraints. In A. Rajabifard, J. Crompvoets, M. Kalantari, & B. Kok (Eds.), *Spatially Enabling Society: Research, emerging trends and critical assessments* (pp. 131–147). Leuven University Press.

Department of Defence (DoD). (2005). *NCW Roadmap*. Canberra, ACT: Defence Publishing Service.

Department of Defence (DoD). (2009). *NCW Integration and Implementation Strategy 2010*. Canberra, ACT: Defence Publishing Service.

Fleischhauer, M., Greiving, S., & Wanczura, S. (2005). Spatial planning in the focus of hazard and risk assessment/management in Europe. EURO-RIOB Conference, Wroclaw, Poland.

Greiving, S., Fleischhauer, M., & Wanczura, S. (2006). Management of natural hazards in Europe: The role of spatial planning in selected EU member states. *Journal of Environmental Planning and Management*, *49*(5), 739–757. doi:10.1080/09640560600850044

Lindsay, B. R. (2011). *Social Media and Disasters: Current Uses, Future Options, and Policy considerations*. Washington, DC: Congressional Research Service.

Mener, A. S. (2007). *Disaster response in the United States of America: An analysis of the bureaucratic and political history of a failing system* (PhD thesis). University of Pennsylvania.

NSDR. (2011). *National Strategy for Disaster Resilience*, retrieved from http://www.em.gov.au/Documents/1National%20Strategy%20for%20Disaster%20Resilience%20-%20pdf.PDF

Price-Robertson, R., & Knight, K. (2012). *Natural disasters and community resilience: A framework for support, Communities and Families.* CFCA paper no. 3. Retrieved from http://www.aifs.gov.au/cfca/topics/communities.php

Rajabifard, A., Crompvoets, J., Kalantari, M., & Kok, B. (2010b). Spatially enabled societies. In A. Rajabifard, J. Crompvoets, M. Kalantari, & B. Kok (Eds.), *Spatially Enabling Society: Research, emerging trends and critical assessments* (pp. 15–25). Leuven University Press.

Rajabifard, A., Williamson, I., & Vaez, S. (2010a). Spatially enabled societies: Asia and the Pacific. In A. Rajabifard, J. Crompvoets, M. Kalantari, & B. Kok (Eds.), *Spatially Enabling Society: Research, emerging trends and critical assessments* (pp. 87–94). Leuven University Press.

ST060088. (2013). *AISRF additional report: Spatially enabled nations: A meta-integrative framework for improving disaster management practices in India and Australia.* A report submitted to the Governments of Australia and India.

So, M. M. L., Mote, T. I., Pappin, J. W., Lau, V. S. Y., & Yim, I. P. H. (2012). 3D GIS technique for seismic hazard assessment for a subduction zone region. *Proc. of. The Annual Technical Conference: Australian Earthquake Engineering Society.*

Sutanta, H., Rajabifard, A., & Bishop, I. D. (2010). Integrating spatial planning and disaster risk reduction at the local level in the context of spatially enabled government. In A. Rajabifard, J. Crompvoets, M. Kalantari, & B. Kok (Eds.), *Spatially Enabling Society: Research, emerging trends and critical assessments* (pp. 205–218). Leuven University Press.

Chapter 4
Mitigation of Liquefaction Hazard Using Granular Piles

A. Murali Krishna
Indian Institute of Technology Guwahati, India

ABSTRACT

Ground improvement techniques are employed to mitigate liquefaction hazards. Granular piles are the most widely preferred alternative all over the world, due to several advantages associated with them. Different mechanisms operate in the function of stone columns/granular piles in liquefaction mitigation like drainage, storage, dilation, densification, and reinforcement. This chapter presents an overview of the use of granular piles as a liquefaction remedial measure for sand deposits. A brief description on the phenomenon of liquefaction and the associated features has been presented. A short discussion on various ground improvement methods available for liquefaction mitigation is presented in light of importance of granular piles. Different installation methods and design concepts for granular piles are presented. Various mechanisms of granular piles in mitigating the liquefaction potential of loose sand deposits are discussed and quantified in detail proving their effectiveness in hazard mitigation.

INTRODUCTION

Amongst the various natural disasters, earthquakes are probably the most disastrous ones because they cannot be predicted in advance and catch people unaware. The amount of energy released is extremely large and the area under its influence could be very extensive. Seismic hazards date from many centuries. Earthquakes are constantly posing risk to life and infrastructure facilities. The recent Bhuj/Kuchch

DOI: 10.4018/978-1-5225-6948-0.ch004

and Sumatra earthquakes have shaken the geotechnical community in South and Southeast Asia both physically and professionally. Large areas have suffered with building and infrastructural damages as well as liquefaction.

Seismic hazards can be described by ground shaking, structural hazards, liquefaction, landslides, failure of earth structures, lifeline hazards and tsunami and seiches (Kramer 1996). One of the most dramatic causes of damage to structures during earthquakes has been the development of liquefaction in saturated loose sand deposits, manifested either by the formation of sand boils and mud-spouts at the ground surface, by seepage of water through ground cracks or in some cases by the development of quicksand conditions over substantial areas (Seed & Idriss 1982). Liquefaction, the state under which soil deposit loses its strength and flows as a fluid is a major cause for damage during earthquakes.

Ground improvement techniques are commonly employed to mitigate liquefaction hazards. Granular Drains/Piles (GP) are the most widely preferred alternative all over the world, due to several advantages associated with them. Granular piles function as drains and permit rapid dissipation of earthquake induced pore pressures by virtue of their high permeability. As an additional advantage, they tend to dilate as they get sheared during an earthquake event. Seismic forces which tend to generate positive pore pressures in these deposits cause an opposite effect of dilation in the dense granular piles. One of the chief benefits of ground treatment with granular piles is the densification of in situ ground by which the in-situ properties of the ground will be modified to mitigate liquefaction potential. Further, the very high deformation modulus and stiffness of the granular pile material provide reinforcement for the in situ soil and offer another mechanism to mitigate liquefaction. Different mechanisms operate in the function of stone columns/granular piles in liquefaction mitigation. These mechanisms can be stated as Drainage, Storage, Dilation, Densification and Reinforcement.

This paper presents an overview of the use of granular piles as a liquefaction remedial measure for sand deposits. Brief descriptions about liquefaction and its evaluation, general liquefaction mitigation methods in practice were presented. Use of granular piles for ground improvement method, their installation and design aspects are discussed in detail. Various mechanisms that function for granular piles as liquefaction remediation are discussed along with some case histories.

LIQUEFACTION, ITS EFFECTS, AND EVALUATION

One of the major consequences of earthquakes is the phenomenon of 'liquefaction'. The concept of liquefaction gathered worldwide attention in the 1960's, when in 1964 large magnitude earthquakes located near Anchorage, Alaska and Niigata, Japan caused massive structural damage through ground failure. Significant amount of work on this topic has been performed in the last few decades since these earthquakes, resulting in several state-of-the-art papers relating to the study, evaluation and remediation of liquefaction (Martin et al. 1975; Seed, 1979; Ishihara, 1993; Seed et al. 2001; Youd et al. 2001; Idriss & Boulanger, 2005; Sawicki & Mierczynski, 2006; Madabhushi 2007 etc.).

Liquefaction occurs in saturated loose sandy deposits if subjected to seismic forces. Deposits of loose granular soils get affected by the ground vibrations induced by an earthquake, resulting in large total and differential settlements of the ground surface. In cases where the deposit consists high water table, the tendency to get densified may result in the development of excess hydrostatic pore water pressures of sufficient magnitude to cause liquefaction of the soil, resulting in settlements and tilting of structures. The pore pressure in the soil increases under repeated earthquake forces. Consequently, the effective stress decreases and if its value approaches zero, the deposit loses its strength completely. Even if the effective stresses do not reduce to zero, the ground becomes very soft and large strains, deformations and lateral flows result.

Castro (1975) distinguishes two different phenomena that were referred to as 'liquefaction' and 'cyclic mobility'. Liquefaction was referred to as complete loss of shear strength and can occur only in loose sands. Cyclic mobility is the gradual increase of cyclic strains without complete loss of shear strength. It could occur in loose to medium dense sands. The different types of ground failures caused by liquefaction may be defined or identified, and vary in intensity, placement in environment, and consequence. They are sand boils, lateral spreads, flow failures, loss of bearing strength and ground oscillations (Ishihara, 1993).

The liquefaction potential of any given soil deposit is determined by a combination of the properties of soils, environmental factors and characteristics of the earthquake to which they may be subjected to. Simplified techniques based on an in situ testing measurement index are commonly used to assess seismic liquefaction potential. Liquefaction potential can be evaluated based on various in-situ tests like standard penetration tests, cone penetration tests, shear-wave velocity measurements, and Becker penetration test for gravely soils. Most of the simplified charts or equations rely on the analyses of liquefaction case histories. Using empirical, simple regression, or statistical methods, a boundary (liquefaction curve) or classification technique is used to separate the occurrence or non-occurrence of liquefaction. There are different probabilistic and deterministic approaches also in practice for liquefaction hazard assessment. A detailed summary on the different liquefaction evaluation methods and information on magnitude scaling factors, correction factors for overburden pressure, slope of the ground, and input values for earthquake magnitude and peak acceleration etc. are presented by Youd et al. (2001) and Seed et al. (2001).

LIQUEFACTION MITIGATION METHODS

One redeeming feature of earthquakes is that with proper planning and implementation of the appropriate measures using the available technical expertise most of the detrimental effects can either completely be prevented or to a very large extent mitigated.

The basic approach for earthquake disaster mitigation can be broadly classified into two major categories, viz., 1. Prevention and 2. Minimization of damage. The former can further be divided into measures that will improve (a) ground characteristics and (b) condition of stress, deformation and pore pressure in the ground. The ground can be engineered to withstand the expected intensity of earthquake by (i) densification, (ii) solidification, (iii) replacement of part of the soil or (iv) by lowering of the ground water table. The in situ stresses, formation of porewater pressure conditions can be controlled by (v) increasing the effective stress, (vi) permit dissipation of excess pore pressure or (vii) restrict shear deformations (PHRI 1997, JGS 1998).

If liquefaction cannot be prevented, its effect on damaging the structures founded on ground that possibly gets liquefied can be minimized. Basically, there are three possibilities to reduce liquefaction hazards when designing and constructing new structures. The first one is to avoid construction on liquefaction susceptible soils; the second one is to build liquefaction resistant structures. The third possibility is to strengthen the ground, by improving the strength, density, and/or drainage characteristics of the soils, using a variety of ground improvement techniques. The third one is considered to be the best strategy and is the most preferred choice. Resistance to liquefaction can be improved by increasing the density, modifying the grain size distribution, stabilizing the soil fabric, reducing the degree of saturation, dissipation of the excess pore pressures generated and intercepting the propagation of excess pore pressures, etc.

Ground improvement techniques are employed to mitigate liquefaction hazards. Most common methods to improve the engineering properties of the soils can be classified as densification, reinforcement, grouting/mixing and drainage. Out of the various ground treatment methods, Granular Drains/Piles (GP) are the most widely preferred alternative all over the world, due to technical feasibility, low energy utilization and cost effectiveness. Provision of granular piles/stone columns is the most commonly adopted ground treatment methodology for liquefaction mitigation which has proved its effectiveness in many instances (Mitchell & Wentz, 1991). GP improve the ground by reinforcement, densification of the surrounding soil and by providing drainage.

GRANULAR PILES, INSTALLATION, DESIGN METHODOLOGIES

Columnar inclusions such as stone columns/granular piles have been used as a ground improvement technique to increase bearing capacity, reduce settlement, increase the time rate of consolidation, improve stability and resistance to liquefaction of soft ground. Ground improvement by means of granular piles/stone columns/geopiers, which is associated with partial substitution of the in–situ soil, originated in sixties. Stone columns generally use gravel or crushed stone as backfill. Theoretical background, analysis, design aspects and installation techniques were being developed since 1970s by various researchers and practicing engineers all over the world (Hughes & Withers, 1974; Engelhardt & Golding, 1975; Madhav et al. 1979; Datye & Nagaraju, 1981; Priebe 1998 etc.). Abhishek et al. (2016) critical review of developments on various aspects related to granular piles and their seismic responses along with case histories to illustrate their versatility.

Various techniques of installation have been conceived for various types of columnar inclusions in a wide variety of soils such as loose sandy to soft compressible soils depending on technical ability, efficiency and local conditions. Effect of method of installation, cased and uncased holes, number of lifts and magnitude of compactive energy per lift given to granular piles and pile spacing were discussed by Madhav and Thiruselvam (1988).

Granular piles are installed by vibro-compaction, vibro-replacement, cased bore hole (rammed stone columns/RGP) or by simple auger boring methods (Datye & Nagaraju, 1981; Balaam & Booker, 1981; Barksdale & Bachus, 1983; JGS, 1998). The vibro-compaction and vibro-replacement methods utilize a vibroflot to densify cohesionless granular deposits. The vibroflot sinks into the ground under its own self-weight aided by vibrations and water jets. The vibrator in the vibroflot induces vibrations into the ground through which the loose deposits get densified. In case of fine-grained soils, the densification effect is negligible and the cavity formed by the vibroflot is filled with granular material and thus the original soil is replaced with granular material/crushed stone material. In the vibro-replacement method cavity created by the vibrofloat is replaced with crushed stone material. There are mainly two types of construction methods in practice for vibro stone columns, top feed method (wet method) and bottom feed method (dry method). Figure 1 shows the typical construction sequence of the top feed method (Hayward Baker, 2004). Vibro-compaction method is applicable for soils with less than 18 to 20% fines content. The range of soil types treatable by different vibro methods are shown in Figure 2 (Sondermann & Wehr, 2004). Vibroflotation technique was widely used in Europe and U.S.A for stone columns installation (Engelhardt & Golding 1975; Barksdale and Bachus 1983). Vibro-compozer method was used in Japan for the installation of granular piles (Aboshi et al. 1979). In India, cased borehole method

(Figure 3) or ramming stone column method (Datye & Nagaraju, 1981) has been used for the construction of granular piles. In this method, RGP are installed into the ground by partial or full displacement methods and by ramming in stages, using a heavy falling weight, within a 'pre-bored casing' or 'driven closed end casing', retracting the casing pipe stepwise. In the latter case, driving of closed end tube itself densifies the surrounding soil. Ramming of granular piles further densifies and reinforces the ground. Granular material is dumped into the hole and rammed by a 15 to 20 kN rammer in lifts of 1.0 to 1.5 m keeping the drop height of the rammer is 1.5 to 2.0 m. After the formation of the full length of the granular pile, it is further rammed with a heavier weight of about 40 kN to recompress the same.

Stone columns/granular piles can be adapted to a wide variety of soils as ground improvement method. Loose sandy soils, silty or clayey sands and soils with undrained shear strength within the range 7 to 50 kPa can use stone columns. However, clays and silts with high sensitivity (sensitivity ≥ 4) which lose strength when subjected to vibrations are not suited for stone column improvement (IS 15284).

Design of granular piles/stone columns involves the consideration of different mechanisms that are expected to function as per the improvement programme. The primary mechanisms to consider for the design are densification, reinforcement and drainage. The diameter of the granular pile, spacing between the granular piles, pattern of the piles installation, information about the length of the pile (termination depth) and gradation details of the granular material etc. need to be defined in the design process. Barksdale and Bachus (1983) presented various design aspects to be

Figure 1. Vibro-replacement method of construction
After Hayward Baker 2004.

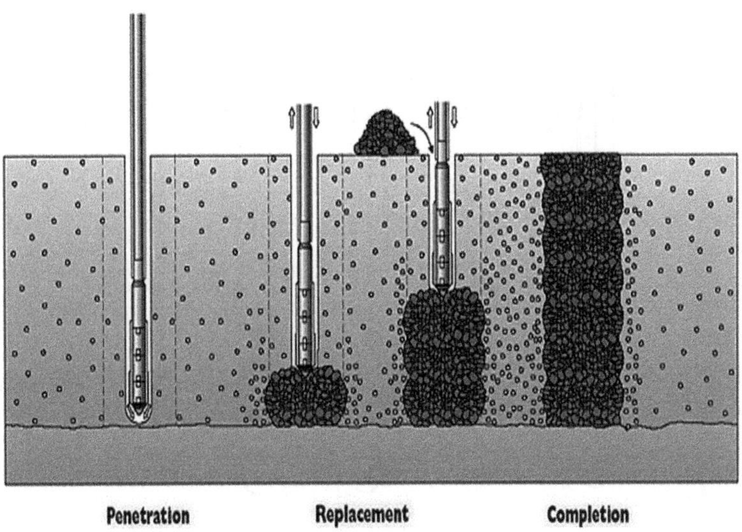

Figure 2. Range of soils treatable by different vibro methods
After Sondermann and Wehr 2004.

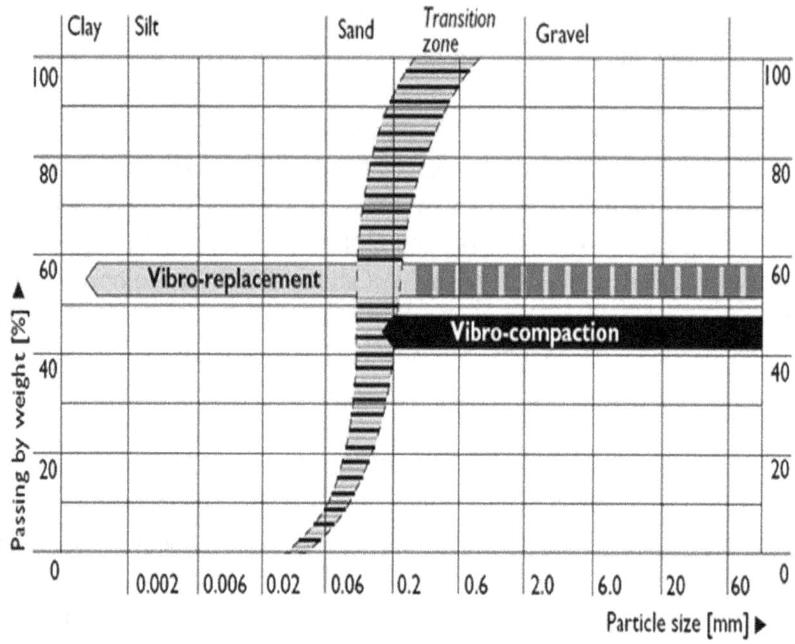

Figure 3. Installation of stone column using bored piling equipment and ramming
After Datye and Nagaraju 1981.

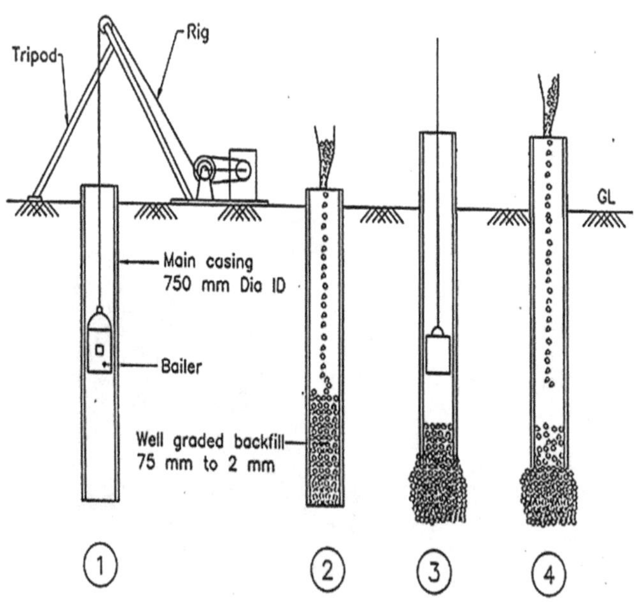

considered while designing the stone columns. Priebe (1995, 1998) demonstrated the design concepts for vibro replacement. Priebe (1998) proposed a relatively simple procedure to evaluate the improved liquefaction resistant capacity of vibro columns installed ground, based on stress transfer mechanism. The method did not consider the drainage mechanism of vibro columns.

IS 15284 and JGS 1998 are the other two good resources to have the concepts for granular piles design. Design curves developed by Seed and Booker (1977) are used in general for sizing the granular piles as a liquefaction remediation measure (Fig. 4) considering the drainage mechanism. To consider the densification effect in connection to liquefaction mitigation, improved SPT N values or cone tip resistance values could be correlated to cyclic resistance ratio and design accordingly (Baez, 1995). Selection of gravel material for granular pile construction is very important as the performance of improved system governed by the granular material. In general, coarse, open-graded stone is used, of size varying from about 12 to 75 mm (Barksdale & Bachus, 1983). However, the actual size of the granular material depends mainly on the gradation of in-situ soil. Saito et al. (1987) proposed a formula for selection of the size of the material for stone column in relation to grain size of the surround in-situ soil as:

$$20D_{s15} < D_{G15} < 9D_{s85}$$

where D_{s15} and D_{G15} are the size/diameter (mm) of soil particles and gravel material, respectively, passing 15 percent and D_{s85} is the diameter of soil passing 85 percent in a grain size analysis test.

GRANULAR PILES AS LIQUEFACTION COUNTER MEASURE, MECHANISMS

Numerous publications (Barksdale & Bachus, 1983; Sasaki & Taniguchi, 1982; Munfakh et al., 1987; Iai et al. 1988, Baez & Martin, 1992; Lopez & Hayden, 1992; Brennan & Madabhushi, 2002; Adalier & Elgamal, 2004 & Murali Krishna et al. 2006; Madabhushi 2007 etc.) describe the use of stone columns for ground reinforcement and their potential to mitigate the liquefaction. Liquefied and non-liquefied subsoil conditions of two reclaimed islands in Kobe City after the 1995 Hyogoken-Nambu earthquake were investigated by Yasuda et al. (1996) who identified that subsoils treated with granular inclusions did not liquefy and nor subside even though the earthquake shaking was very strong.

Figure 4. Effect of drain diameter and drain spacing on maximum pore pressure ratio
Adapted from Seed and Booker 1977.

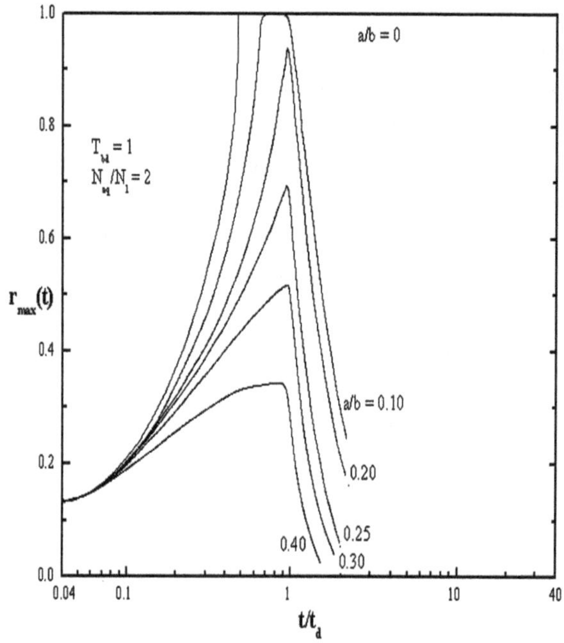

Adopting ground treatment by stone columns as a method to determine the stability of potentially liquefiable sand deposits was initiated originally by Seed and Booker (1977). Granular pile inclusions improve the deformation properties of the ambient soil. Granular piles help in mitigating earthquake induced liquefaction effects through one or more of the following functions.

1. Granular piles function as drains and permit rapid dissipation of earthquake induced pore pressures by virtue of their high permeability with the additional advantage that they tend to dilate as they get sheared during an earthquake event.
2. Granular piles densify and reinforce the in-situ soil;
3. Granular piles, installed into a very dense state, are not prone to liquefaction and replace a significant quantity of in-situ liquefiable soil; and
4. Granular piles modify the nature of earthquake experienced by the in-situ soil.

The following sections discuss each of these mechanisms in detail.

Drainage Mechanism of Stone Columns

Due to installation of stone columns, the generated porewater pressure due to repeated loading may be dissipated almost as fast as they are generated Seed and Booker (1977). In cases of high liquefaction potential the installation of stone columns may well provide an efficient method for preventing the development of excessively high pore water pressures. Generally, the horizontal permeability of a sand/gravel is several times greater than its vertical permeability. As such the spacing between vertical drains can be made less than the distance required for water to drain vertically to a free surface. Hence, the drainage mechanism by stone columns is attributed to their high permeability and the ability to reduce the drainage path significantly so that the developed excess pore water pressures are dissipated almost as fast as they are generated.

Seed and Booker (1977) applied the one-dimensional theory of porewater pressure generation and dissipation developed by Seed et al. (1975) to the analysis of columnar gravel drains under a variety of earthquake conditions and proposed a simple radial consolidation analytical model. For flow into a stone column, assuming pure radial flow, and constant coefficients of permeability (k_h) and volume compressibility (m_v), the governing equation for the phenomenon can be written as (Seed & Booker, 1977):

$$\frac{k_h}{\gamma_w \cdot m_v}\left(\frac{1}{r}\frac{\partial u}{\partial r}+\frac{\partial^2 u}{\partial r^2}\right)=\frac{\partial u}{\partial t}-\frac{\partial u_g}{\partial N}\cdot\frac{\partial N}{\partial t} \tag{1}$$

Seed and Booker (1977) presented the charts (Figure 4), for the variation of maximum pore pressure ratio against time during an earthquake for different area ratios and other earthquake parameters, to aid in the design of stone columns for liquefaction mitigation. These charts are based on the assumption that the drains are infinitely permeable so that the excess pore water pressure in the drain is effectively zero and the coefficient of permeability of the ambient soil is constant.

Tokimatsu and Yoshimi (1980), Sasaki and Taniguchi (1982) and Onoue (1988), report results similar to those of Seed and Booker (1977) taking into consideration additional factors such as well resistance (finite permeability of stone column) and column slenderness ratio (slenderness ratio: *L*/*a*, where *L* is the length, and *a* is the radius of the stone column). Iai and Koizumi (1986), Onoue et al. (1987) and Iai et al. (1988), present design procedures for stone columns derived analytically and verified by model or in situ tests. All of the above studies considered only the drainage effect of stone columns (Adalier & Elgamal, 2004).

Poorooshasb et al. (2000) demonstrated the effectiveness of inclusion of stone columns in reducing the risk of liquefaction of very loose to loose sandy and silty sand layers using the concept of equivalent permeability. Based on the radial consolidation theory as applicable to granular pile treated ground, Poorooshasb et al. (2000) proposed an equivalent coefficient permeability, k_{eq}, for the treated soil in terms of the permeability, k_h, of untreated ground, as $k_{eq} = k_h . t_{50(untr)} / t_{50(tr)}$ where $t_{50(untr)}$ and $t_{50(tr)}$ are the times for the untreated and granular pile treated ground for 50% degree of consolidation, respectively, based on one dimensional and radial consolidation theories. The ratio, r ($=k_{eq}/k_h$) is derived (Figure 5) in terms of the spacing ratio, n ($=S/d$) where S and d are respectively the spacing between and diameter of granular piles. The results presented were for triangular arrangement of drains but can easily be modified for square arrangement. Krishna et al. (2018) presented short summary of recent developments in the design of stone columns for liquefaction mitigation considering drainage along with two case studies to demonstrate the effectiveness of vibro stone columns for improving bearing capacity and liquefaction potential mitigation.

Thus, the principal mechanism of granular drains/stone columns in mitigating the liquefaction potential is their drainage effect.

Figure 5. Equivalent permeability concept
After Poorooshasb et al. 2000.

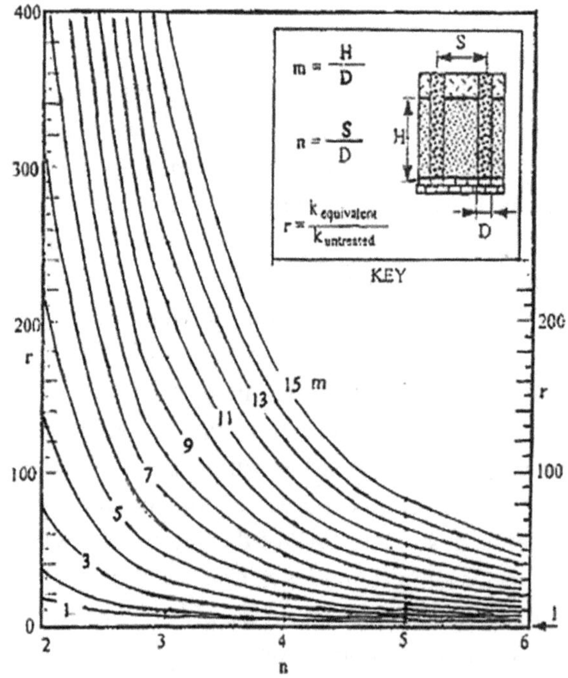

Storage Effect of Stone Columns

The effect of storage capacity on vertical drain performance in liquefiable sand deposits is presented by Pestana et al. (1998). Generally, in most practical cases, the water level is at some depth below ground surface. During an earthquake, excess pore water pressure is dissipated by water discharging into the vertical drains, causing a rise in the water level within the drains. This rise in water level results in an increased "static" excess pore pressure or backpressure inside the drain elements, which in turn increases the excess pore pressure in the soil deposit. However, the accumulated water within the drain can lead to a significant retardation of flow into the drain. Pestana et al. (1997) recommended providing a larger diameter of the drain, filled with crushed rock or gravel, near the ground surface to act as a reservoir for flow rising out of the drain during earthquake. A reservoir (natural or man-made) with additional storage capacity above the original water table provides a buffer storage capacity and thus reduces the rate of rise of water level, thus minimizing backpressure effect in the drain (Pestana et al. 1998).

Pestana et al. (1998) considered a two-layer soil system with very low permeability clay overlying the liquefiable sand layer as shown in the Figure 6. Figure 7 shows average pore pressure ratios, $Ru_z(max)$, versus cycle ratio, r_N, for different s/d ratio (s is diameter of the unit cell and d is diameter of the granular pile) for a perfect drain with and without the storage capacity. Here the perfect drain implied the drain without any well resistance as that of the Seed and Booker (1977) model. The well resistance is a phenomenon in which limited permeability in the drain and in the close vicinity of the drain hinders, to some extent, the dissipation of excess pore water pressure. If the water table in the system is not at the ground surface, there is room within the drain for water to rise that represents the storage capacity. In the figure for all storage cases (*storage*, in the legend), the water level in the drain was initially at 1 m depth. Ru_z is the average pore pressure ratio at depth, z, in the liquefiable layer. As the upper soil layer acts as a barrier to vertical flow, the maximum value of Ru_z, $Ru_z(max)$, occurs at the interface between the upper and lower soil layers. The cyclic ratio, r_N, is the ratio of the number of uniform stress cycles in an earthquake versus the number of uniform stress cycles required to cause liquefaction in the soil layer in question under undrained conditions. The figure shows that at a small drain spacing (s/d =4), the rise in the water level in the drain forces a steady rise in the pore pressure ratios in the drain compared to the case with no storage. Similarly, in the other cases (s/d =6, 8 and 10), the effect of storage is to increase the maximum pore pressure ratios.

Pestana et al. (1997, 1998) analysed the use of providing the reservoir in reducing the depth of water levels within the drain and also to minimise the drain resistance to flow into the drain. The reservoir may be formed by using a composite drainage

Figure 6. Unit cell with EQ drain
After Pestana et al. 1998

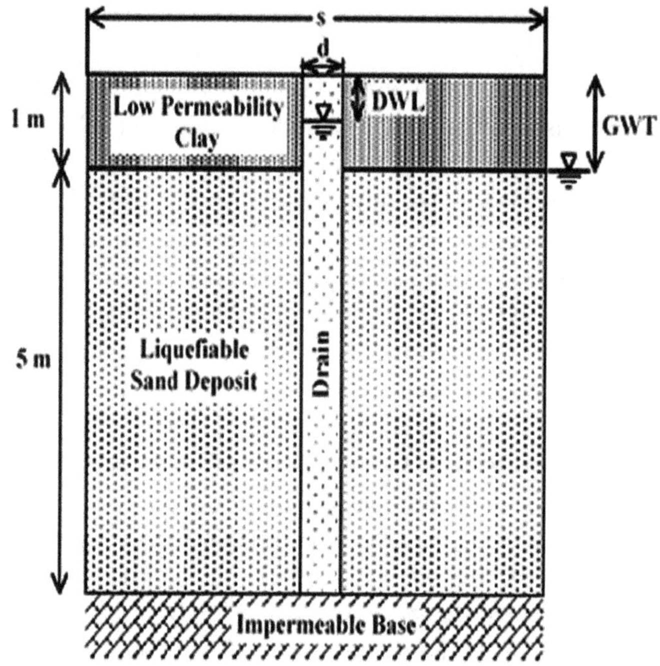

Figure 7. Pore pressure ratio variation with drain spacing with and without storage
After Pestana et al. 1998.

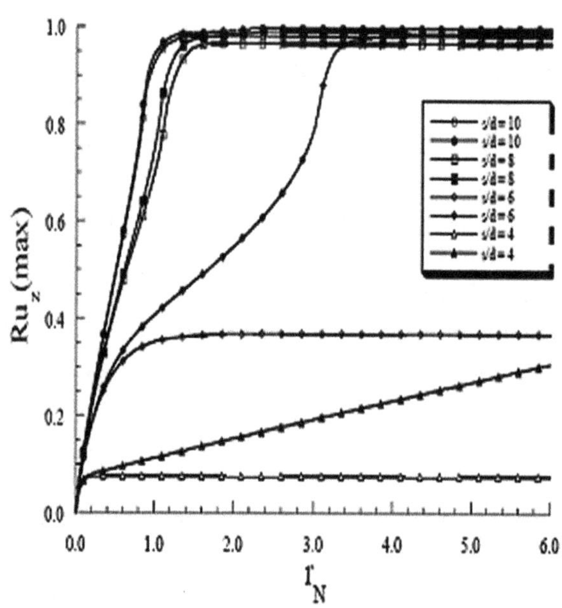

product consisting of a perforated geopipe surrounded by a filler fabric to prevent clogging of the orifices (Pestana et al. 1998). This could achieved by a technique of auguring a hole larger than the diameter of the drain near the ground surface and filling it with crushed rock or gravel. This creates a reservoir for flow rising out of the drain during an earthquake, and at the same time, provides a means for reducing the rise in water level within the drain as the same volume of water is now spread out over a lager area (Figure 8) as the equivalent area of the reservoir is larger than that of the drain.

Dilation Effect of Stone Columns

It has been generally recognized that the susceptibility of a given soils to liquefaction is to a high degree governed by its void ratio or relative density. For any given earthquake loose sands are very likely to liquefy but the same granular material in a denser condition would not (Seed & Idriss, 1971). Shearing of dense dilative soils may generate some small positive pore pressures at small strains. However, at larger strains, the pore pressures decrease and become negative as the soil grains move up over one another, tending to cause an increase in soil volume (dilation). For dense, saturated sands when sheared without drainage, the tendency for dilation or volume increase results in generation of negative pore pressure, with consequent increases in the effective stress and the shear strength of the granular material. While positive pore pressures are generated in loose sands, generation of very high negative pore pressures can be observed due to suppression of the tendency for dilation in medium dense and dense sands. Loose samples undergo volume decrease while dense samples

Figure 8. Rise in water level inside the drain and reservoir
After Pestana et al. 1997.

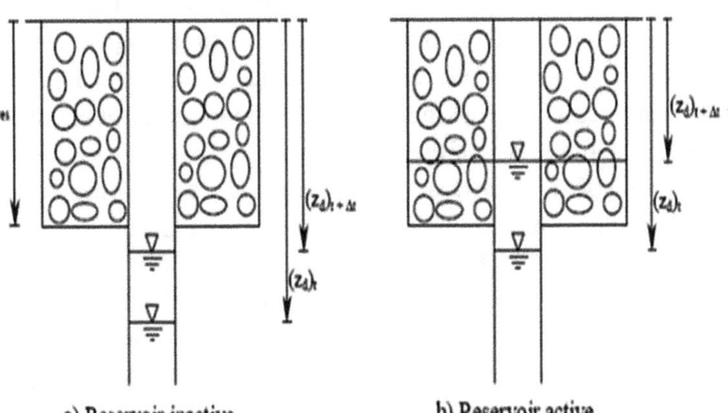

experience volume increase (dilation) during shearing (Leonards, 1962; Vaid et al. 1981). The rate of dilation increases with relative density.

Dilation effect on the drainage function of granular piles was studied by Madhav and Arlekar (2000) by extending the Seed and Booker (1977) model. Dilation effect is included in the analysis by adopting a dilation coefficient (d_c). The dilation effect on pore pressure dissipation by granular piles for the range of parameters (d_c = 0 (no dilation), 2 and 5) considered is shown in Figure 9. Granular piles installed in loose sand deposits are often compacted at a relatively high density. Seismic forces which tend to generate positive pore pressures in these deposits cause an opposite effect of dilation in the dense granular piles. During the seismic event, negative pore pressures that tend to get generated therein increase the gradient and permit rapid rates of drainage than otherwise. Madhav and Arlekar (2000) quantify this effect of dilation in granular pile on mitigating the liquefaction damage. The variation of maximum pore pressure ratio, W_{max}, with normalized time (T, t/t_d) for different rates of dilation is depicted in Figure 10. The figure shows the plots different T_{bd} values, a dimensionless time factor which is a function of flow parameters of soil and duration of earthquake, t_d (Equation 2). The effect of dilation rates in reducing the peak values of the pore pressure ratios for different T_{bd} values can be clearly seen in the figure.

$$T_{bd} = \left(\frac{k_h}{\gamma_w}\frac{t_d}{m_v}\right) \cdot \frac{1}{b^2} \tag{2}$$

Densification Effect of Stone Columns

One of the chief benefits of ground treatment with granular piles is the densification of in situ ground. The densification effect can be easily but indirectly quantified by in situ tests. The effect of densification is manifested through an increase in the coefficient of earth pressure at rest and in the values of modulus of deformation of the soil. The densification is maximum close to the granular pile and least at the farthest point, i.e. at the center of the grid points.

Ramming action of the vibrofloat/falling weight tends to ram the stone into sides of the hole, densifying while reinforcing the ground in addition to compacting the stone column substantially. Thus, densification and reinforcement effects cause modifications in the properties of the in situ soil. This densification and modification effects on the soil parameters of the ground are not uniform over the entire zone of surrounding ground but are functions of the distance from the point of densification. During the process of installation of GP, the soil adjacent to and in the vicinity of

Figure 9. Effect of dilation of granular material on pore pressure generation and dissipation
After Madhav and Arlekar 2000.

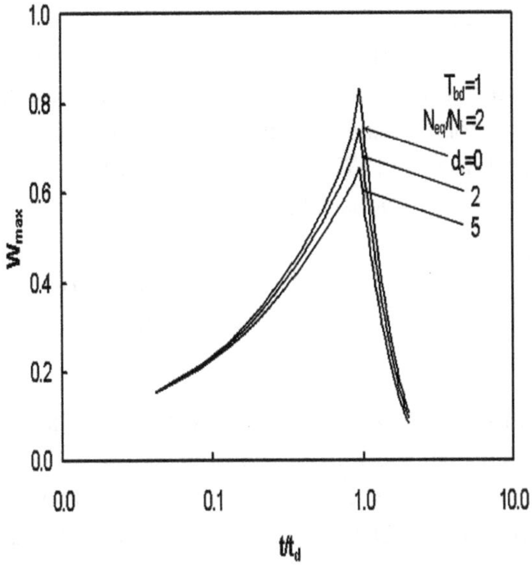

Figure 10. Maximum pore pressure versus time: Effect of dilation and T_{bd}
After Madhav and Arlekar 2000.

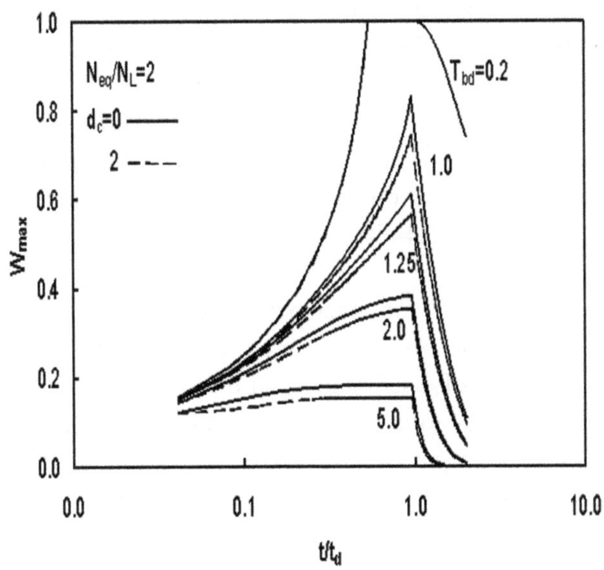

the point of treatment gets densified most. This densification effect decreases with the distance from the point of densification. Densification by RGP causes increase in deformation moduli and decrease in the coefficients of permeability and of volume change (Murali Krishna et al. 2007).

Murali Krishna et al. (2006) incorporated the densification effect of granular piles, with respect to the variation of flow parameters from the centre of the granular pile, in the analysis of pore pressure generation and dissipation that was originally developed by Seed and Booker (1977). The modified form of the governing Equation 1 with the inclusion effects of densification is:

$$\frac{k_h(r)}{\gamma_w \cdot m_v(r)}\left(\frac{1}{r}\frac{\partial u}{\partial r}+\frac{\partial^2 u}{\partial r^2}\right)+\frac{1}{\gamma_w \cdot m_v(r)}\cdot\frac{\partial(k_h(r))}{\partial r}\cdot\frac{\partial u}{\partial r}=\frac{\partial u}{\partial t}-\frac{\partial u_g}{\partial N}\cdot\frac{\partial N}{\partial t} \tag{3}$$

In this case coefficients of permeability and volume change, $k_h(r)$ and $m_v(r)$, are functions of radial distance r from the point of densification and degree of densification. Murali Krishna et al. (2006) studied the densification effect with respect to the coefficients of permeability and volume change at the near and at the farthest ends of the granular pile, individually and together, on maximum pore pressure variations during an earthquake event. Figures 11 and 12 show the densification effect on maximum pore pressure ratio with respect to coefficient of volume change at the near (R_{ma}) and farthest (R_{mb}) ends respectively. R_{ma} ($=m_v(a)/m_v$) and R_{mb} ($=m_v(b)/m_v$)are the normalized coefficients of volume change due to densification at the near and farthest ends respectively. Coefficients of volume change at the near and far end, after the treatment, are normalized by the in-situ coefficient of volume change, m_v (i.e., $R_{ma}=m_v(a)/m_v$ *and* $R_{mb}=m_v(b)/m_v$). Figures show that densification effect minimises the pore pressure ratios. Figures 13 and 14 show the densification effect on maximum pore pressure ratio with respect to both the coefficients of permeability and volume change at the near and farthest ends respectively.

Krishna and Madhav (2008) combined both the densification and dilation effects and incorporated them in the analysis of pore pressure generation and dissipation. Figure 15 shows the effect of densification with respect to flow parameters at the near end in addition to the dilation effect. It is seen from the figure that the dilation effect ($d_c = 2$) reduces the negative effect of the densification effect.

Densification effect on the coefficient of volume change is positive in that the maximum induced pore water pressure ratios get reduced. Densification effect, on the coefficient of permeability alone or in addition to effect on coefficient of volume change, increases the maximum pore water pressure ratios giving a negative effect.

Figure 11. Effect of densification at near end (R_{ma}) on W_{max}
After Murali Krishna et al. 2006.

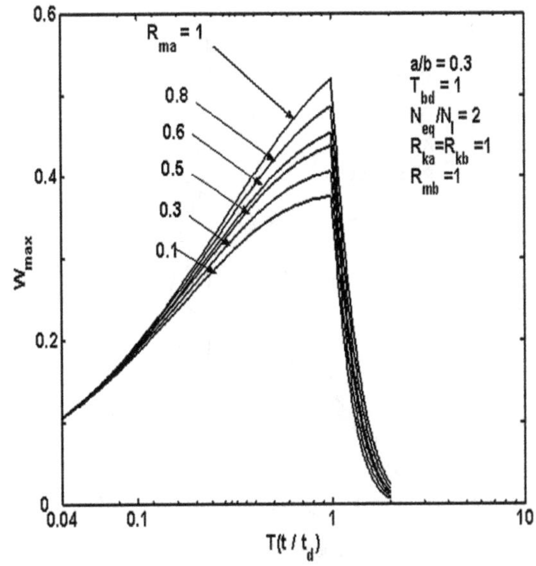

Figure 12. Effect of densification at farthest end (R_{mb}) on W_{max}
After Murali Krishna et al. 2006.

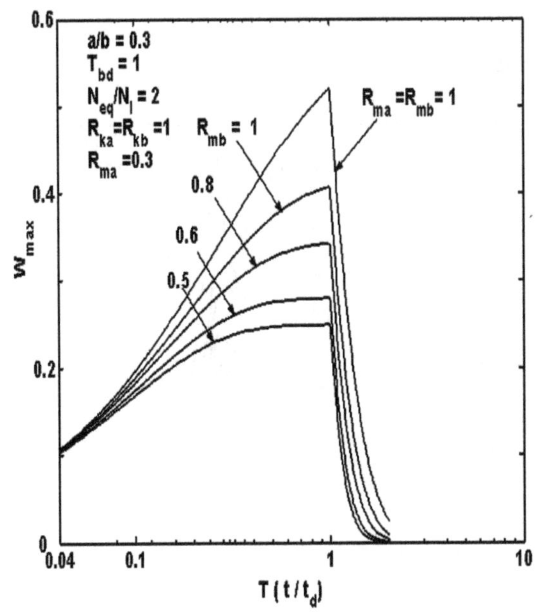

Figure 13. Effects of R_{ma} and R_{ka} on W_{max}
After Murali Krishna et al. 2006.

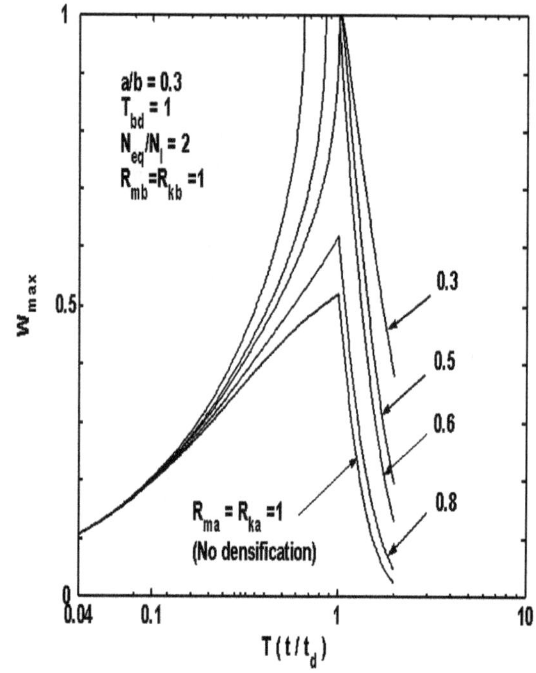

Figure 14. Effects of R_{mb} and R_{kb} on W_{max}
After Murali Krishna et al. 2006.

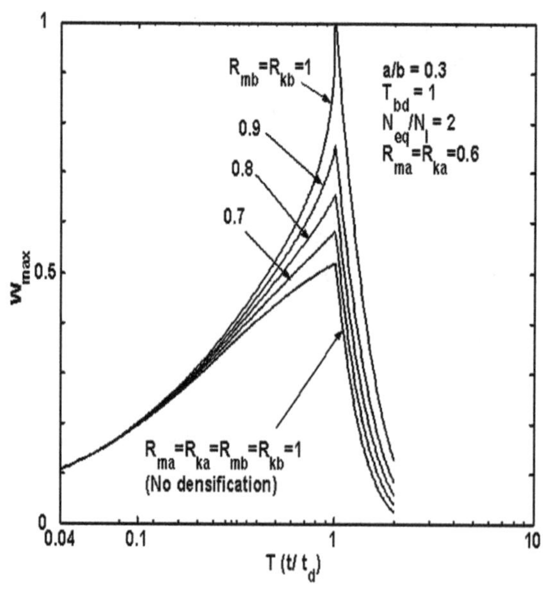

Densification effect on both coefficients of permeability and volume change result in a either slightly negative or positive effect depending on the degree of densification.

The negative effect of the densification (reduction in permeability) is offset by the dilation effect thus proving the effectiveness of granular piles in liquefaction mitigation. It is recommended that the densification and dilation effects should be considered while designing granular pile/stone column treatment for liquefaction mitigation.

Bouckovalas et al. (2009, 2011) considered sand fabric evolution effects on drain design for liquefaction mitigation. They proposed that overlooking the shakedown effects of fabric evolution during cyclic loading underestimates the effectiveness of gravel drains. Krishna et al. (2014) combined the soil fabric evolution effect and the densification effect in analyzing the pore pressures for the granular pile reinforced ground.

Reinforcement Effect

Granular piles, installed into a very dense state, are not prone to liquefaction and replace a significant quantity of in situ liquefiable soil and reinforce the same. The very high deformation modulus and stiffness of the granular pile material provide reinforcement for the in situ soil and offer another mechanism to mitigate liquefaction. Stone columns installed in loose sands are effective in damage control as they are installed to a relative density (more rigid/stiff) well above the density of the surrounding soil and hence increase the ability of the layer to withstand the earthquake forces more efficiently.

Poorooshasb et al. (2006) and Noorzad et al. (2007) demonstrated the reinforcement effect of stone columns while analyzing their performance during an earthquake. They proposed that the seismic load imposed on the soil is shared between the stone column and the surrounding ground and stone column carries the major load. The geometry of the stone column installation and the free body diagram of the load distribution as a result of reinforcement are shown in the Figure 16. It was concluded that the increase in diameter of the columns shows dramatic effect on the performance of system in terms of the development of the excess pore water pressure and the deformations get reduced significantly. Noorzad et al. (2007) presented the performance of partially penetrating stone columns during earthquake on similar lines.

Figure 15. Effect of densification with respect to R_{ka} & R_{ma} and dilation on W_{max}
After Murali Krishna and Madhav 2008.

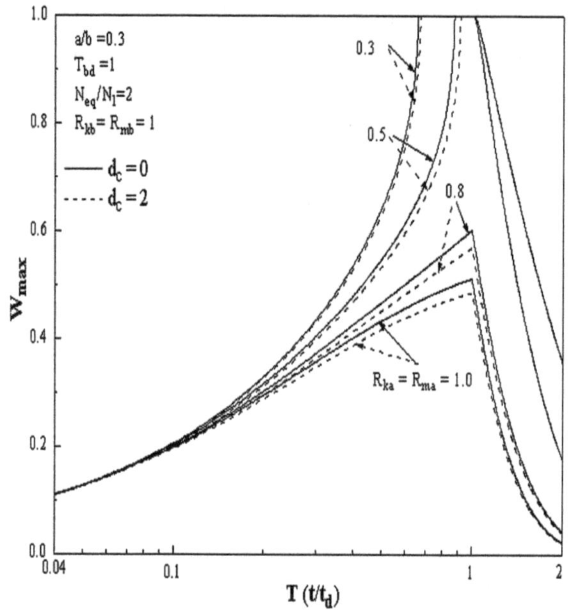

Figure 16. Stone column configuration considered for reinforcement effect a) Definition sketch, b) FBD of the Unit Cell
After Poorooshasb et al. 2006.

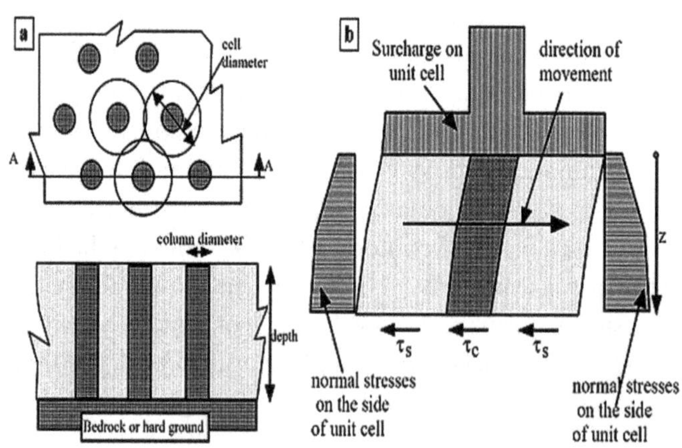

CASE HISTORIES OF GRANULAR PILES AS LIQUEFACTION COUNTER MEASURE

Many case histories are reported in the literature on the performance of granular piles as liquefaction counter measure (Englehardt & Golding, 1975; Barksdale & Bachus, 1983; Dobson, 1987; Mithcel & Wntz, 1991; Andrus & Chung, 1995; Kerwin & Stone, 1997; Luehring et al. 2001; Wijewickreme & Atukorala, 2005). Mitchell and Wentz (1991) evaluated twelve sites treated with different improvement methods prior to the 1989 Loma Prieta earthquake and concluded that the ground improvement by granular columns was very effective in mitigating liquefaction. Adalier and Elgamal (2004) summarized some of the field case histories on the use of stone columns as liquefaction counter measure in tabular form.

To reduce overall settlement and to mitigate liquefaction potential in a possible seismic event for Hazira liquefied natural gas (LNG) terminal, located at an estuary on the west coast of the Khambhat Gulf in India, vibro stone columns were used (Raju & Sondermann, 2005). Figure 17 shows the two LNG tanks of diameter 84 m each founded on stone columns. The subsoil at site consists of loose silty sands to a depth of 16 m followed by very dense sandy layers with SPT N more than 50. Stone columns of diameter 1m and 16 m length were installed using the wet top feed method.

Mejia (2005) discussed about the use of stone columns for seismic retrofit against liquefaction for Tolt Regulating Basin South Dam (Fig. 18). The Tolt Regulating Basin South Dam is an 11-m-high embankment dam located about 50 km east of Seattle in western Washington that was build in the early 1960's. The embankment is founded on a 6-m-thick mat of medium dense to dense sandy gravel, placed to spread the load over a deposit of soft to stiff silty clay up to 12 m thick. It was concluded that, based on deterministic and probabilistic analysis of the seismic hazard at the site, an earthquake of magnitude 6½ at a distance of 10 km could affect the dam safety. Hence, the seismic retrofit consisted of placing a buttress on the downstream slope of the dam and strengthening the foundation beneath the buttress with stone columns was designed and successfully completed in 1995.

Al-Homoud and Degen (2006) presented a case history (Figure 19) demonstrating the use of marine stone columns for liquefaction remediation at the Patras Harbor extension project. At the Petras Harbor breakwater, the depth to seabed is 34 m that is underlain by 14 to 16 m of soft soils. In the neighborhood of the new quay and seawall, a similar structure existed in the past but was destroyed by a slope failure occurred within the upper 20 m of loose silty sand and sandy silt during due to soil

Figure 17. Use of vibro stone columns to mitigate liquefaction potential at Hazira LNG terminal, India
After Raju and Sondermann 2005.

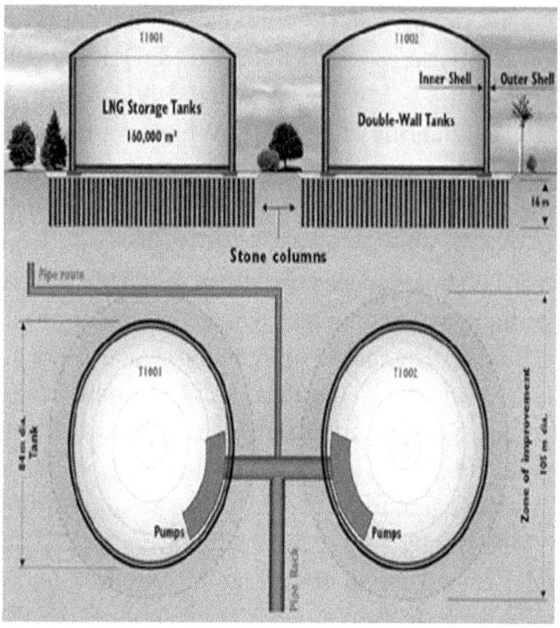

Figure 18. Use of stone columns for seismic retrofit against liquefaction for Tolt Regulating Basin South Dam
After Mejia 2005.

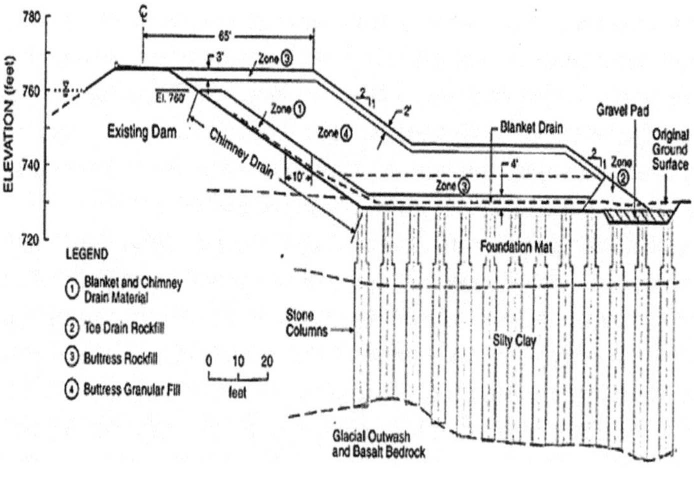

Figure 19. Use of stone columns for Patras Harbor extension project
After Al-Homoud and Degen 2006.

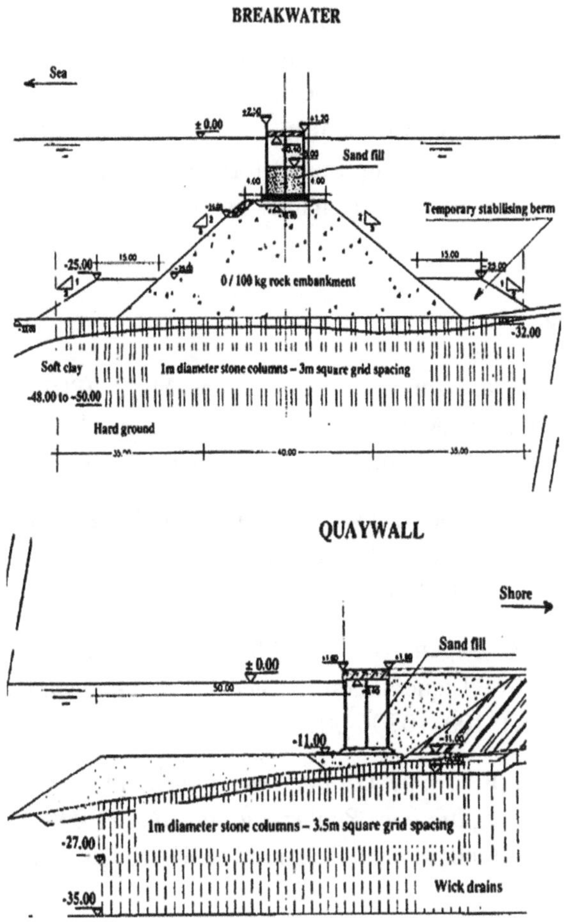

liquefaction developed during an earthquake. Based on this past experience, the new design incorporated soil improvement by a combination of wick drains and stone columns. Stone columns of 1 m diameter in a 2.7 to 3.3 m grid that extend 20 m into the soft silty and clayey marine sediments were constructed for improving the liquefaction mitigation.

CONCLUSION

Liquefaction is the most hazardous damage during an earthquake. Among various remedial measures available, installation of stone columns/granular piles is the most widely adopted method for liquefaction mitigation. Granular piles provide drainage facility, densify the surrounding soil and reinforce the in-situ ground. Different mechanisms operate in the functioning of the stone columns during the process of liquefaction mitigation. The major fundamental mechanism is drainage by which developed excess pore water pressures get dissipated almost as fast as they get generated. The other mechanisms include storage, dilation, densification and reinforcement. While studying the performance of the gavel drain system all the mechanisms along with their mechanics are to be considered.

REFERENCES

Abhishek, S. V., Rajyalakshmi, K., & Madhav, M. R. (2016). Engineering of ground with granular piles: A critical review. *International Journal of Geotechnical Engineering*, *10*(4), 337–357. doi:10.1080/19386362.2016.1145942

Adalier, K., & Elgamal, A. (2004). Mitigation of liquefaction and associated ground deformations by stone columns. *Engineering Geology*, *72*(3-4), 275–291. doi:10.1016/j.enggeo.2003.11.001

Al-Homoud, A. S., & Degen, W. S. (2006). Marine Stone Columns to Prevent Earthquake Induced Soil Liquefaction. *Geotechnical and Geological Engineering*, *24*(3), 775–790. doi:10.100710706-005-2783-5

Andrus, R., & Chung, R. (1995). *Ground improvement techniques for liquefaction remediation near existing lifelines. Report, Building and Fire Research Laboratory*. National Institute of Standards and Technology, U.S. Department of Commerce. doi:10.6028/NIST.IR.5714

Baez, J. I. (1995). *A design model for the reduction of soil liquefaction by vibro-stone columns* (PhD thesis). University of Southern California, Los Angeles, CA.

Baez, J. I., & Martin, G. R. (1992). Quantitative evaluation of stone column technique for earthquake liquefaction mitigation. *Proc. 10th World Conference on Earthquake Engineering*, 1477-1483.

Balaam, N. P., & Booker, J. R. (1981). Analysis of rigid raft supported by granular piles. *International Journal for Numerical and Analytical Methods in Geomechanics*, *5*(4), 379–403. doi:10.1002/nag.1610050405

Barksdale, R. D., & Bachus, R. C. (1983). *Design and Construction of Stone Columns*. Report No. FHWA/RD-83/026, U. S. Department of Transportation, Federal Highway Administration, Washington, DC.

Bouckovalas, G. D., Papadimitriou, A. G., & Niarchos, D. (2009). Gravel drains for the remediation of liquefiable sites: The Seed & Booker (1977) approach revisited. *Proceedings of the keynote lecture.*

Bouckovalas, G. D., Papadimitriou, A. G., Niarchos, D., & Tsiapas, Y. Z. (2011). Sand fabric evolution effects on drain design for liquefaction mitigation. *Soil Dynamics and Earthquake Engineering*, *31*(10), 1426–1439. doi:10.1016/j.soildyn.2011.05.019

Brennan, A. J., & Madabhushi, S. P. G. (2002). Effectiveness of vertical drains in mitigation of liquefaction. *Soil Dynamics and Earthquake Engineering*, *22*(9), 1059–1065. doi:10.1016/S0267-7261(02)00131-8

Castro, G. (1975). Liquefaction and cyclic mobility of saturated sands. *J. of the Geotechnical Engineering Division, ASCE.*, *101*(GT6), 551–569.

Datye, K. R., & Nagaraju, S. S. (1981). Design approach and field control for stone columns, *Proc.10th International Conf. on Soil Mechanics and Foundation Engineering*, *3*, 637-640.

Dobson, T. (1987). Case Histories of the Vibro Systems to Minimize the Risk of Liquefaction, Soil Improvement–A Ten Year Update. *ASCE Symposium*, 167-183.

Engelhardt, K., & Golding, H. C. (1975). Field Testing to Evaluate Stone column Performance in a Seismic Area. *Geotechnique*, *25*(1), 61–69. doi:10.1680/geot.1975.25.1.61

Hayward Baker. (2004). *Vibro systems*. Hayward Baker Inc. Retrieved from www.HaywardBaker.com

Hughes, J. M. O., & Withers, N. J. (1974). Reinforcing of Soft Cohesive Soils with Stone Columns. *Ground Engineering*, *7*(3), 42–49.

Iai, S., & Koizumi, K. (1986). Estimation of earthquake induced excess pore water pressure for gravel drains. *Proc. 7th Japan Earthquake Engg. Symposium*, 679-684.

Iai, S., Koizumi, K., Noda, S., & Tsuchida, H. (1988). Large scale model tests and analysis of gravel drains. *Proc., 9th World Conference on Earthquake. Engg.*, *3*.

Idriss, I. M., & Boulanger, R. W. (2005). Evaluation of Liquefaction Potential, Consequences and Mitigation. *Proc. Indian Geotechnical Conference*, 3-25.

IS 15284 (Part 1). (2003). *Design and construction for ground improvement — Guidelines*. Part 1 Stone Columns, analysis. Bureau of Indian Standards, New Delhi.

Ishihara, K. (1993). Liquefaction and flow failure during earthquakes. *Geotechnique*, *42*(3), 351–415. doi:10.1680/geot.1993.43.3.351

JGS. (1998). Remedial Measures against Soil Liquefaction. The Japanese Geotechnical Society.

Kerwin, S., & Stone, J. (1997). Liquefaction failure and remediation: King Harbor Redondo Beach, California. *Journal of Geotechnical and Geoenvironmental Engineering*, *123*(8), 760–769. doi:10.1061/(ASCE)1090-0241(1997)123:8(760)

Kramer, S. L. (1996). *Geotechnical Earthquake Engineering*. Upper Saddle River, NJ: Prentice Hall.

Krishna, A. M., Kumar, A. M., & Baruah, A. K. (2018). Stone Columns/Granular Piles for Improving Liquefiable Sites: Case studies. *Geotechnical Engineering Journal of the SEAGS & AGSSEA*, *49*(1), 136–142.

Krishna, A. M., & Madhav, M. R. (2008). Densification and dilation effects of granular piles in liquefaction mitigation. *Indian Geotechnical Journal*, *38*(3), 295–316.

Krishna, A. M., Madhav, M. R., & Kumar, K. (2014). Ground Engineering with Granular Inclusions for Loose Saturated Sands subjected to Seismic Loadings. *Indian Geotechnical Journal*, *44*(2), 205–217. doi:10.100740098-013-0085-z

Krishna, A. M., Madhav, M. R., & Latha, G. M. (2006). Liquefaction mitigation of ground treated with granular piles: Densification effect. *ISET Journal of Earthquake Technology*, *43*(4), 105–120.

Krishna, A. M., Madhav, M. R., & Latha, G. M. (2007). Densification effect of granular piles on settlement response of treated ground. *Ground Improvement*, *11*(3), 127–136. doi:10.1680/grim.2007.11.3.127

Leonards, G. A. (Ed.). (1962). *Foundation Engineering*. New York: McGraw-Hill.

Lopez, R. A., & Hayden, R. F. (1992). The Use of Vibro Systems in Seismic Design. Grouting, Soil Improvement and Geosynthetics, 30(2), 1433-1445.

Luehring, R., Snorteland, N., Mejia, L., & Stevens, M. (2001). Liquefaction Mitigation of a Silty Dam Foundation Using Vibro-stone Columns and Drainage Wicks: A Case History at Salmon Lake Dam. *Proc. 21st USSD annual meeting and lecture*.

Madabhushi, S. P. G. (2007). Ground improvement methods for liquefaction remediation. *Ground Improvement*, *11*(4), 195–206. doi:10.1680/grim.2007.11.4.195

Madhav, M. R., & Arlekar, J. N. (2000).Dilation of granular piles in mitigating liquefaction of sand deposits. *Proc. 12th World Conference Earthquake Engineering*.

Madhav, M. R., Iyengar, N. G. R., Vitkar, R. P., & Nandia, A. (1979). Increased Bearing Capacity and Reduced Settlements Due to Inclusions in Soil. *Proc. International Conference on Soil Reinforcement: Reinforced Earth and Other Techniques*, *2*, 329-333.

Madhav, M. R., & Thiruselvam, K. (1988). Effect of Installation Methods on Granular Pile Response. *Proc. Int. Symp. on Theory and Practice of Earth Reinforcement*, 221-226.

Martin, G. R., Finn, W. D. L., & Seed, H. B. (1975). Fundamentals of liquefaction under cyclic loading. *J. of the Geotechnical Engineering Division, ASCE*, *101*(GT5), 425–438.

Mejia, L. H. (2005). Seismic Retrofit of Embankment Dams to Mitigate Liquefaction: Selected Case Histories. In *Proc. Geo-Frontiers 2005 Congress*. ASCE Geotechnical Special Publication.

Mitchell, J. K., & Wentz, F. K. (1991). *Performance of improved ground during Loma Prieta earthquake*. Report No. EERC91/12. Earthquake Engineering Research Center, University of California, Berkeley, CA.

Munfakh, G. A., Abramson, L. W., Barksdale, R. D., & Juran, I. (1987). In-Situ Ground Reinforcement. In J. P. Welsh (Ed.), Geotechnical Special Publication: Vol. 12. *Soil Improvement – A Ten Year Update* (pp. 1–17). ASCE.

Noorzad, A., Poorooshasb, H. B., & Madhav, M. R. (2007). Performance of partially penetrating stone columns during an earthquake. *Proc. Tenth Symposium on Numerical models in Geomechanics (NUMOG X)*, 503-508.

Onoue, A. (1988). Diagrams considering well resistance for designing spacing ratio of gravel drains. *Soil and Foundation*, *28*(3), 160–168. doi:10.3208andf1972.28.3_160

Onoue, A., Mori, N., & Takano, S. (1987). In-situ experiment and analysis on well resistance of gravel drains. *Soil and Foundation*, *27*(2), 42–60. doi:10.3208andf1972.27.2_42

Pestana, J. M., Hunt, C. E., & Goughnour, R. R. (1997). *FEQDrain: a finite element computer program for the analysis of the earthquake generation and dissipation of pore water pressure in layered sand deposits with vertical drains.* Report No. UCB/EERC-97-15.

Pestana, J. M., Hunt, C. E., Goughnour, R. R., & Kammerer, A. M. (1998). Effect of storage capacity on vertical drain performance in liquefiable sand deposits. *Proc. Second International Conference on Ground Improvement Techniques*, 373-380.

PHRI. (1997). Handbook on liquefaction remediation of reclaimed land. Port and Harbor Research Institute.

Poorooshasb, H. B., Madhav, M. R., & Noorzad, A. (2006). Performance of Stone Columns subjected to a seismic Base Excitation. *Proc. Int. Symp. on Lowland Technology*, 1189-1194.

Poorooshasb, H. B., & Noorzad, A. (2000). Miura, N. and Madhav, M. R. Prevention of Earthquake Induced Liquefaction of Sandy Deposits using Stone Columns. *Proc. Int. Symp. on Lowland Technology*, 213-220.

Priebe, H. J. (1995). Design of vibro replacement. *Ground Engineering*, *28*(10), 31–37.

Priebe, H. J. (1998). Vibro Replacement to Prevent Earthquake Induced Liquefaction. *Proceedings of the 5th Geotechnique-Colloquium.*

Raju, V. R., & Sondermann, W. (2005). Ground Improvement using Deep Vibro Techniques. In Ground Improvement — Case Histories. Elsevier Publishers (UK).

Saito, A., Taghawa, K., Tamura, T., Oishi, H., Nagayama, H., & Shimaoka, H. (1987). A countermeasure for sand liquefaction: gravel drain method. Nippon Kokan Technical Report, Overseas No. 51.

Sasaki, Y., & Taniguchi, E. (1982). Shaking table tests on gravel drains to prevent liquefaction of sand deposits. *Soil and Foundation*, *22*(3), 1–14. doi:10.3208andf1972.22.3_1

Sawicki, A., & Mierczynski, J. (2006). Developments in Modeling Liquefaction of Granular Soils, Caused by Cyclic Loads. *Applied Mechanics Reviews, Transactions of the ASME*, *59*(3), 91-106.

Seed, H. B. (1979). Soil liquefaction and cyclic mobility evaluation for level ground during earthquakes. *J. of the Geotechnical Engineering Division, ASCE*, *105*(2), 201–255.

Seed, H. B., & Booker, J. R. (1977). Stabilization of potentially liquefiable sand deposits using gravel drains. *J. of the Geotechnical Engineering Division, ASCE*, *103*(7), 757–768.

Seed, H. B., & Idriss, I. M. (1971). Simplified procedure for evaluating soil liquefaction potential. *Journal of the Soil Mechanics and Foundations Division*, *97*(9), 1249-1273.

Seed, H. B., & Idriss, I. M. (1982). *Ground Motions and Soil Liquefaction during Earthquakes*. Oakland, CA: Earthquake Engineering Research Institute.

Seed, H. B., Idriss, I. M., Makdisi, F., & Bannerjee, N. (1975). *Representation of irregular stress time histories by equivalent uniform stress series in liquefaction analyses*. Report No. UCB/EERC/75-29, Earthquake Engineering Research Center, University of California, Berkeley, CA.

Seed, R. B., Cetin, K. O., Moss, R. E. S., Kammerer, A. M., Wu, J., Pestana, J. M., & Riemer, M. F. (2001). Recent Advances in Soil Liquefaction Engineering and Seismic Site Response Evaluation. *Proc. 4th Int. Conf. Recent Advances in Geotechnical Earthquake Engineering and Soil Dynamics*, 1-45.

Sondermann, W., & Wehr, W. (2004). Deep vibro techniques. In Ground Improvement (2nd ed.). Spon Press, Taylor & Francis Group.

Tokimatsu, K., & Yoshimi, Y. (1980). Effects of vertical drains on the bearing capacity of saturated sand during earthquakes. *Proc., International Conference on Engineering for Protection from Natural Disasters*, 643–655.

Vaid, Y. P., Byrne, P. M., & Hughes, J. M. O. (1981). Dilation angle and liquefaction potential. *Journal of the Geotechnical Engineering Division. Proceedings of the ASCE, 107*(GT7), 1003–1008.

Wijewickreme, D., & Atukorala, U. (2005). Ground improvement for mitigating liquefaction-induced geotechnical hazards. In *Ground Improvement— Case Histories*. UK: Elsevier Publishers.

Yasuda, S., Ishihara, K., Harada, K., & Shinkawa, N. (1996). Effect of soil improvement on ground subsidence due to liquefaction. *Soils and Foundations*, 99-107.

Youd, T. L., Idriss, I. M., Ronald, D. A., Arango, I., Castro, G., Christian, J. T., ... Stokoe, K. H. II. (2001). Liquefaction Resistance of Soils: Summary Report from the 1996 NCEER and 1998 NCEER/NSF Workshops on Evaluation of Liquefaction Resistance of Soils. *Journal of Geotechnical and Geoenvironmental Engineering, 124*(10), 817–833. doi:10.1061/(ASCE)1090-0241(2001)127:10(817)

APPENDIX

Notation

a: radius of the granular pile
b: radius of the unit cell
d, s: diameter of granular pile
d_c: dilation coefficient
GP: granular pile
k_h: horizontal permeability of untreated ground
$k_h(r)$: horizontal permeability of treated ground
L: length of pile
m_v: coefficient of volume change of untreated ground
$m_v(r)$: coefficient of volume compressibility/volume change of treated ground
N_{eq}: equivalent number of uniform stress cycles induced by earthquake
N_l: number of uniform stress cycles required to cause liquefaction
R: non-dimensionalized radial distance, r/b
R_{ma}: normalized coefficient of volume change at near end, $m_v(a)/m_v$
R_{mb}: normalized coefficient of volume change at farthest end, $m_v(b)/m_v$
R_{ka}: normalized coefficient of permeability at near end, $k_h(a)/k_h$
R_{kb}: normalized coefficient of permeability at farthest end, $k_h(b)/k_h$
r: radial distance measured from the center of granular pile
r_N: cyclic ratio
s: diameter of unit cell
S: spacing between the granular piles
SPT N: standard penetration test number
T: normalized time, t/t_d
t: time
$T_{bd} = \left(\frac{k_h}{\gamma_w}\frac{t_d}{m_v}\right)\cdot\frac{1}{b^2}$: dimensionless time factor
t_d: duration of earthquake
u: excess hydrostatic pressure
u_g: excess hydrostatic pressure generated by earthquake shaking
W or r_u or Ru_z: pore pressure ratio (= u/σ'_{v0})
W_{max}: maximum pore pressure ratio W throughout the layer at a given T

Chapter 5
Mitigation of Seismic Accelerations by Soft Materials Embedded in the Ground

Andrew J. Brennan
University of Dundee, UK

Assaf Klar
Technical University of Denmark, Denmark

S. P. Gopal Madabhushi
University of Cambridge, UK

ABSTRACT

Seismically induced accelerations are greatly affected by the transmission through soil layers. The stiffness and damping of the material beneath a structure can therefore influence the seismic loads experienced in the structure. Recently, there has been interest in the use of very low shear stiffness materials in subsoil beneath structures in order to act as a filter to seismic motions, but there is significant skepticism as to its practicability. In this chapter, some notable contributions are discussed, and some work showing the physical modelling and numerical modelling of a soft-material scheme that showed initial promise is presented. This is seen to be readily modelled as a simple mass-spring system, which could be applied to other similar cases. Future challenges include development of a suitable material for the purpose, and reliably proving static stability.

DOI: 10.4018/978-1-5225-6948-0.ch005

BACKGROUND

The authors' interest in this subject began in early 2000s whilst working together at the University of Cambridge. During the EU FP5 project NEMISREF (NEw Methods In Seismic Remediation of Existing Foundations), it became apparent that there was curiosity regarding the possibility of moving standard structural base-isolation into the soil, through targeted ground improvement involving some low stiffness material. Colleagues at a partner organisation, a major contractor, had developed such a material and were eager to investigate whether it could be beneficial in reducing structural accelerations (and, hence, damage), and if so what the conditions were for benefits to be derived. Thus, we investigated the problem using physical modelling on the centrifuge, supported by validated numerical models. This led to the realisation that the problem could be readily simplified, and the basic mass-spring model presented below was hence developed. In the years since the original work, a number of other authors have also looked at the problem, and some of these contributions are mentioned below.

One drawback with the prototype material in its original form was that its bitumen-base made it difficult to handle, and the manufacturers were unsure at the time of the potential for time-dependent degradation, or drying out in the dry sand of a physical model. For that reason, an alternative material was sought which led to the unusual selection of concentrated jelly, a standard dessert material in the UK. Jelly's mechanical properties proved very similar to the target values indicated for the prototype material, which made it suitable. In revisiting this work, it has been interesting to see the new work that has been carried out, but we have yet to see a definitive material developed for practical application, to the authors' knowledge.

SEISMIC LOADING AND SOFT MATERIALS

Seismically-induced accelerations create many problems for structures on the ground surface. Excessive shaking can cause many types of failure, as summarised by, e.g., Wiegel (1970). Ordinarily, seismic structural design is performed by generating a surface motion for the site soil profile, using a site response code such as SHAKE (e.g. Schnabel et al., 1972), and then using this as an input motion for further seismic analysis. Soil would normally be expected to amplify the accelerations as they are transmitted from bedrock to surface, depending on its natural period (e.g. Iwasaki et al., 1978). Response spectra are commonly used to quantify the effects of soil on the response of single degree-of-freedom oscillators at the surface, for particular earthquake motions. Whether this motion is affected by interaction with the soil would not normally be considered, although recent evidence suggests that such soil-structure interaction can significantly affect structure response, either positively or negatively (Gazetas & Mylonakis, 1998; Ghosh & Madabhushi, 2005; Mylonakis & Gazetas, 2000).

Earthquake loading in general contains pressure wave, shear wave and surface wave components, of which the shear wave component is usually the most damaging for structures. Softened liquefied soil is seen to reduce shear accelerations felt by surface structures (e.g. Tokimatsu et al., 1996). The effect of underground liners that slip to prevent shear wave transmission has also been found to reduce seismic shear stresses experienced at the ground surface (Yegian & Catan, 2005; Yegian & Kadakal, 2005). An idea suggested has been to incorporate a soft layer horizontally in the ground beneath a structure such as in Figure 1a, but unless this is very wide then shear waves still diffract around the ends, and there is no protection from surface waves such as Rayleigh or Stoneley waves. Open trenches have been studied as a barrier to surface wave propagation whether seismically-induced (May & Bolt, 1982; Wang et al., 2009), rail/traffic-induced (Ahmed & Al Hussaini, 1991; Beskos et al., 1986; Yang & Hung, 1997), machine-induced (Celebi et al., 2009) or construction induced (Ye & Tan, 2011). Alternative solutions could be a combination method involving horizontal and vertical layers acting together as a box, and the parametric work by our co-investigators in the NEMISREF project Kirtas & Pitilakis (2009) showed that this was the most effective way of modifying seismic motions. This led to the test designs for the paper on which this Chapter is based, Brennan et al. (2013), detailed below. Further numerical investigation has been more recently presented by Lombardi et al. (2014) and Flora et al. (2018), that concentrate mostly on the demands of constitutive models but also identify that caisson-type boxes of the sort shown below are most suited for structures with high natural frequencies – a natural corollary of the original conclusions of the simplified mass-spring models shown below. Related to these fully enclosed systems, Le et al. (2016) solved analytically the equations of motion for the soil structure interaction of a structure sitting on a “bowl” of rigid soil surrounded by rather softer material, equivalent to a hemispherical soft layer beneath the structure. Flora et al. (2018) also investigated an alternative, angled planar soft layers (called “V shaped” due to their appearance in two dimensional view) that also showed promise, but as with the Kirtas & Pitilakis (2009) investigation of a finite-length horizontal barrier, this needed to extend beyond the footprint of the structure base in order to prevent waves coming around the end of the barrier and reaching the structure.

The work presented in this Chapter investigates a combination method, utilizing the filtering effect of a soft layer in the ground, akin to liquefied soil and related to a sliding surface, with soft trenches to prevent diffraction of shear waves around the horizontal layer. Such a configuration is shown in Figure 1b with a basic surface structure. The soft material can be seen to resemble a caisson, impeding wave travel in all directions. For this study, only unidirectional shear loading is considered.

MODELLING TECHNIQUES

In this work, both physical and numerical models of the system are presented. This section presents the basis behind these analysis techniques, firstly the physical modelling, including the physical modelling of soft layers, and secondly the numerical analysis.

Centrifuge Modelling

Centrifuge tests were carried out on the 10m diameter beam centrifuge at Cambridge. Schofield (1980) describes this and gives a thorough discussion of appropriate scaling laws. The purpose of centrifuge modelling is to allow small scale physical models to be tested under the effective stress levels of a larger prototype, in order to correctly model the effective-stress dependent mechanical properties of the soil. Testing is carried out at 50g (i.e. 1/50th length scale) for all tests described here. Values quoted in this paper are in prototype scale (unless otherwise specified).

Earthquake shaking is provided by a mechanical shaker (Madabhushi et al., 1998) capable of transferring single frequency motion for short duration, although being a mechanical system, higher harmonics of the intended shaking frequency are inevitably applied to the model too. This is observed later, for example in Figure 7.

Figure 1. Schematic representation of soft layers impeding seismic motion a) horizontal layer b) "caisson" constructed of layer plus walls

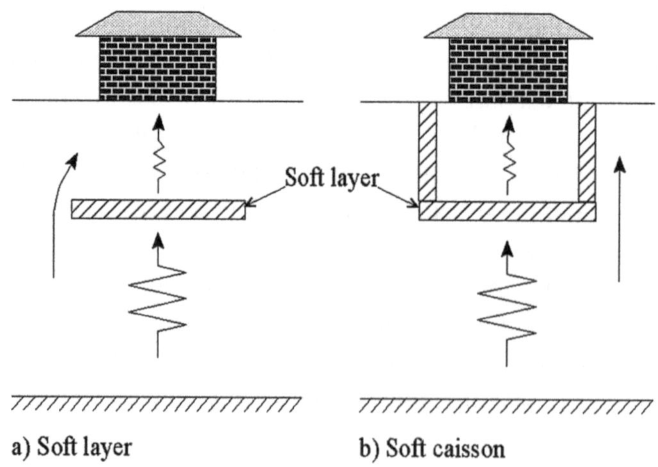

Model containment is highly significant for wave transmission problems. For this centrifuge test, a new laminar model container was used (Brennan et al., 2006). This container consists of a stack of low-mass thin frames separated by roller bearings. Therefore, the soil alone loads the boundary rather than the input motion, in the same manner as the boundaries used in the numerical analysis. Figure 2 shows the layout of the centrifuge test, with some salient instrument positions marked. Dimensions marked in Figure 2 are given in model scale; multiplying dimensions by the scale factor of 50 provide the appropriate prototype dimension.

Soft Materials

Ground improvement is to be achieved in the field using any suitable flexible material. A polymeric solution with stiffness properties as shown in Table 1 has already been developed, and further solutions are also currently under investigation. For small scale modelling, an alternative representative material had to be used. The chosen material was a concentrated dessert jelly, produced in the UK by Rowntrees Ltd. For ease of observation in contrast to the pale yellow sand, the red-coloured strawberry flavour was selected. Elastic properties for this material were obtained by unconfined compression testing (Figure 3), and tabulated in Table 1. In addition to its elastic properties, the jelly possesses approximately 15% hysteretic damping. It is not the primary function of the soft material to dissipate energy through damping, but this will certainly have an effect on measured results, by dissipating energy and reducing the system fundamental frequencies. During model construction, cubes of jelly were built layer by layer into a wall. Heating additional jelly pieces in a microwave oven produced a sticky "mortar" which was used to cement adjacent pieces. Figure

Figure 2. Centrifuge model layout (dimensions at 1/50th scale)

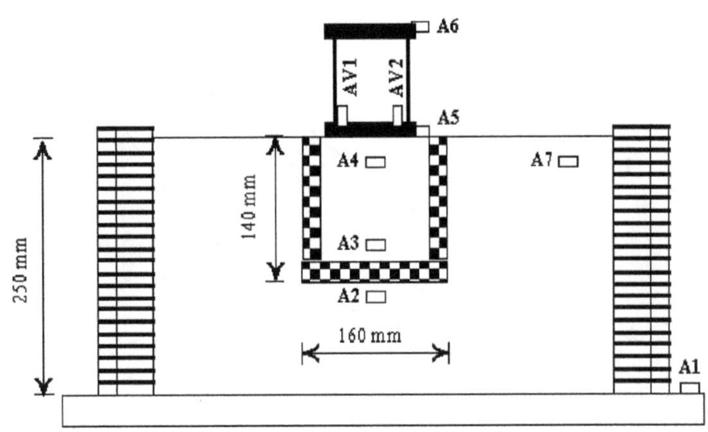

4 shows the model prior to testing, with the soft material clearly visible at the soil surface. No material damage was observed in the jelly following testing.

Model Construction

The soil used throughout the test series was fine Hostun S28 silica sand. This is a uniformly graded sand, with D50 = 350 µm, e_{min} = 0.60 and e_{max} = 1.01. Models are prepared by dry-pluviating sand from a hopper at a controlled rate to achieve the desired relative density of around 80%.

Table 1. Elastic properties of target material and model material. Italic values inferred from non-italic values

	Target Material	Rowntrees Jelly
Young's modulus E (kPa)	54	50
Shear modulus G (kPa)	20	17
Poisson's ratio ν	0.33	0.42
Bulk modulus K (kPa)	50	104

Figure 3. Stress strain behaviour of Rowntree's jelly

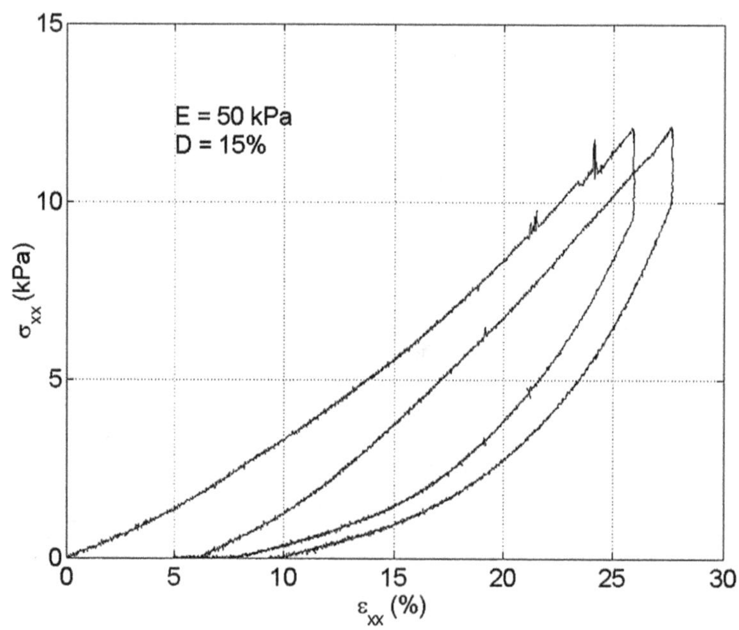

Figure 4. Jelly as a soft caisson after model construction

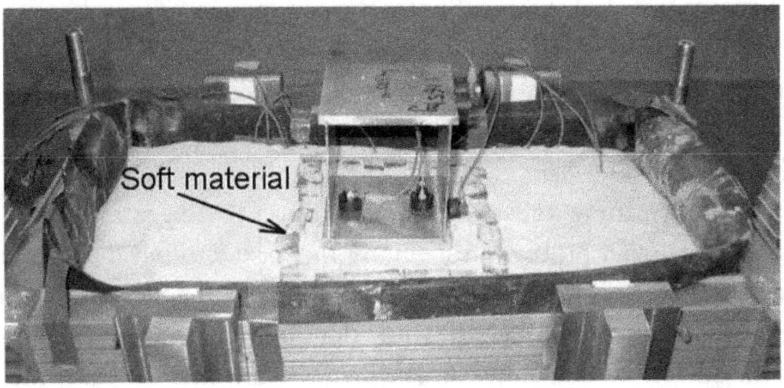

The model structure is constructed from a plate metal base and first floor, with sheet metal walls slotted in on two sides to form a sway frame. Such models have been used before in physical modelling to produce single-degree-of-freedom structures, e.g. Mitrani & Madabhushi (2008). Prototype scale structure height, length and breadth are all approximately 5 m. Structure mass was 65600 kg distributed equally between base and roof, imposing a bearing pressure of 26 kPa.

Numerical Analysis

In order to confirm the benefits of the isolation scheme, the physical model is compared to two computer generated models; one corresponding to a homogenous sand layer and the other to the isolation scheme considered in the physical model.

By assuming the soil stress condition in the problem closely resembles a plain strain condition, it was possible to utilize a plain strain formulation for the soil behaviour. This allowed for the use of the finite difference code FLAC (ITASCA, 2002). The boundary value problem corresponds to the prevailing conditions in the centrifuge model container; the lower boundary is rigid, while the side boundaries are attached to each other, representing the condition that is created in the laminar box by the rigid rings. Figure 5 shows the boundary value problem (in prototype scale) and the discretisation to elements. To sustain an accurate ratio between the fundamental frequencies of the superstructure and the soil in the box, the formulation was modified such that each element corresponded to the thickness of the box (13.15 m in prototype), rather than to a unit length. The modification from unit width formulation was simply conducted by attributing a heavier mass, by a factor of 13.15, to each soil grid point, and multiplying the stresses (from the constitutive model) by the appropriate surface area (i.e. additional factor of 13.15). The superstructure itself retained its true prototype dimensions, mass and stiffness, and was constructed using standard beam elements. The computer model can be visualized as a physical model of the centrifuge box, where the structure is connected at its base to massless rigid rods with length equal to the thickness of the box.

To deal with the important aspect of soil nonlinearity, the equivalent linear approach was chosen, in which an iterative procedure is employed using analysis of linear material as follows.

To define the confining stress, a static condition is first established. It should be noted that the current problem, especially the case without the soft box, is not kinematically constrained. As a result the initial stress condition is practically independent of the constitutive relation, and does not change significantly between iterations. Based on the confining stresses and strain history from the previous iteration, the equivalent stiffness and damping can be defined for the current iteration. The iterative procedure continues until convergence is achieved. The procedure basically follows the one of SHAKE (Schnabel et al., 1972). An important aspect of SHAKE is that it is able to simulate the soil behaviour as frequency independent, as the analysis is in the frequency domain and damping can be regulated as a function of frequency. However, the current analysis is in time domain so the damping is frequency dependent when considering viscoelastic material. Nonetheless, it is believed that the use of Rayleigh damping (which allows for frequency independence in a certain definable range), and the fact that the input motion is dominated by a single frequency, result in a behaviour that is strongly independent of frequency for the current analysis.

The expressions of Seed et al. (1986) and Ishibashi and Zhang (1993) for maximum stiffness and reduction factor are utilized in the current analyses. These expressions are given in Figure 6, which summarizes the iterative procedure described above, and have been shown to be appropriate for the soil tested (Brennan et al., 2005). The use of these expressions is limited to the elements representing the sand. For the soft material, constant values were set according to the large strain behaviour shown in Figure 3.

RESULTS AND COMPARISON

The initial question, before the dynamic performance of this solution can be evaluated, is one of static stability. As gravity is gradually increased during a centrifuge test, the structure will undergo equivalent ultimate settlements to the field case. For simplicity, model construction sequences were not recreated. The ultimate settlements in prototype scale were measured as being 25 mm in the free-field soil, and 150 mm for the structure above the treated ground. Obviously, if soft material is being installed as a retrofit beneath an existing building, care must be taken with

Figure 5. Schematic representation of the boundary value problem

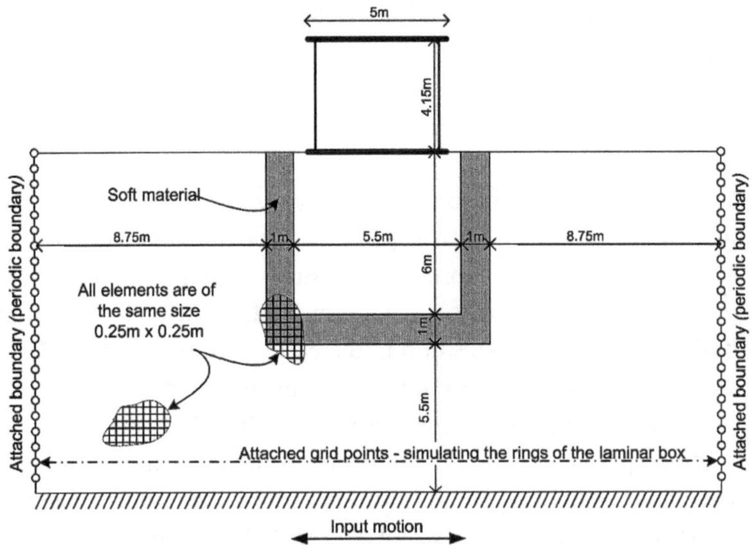

Figure 6. Flow chart of iterative procedure

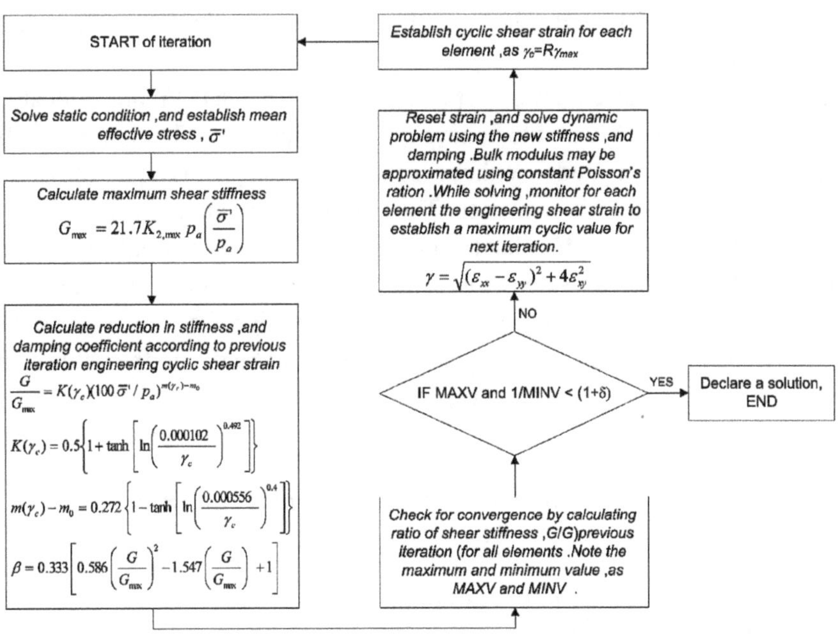

the construction sequence to avoid differential settlements. That issue was beyond the scope of the original work, but its importance to the success of such a scheme must be acknowledged.

Free-field Accelerations

Figure 7 shows the input acceleration as a time trace, together with a corresponding 5% damping response spectrum. It is seen in the time domain that, unlike a "real" earthquake, each cycle of the applied shear stress is approximately uniform in amplitude. It is common to apply a reduction coefficient of 0.65 on maximum strain to obtain the representative strain in the equivalent linear method (Schnabel et al., 1972) in order to account for the variation in peak heights during shaking. If each peak was identical in height, this factor would be 1.0. In this case, the peaks are almost but not quite identical, therefore a compromise value of 0.85 is deemed more appropriate, chosen because it is a little closer to the uniform case (1.0) than the random case (0.65). However, it is not anticipated that this would affect results significantly, as shear modulus degradation is a function of log(shear strain), so such changes will have only minor effects.

Note that although the input motion is nominally at the period 1.0 s, significant energy is present at 0.33 s and 0.20 s also, due to harmonics in the shaking actuator as described above.

Figure 8 shows the accelerations at position A7 (as shown in Figure 2) located far from the structure and assumed to behave as a free-field soil. The figures are, from the top down, computer model without isolation, computer model with isolation and the centrifuge model with isolation. The two computer models correspond very closely to each other, indicating that the presence of the soft material has indeed no significant effect on the free-field soil. The computer and the centrifuge produce results that are similar in magnitude, although less high frequency components have been transmitted in the centrifuge test. This could be due to the value of high frequency damping chosen in the numerical analysis.

Figure 7. Input motion applied at base, and 5% damping response spectrum

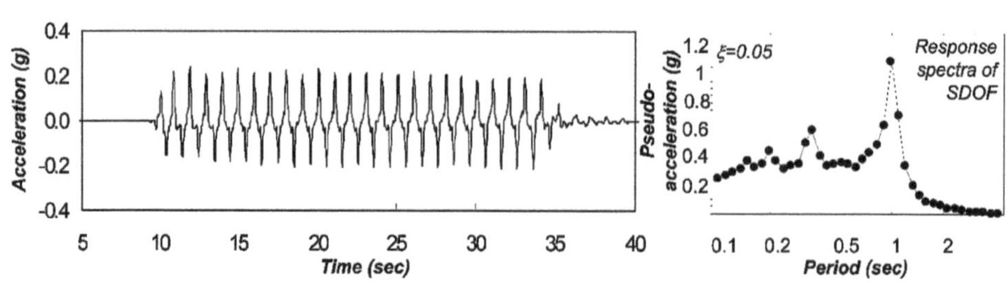

In the absence of ground improvement, a conventional structural analysis ignoring soil-structure interaction would use the acceleration given in Figure 8 (i.e. the free-field acceleration) as the input acceleration to the base of a structure.

Structural Vibration

Figure 9 compares the behaviour at the base of the structure for the two numerical models and the centrifuge test. On the left hand side of the figure, the horizontal acceleration time history of the base of the structure is given (point A5 in Figure 2). The difference between the results of the two computer models indicate that the "soft caisson" significantly reduces the peak accelerations at the base of the structure. It is also evident that the soft caisson has filtered out much of the higher frequencies, leaving the harmonic motion of the fundamental input frequency. The results of the computer model of the soft caisson case agree well with the centrifuge measurements in the overall shape. Although a difference of 20% in acceleration amplitudes is observed, peak amplitude has reduced by a similar factor (2.9 in the numerical case, 2.5 in the centrifuge model) relative to the input acceleration shown

Figure 8. Accelerations at ground surface: Free-field.

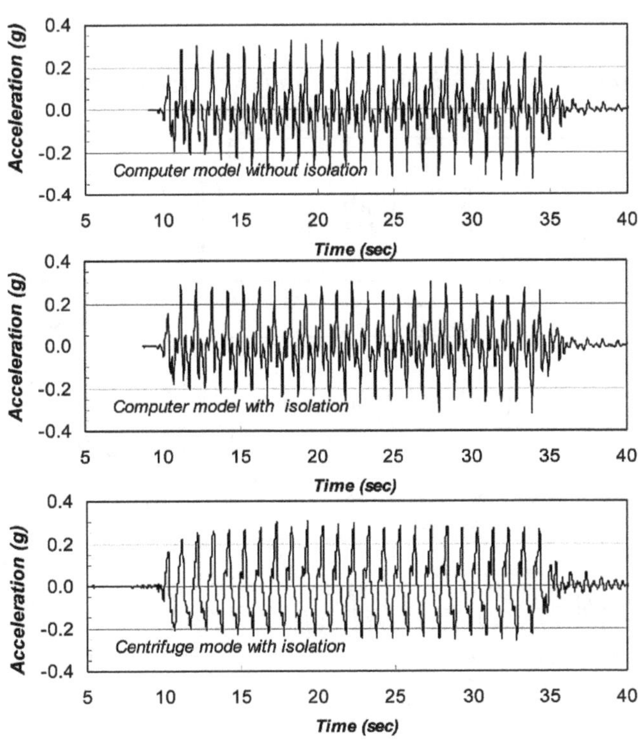

in Figure 7. For a rigid based structure, these motions would ordinarily be considered sufficient to specify as base motion for further structural analysis.

Although the above results seem promising, they do not reveal whether the superstructure itself benefits from the existence of the soft caisson, as frequency changes are apparent as well with the reduction in accelerations. The structural components are only affected by the relative motion between storeys. Using the Duhamel integral method, response spectra for the superstructure's relative motion are presented in Figure 9 to the right of the corresponding time history. The pseudo-acceleration is of great importance from a structural point of view as the horizontal force acting on the walls is the product of the pseudo-acceleration and the top floor mass. Therefore the influence of the base motion on a structure may studied from the response spectra.

The FLAC numerical analysis resulted in a relative acceleration with maximum value of 0.17g in the presence of the soft caisson. Taking the maximum relative displacement recorded during the numerical analysis and multiplying it by $(2\pi/T)2$ results in a pseudo- acceleration of 0.288g for T = 0.43 s, the theoretical period of the structure. This value is plotted on Figure 9, and labelled "FLAC". For the case without the soft caisson the numerical analysis resulted in maximum relative acceleration of 1.01g and a pseudo- acceleration of 1.04. The Duhamel integral values read from Fig. 9 are 0.179 g and 0.517g for the case with and without the soft caisson, respectively. This discrepancy between response spectra values and measured FLAC

Figure 9. Accelerations and response spectra for motion at base of structure

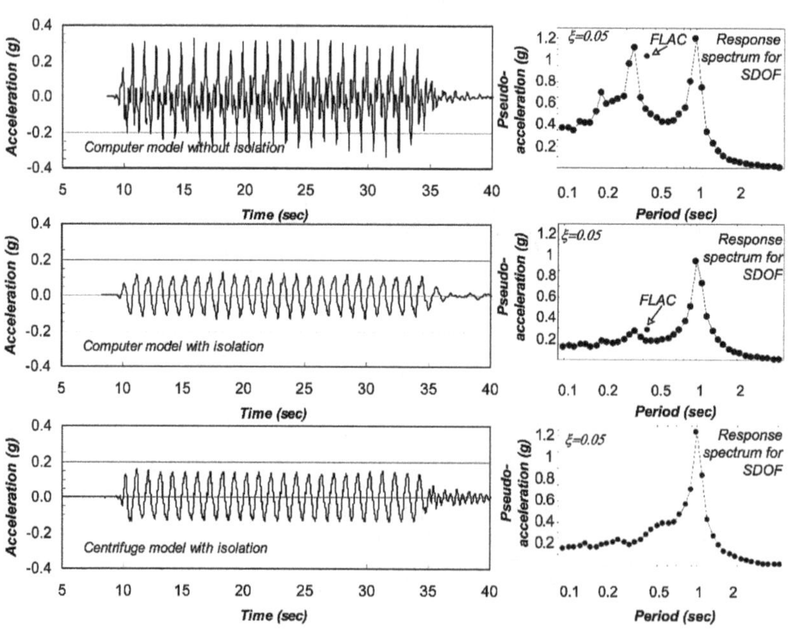

results is an outcome of the simplification of the response spectra, which ignores the rotational degree of freedom. The FLAC relative motion includes some rigid body rotation that does not cause stress in the walls - taking out the rotation component in FLAC changes the pseudo-acceleration from 0.288g to 0.23g. Nonetheless, both calculations suggest a significant improvement in structural performance due to the presence of the soft caisson, and the reduction is approximately a factor of 3 compared to input in each case.

The response spectra of the computer model and the centrifuge in Figure 9 agree well, in that low periods below about 0.8 s are seen to be heavily filtered. The maximum value differs by around 20%, corresponding to the 20% difference noted in the time history, and may be due to over-damping the jelly at this frequency in the numerical model.

Note that response spectra were derived from accelerations that inherently include SSI effects. The influence of SSI on structures other than the one tested (theoretical period 0.43 s) will be slightly different, and consequently slight modifications would be required to the response spectra. However, for the given stiff structure and stiff soil considered here, SSI effects are likely to be much less of an influence than the extremely soft jelly beneath the foundation, so the response spectra can be considered as good approximation for other periods. The spectra also ignore the effect of rotation of the soil-foundation system, the effect of which is discussed in more detail below.

This data therefore suggests that for the given structure and improvement tested, under the given motion, the soft caisson proved to have a significant beneficial effect. It also suggests that the influence of rocking is slight and not particularly harmful to the ultimate limit state. In fact, if the rocking can be made to occur in antiphase to the translational oscillation then accelerations would be further reduced. Now that the feasibility of soft caissons as a means of reducing seismic accelerations present at the surface and improving structural response has been established, a simple model of the situation is presented below for further analysis.

SIMPLE SPRINGS ANALYSIS

In order to generalise the analysis, a simple mass-spring model is created to reproduce the effect of ground improvement by a soft caisson. The structure may be treated as two discrete masses. One is the base floor, moving with the enclosed soil due to its high stiffness relative to the soft material. The other is the top floor, attached to the base by the springy walls as shown in the photograph in Figure 4. This top floor and the walls provide an example single degree of freedom oscillator. The base floor vibrates with the soil and provides the input motion for the structure. In the 3-D configuration of the current test being considered (Figure 2), the participating mass within the caisson is estimated as being 345600 kg soil + 33000 kg structure base = 379×103 kg. Spring stiffness for the soil-caisson system, due to the horizontal soft layer passing shear waves, side soft walls passing shear waves and end soft walls passing compression waves, is estimated as being (Equation 1):

$$k_y = \frac{GLB}{t_{base}} + 2\frac{GLh}{t_{side}} + 2\frac{EBh}{t_{side}} = 612000 + 1224000 + 3600000 = 5.426MNm^{-1} \tag{1}$$

where L = wall dimension in shaking direction, B = wall dimension parallel to shaking direction, h = wall depth and t = thickness of soft material. Equation 1 shows that the dominant term contributing to lateral stiffness in this particular example is the end soft walls in direct compression rather than the shear walls. This is the term that would have to be addressed, perhaps by thicker walls or more flexible material, if lower stiffness was required.

Rotational vibration of the caisson may also occur, with the structure-foundation system rocking about a centre of mass. The moment of inertia for the two elements about a single centre of mass is estimated as being I = 3666860 kgm^2. Stiffness contributions are provided by shear in the end walls and along the system base, plus that from torsion in the side walls. For the 2-D case (not counting side walls), rotational stiffness is estimated as being (Equation 2):

$$k = \frac{GL^2Bh}{2t_{side}} + \frac{GLBh^2}{4t_{base}} = 11016000 + 5508000 = 16.524MNm / rad \tag{2}$$

Including the stiffness based on integrating the torsional shear in the side walls of the jelly produces an additional stiffness of 4.146 MNm/rad. This gives an indication of the potential error between the full 3-D case on the centrifuge and the plane strain assumption in the FLAC analysis.

Model Definition

The situation considered is that shown in side elevation in Figure 10. A structure overlies soil, and around the foundation a certain thickness of soft material has been installed. This extends out of plane to form a box of soil, and is not here treated as plane strain.

Figure 10. System analysed, and a mass-spring representation

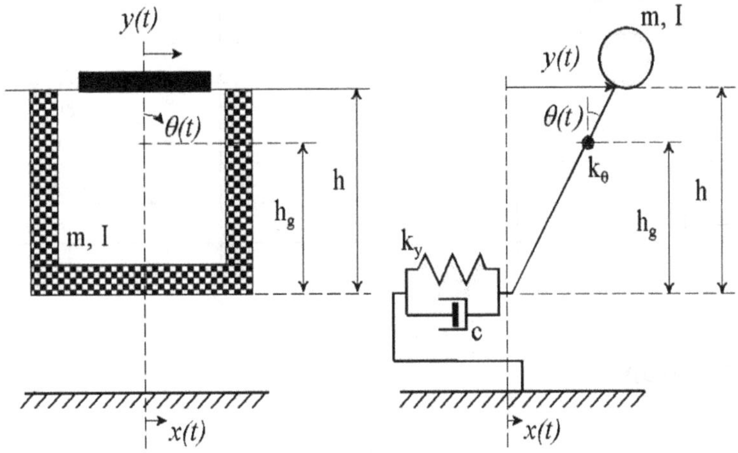

The structure base and enclosed soil combined have a total mass m [kg], centre of mass at a position distance h_g [m] above the lowest edge, and moment of inertia about this point of I [kgm^2]. These quantities are indicated in Table 2. It is assumed that the soil and structure vibrate together, as SSI stiffness will be much greater than that of the soft material. Relative motion between structure floors are also assumed to have a small effect on structure base motion relative to input shaking.

Simple calculation, as in Equation 1 above, enables the translational stiffness of the soft material around the block to be determined in terms of its elastic moduli and thickness. This is termed k_y [N/m]. A parallel dashpot damper is added to provide damping. Integration of shear around the edges, including the sides in torsion not seen in Figure 10, enable a rotational stiffness k_θ [Nm/rad] to be evaluated for the system, as in Equation 2. These parameters are summarised in Table 2, along with values for the configuration modelled.

Figure 10 also shows the variables considered in this simple hand calculation. Coordinate x is the translation of the soil beneath the entire system away from a single equilibrium position. Coordinate θ is the rotation of the system about the centre of mass. Coordinate y is the translation of the base of the structure from the same equilibrium as x. Thus, the net extension of spring k_y, at the base of the caisson as shown in Figure 10, is $(y - \theta h - x)$. For simplicity, damping has not been considered in this analysis. As a small amount of damping should not change resonant frequency significantly then the undamped results will be conservative, and peak values always larger than if damping had been present.

Table 2. Parameter values for centrifuge test: prototype scale

Property	Value	Unit
M	378600	kg
I	3666860	kgm^2
k_y	5426000	N/m
k_θ	20670000	Nm/rad
H	7	m
h_g	4.88	m

Model Solution

Allowing the soil and structure to vibrate as a single unit of mass m and lateral acceleration ÿ, the horizontal force balance is (Equation 3).

$$m\ddot{y} = -k_y\left(y - h\theta - x\right) - c\left(\dot{y} - h\dot{\theta} - \dot{x}\right) \tag{3}$$

Allowing one mode of oscillation, taking moments clockwise about the centre of mass gives (Equation 4).

$$I\ddot{\theta} = -k_\theta\theta + h_g \times k_y\left(y - h\theta - x\right) + h_g \times c\left(\dot{y} - h\dot{\theta} - \dot{x}\right) \tag{4}$$

Assembling these in matrix form produces (Equation 5).

$$\begin{bmatrix} m & 0 \\ 0 & I \end{bmatrix}\begin{pmatrix} \ddot{y} \\ \ddot{\theta} \end{pmatrix} + \begin{bmatrix} c & -ch \\ -ch_g & chh_g \end{bmatrix}\begin{pmatrix} \dot{y} \\ \dot{\theta} \end{pmatrix} + \begin{bmatrix} k_y & -hk_y \\ -k_y h_g & k_\theta + h_g hk_y \end{bmatrix}\begin{pmatrix} y \\ \theta \end{pmatrix} = \begin{pmatrix} k_y \\ -h_g k_y \end{pmatrix} x + \begin{pmatrix} c \\ -h_g c \end{pmatrix}\dot{x} \tag{5}$$

Undamped free vibrations of the system can be analysed from eigenvalues/ vectors of the inverted mass times stiffness matrix in Equation 5. For the properties tabulated in Table 2, which have been evaluated for the configuration modelled above, the natural modes are:

$$f_1 = 0.17 \text{ Hz}, \begin{pmatrix} y \\ \theta \end{pmatrix} = \begin{pmatrix} -0.99 \\ -0.13 \end{pmatrix}$$

and

$$f_2 = 1.33 \text{ Hz}, \begin{pmatrix} y \\ \theta \end{pmatrix} = \begin{pmatrix} 0.87 \\ -0.48 \end{pmatrix}$$

The in-phase mode has a period of 5.82 s, significantly greater than any component of the input to the models. The antiphase mode, of period 0.75 s, is significantly closer to the periods in the applied motion, which are predominantly around 1.00 s, and should therefore be the dominant mode of vibration during the applied shaking. This tallies with the numerical analysis finding a reduction in maximum acceleration when rocking was allowed – the antiphase response of the caisson has been stimulated.

The mass-spring system is now subjected to the input motion applied to the more sophisticated models, as shown in Figure 11, evaluated using a simple finite-difference time stepping algorithm. A nominal 1% damping is incorporated. Also plotted for comparison is the corresponding response of the centrifuge model (at position A5 in Figure 2). In the time domain, the spring model shows a combination of the antiphase mode and the dominant frequency of applied loading, and there is also "ringing" at this mode after shaking. The modal response is so much more evident in the springs model than the centrifuge model because of the inherent nonlinearities of the centrifuge model.

Figure 11. Response of spring model to input motion, compared to centrifuge measured response

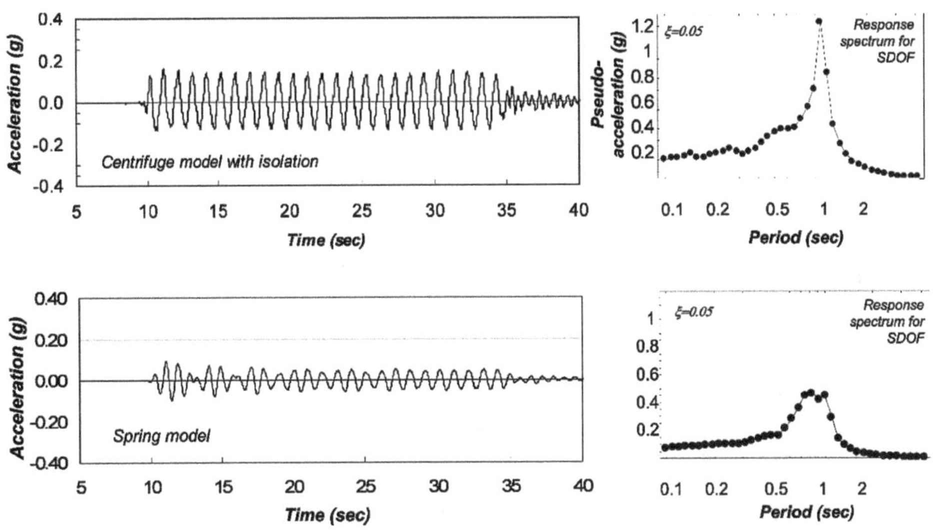

Generalised Results

The above results have demonstrated the effect of the given improvement system under the synthetic motion of the available centrifuge earthquake actuator. This represents the logical testing method for research, as only certain discrete frequencies are applied to the system, making analysis simpler than in the case of the multiple-frequency input of a "real" input motion. To investigate the response of the given system under other frequencies, the time scale of the (approximately) harmonic input motion of the centrifuge actuator was scaled to produce motion at different frequencies and used as input for a succession of spring analyses, with parameters as Table 2. Figure 12 plots the peak surface acceleration generated by each of these synthetic earthquakes at the structure base, against the fundamental period of the applied earthquake. Acceleration axis is normalised by peak input acceleration, and the earthquake period axis is normalised by T_0, the fundamental period of the caisson. Both vibrational modes are seen, the antiphase response at $T_{eq}/T_0 = 1$ (by definition) and the stronger in-phase response at the expected $T_{eq}/T_0 = 7.8$ following normalisation.

For the narrow-band earthquake applied, we would expect the desired attenuation from the caisson as long as its natural periods were a) more than 10% greater than that of the earthquake or, more generally, b) more than 10% different to the applied motion's period. The system tested on the centrifuge was in category (b), and is also plotted as a point on Figure 11. This is more risky than category (a), as a wider bandwidth process might contain frequencies that could stimulate a resonant response. These are here represented by the higher harmonics present in the applied motion, at $T_{eq} = 2 \times T_0$, etc, which are seen to stimulate a response in Figure 11. Including damping into the model could well reduce the response at these higher

Figure 12. Response of particular caisson to different frequency motions

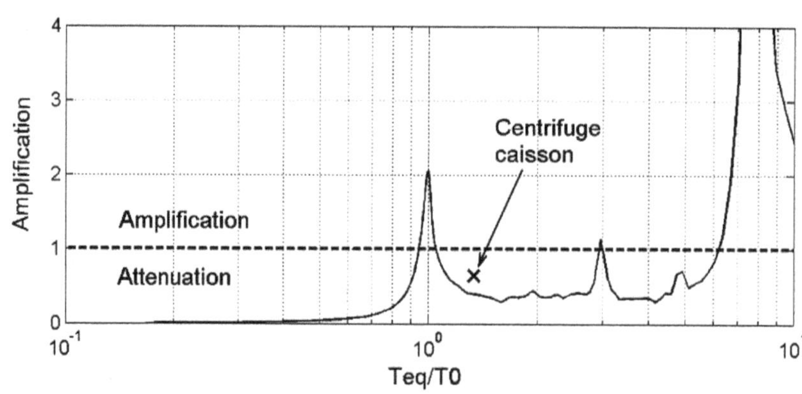

frequencies. Better design would aim to be to the left of the resonant peak (category a), which is less likely to be stimulated but can also achieve significantly greater attenuations. The difficulty is that this would involve using even softer material, which poses its own engineering challenges such as finding a suitable material and ensuring static stability.

Finally, to test response against broad-band motion, the simplified model (retaining parameters from Table 2) is subjected to recorded strong motion data. The motions chosen are from the Tokachi-Oki earthquake (Urahoro station, gravel/fill site, E-W component, 119 km from epicentre, data provided by K-NET, National Research Institute for Earth Science and Disaster Prevention, Japan) and the Northridge earthquake (Century City station, alluvial deposits, E-W component, 20 km from epicentre), selected to contrast with the narrow-band motion applied above. Figure 13 shows the response spectra of the original motions superimposed on the response spectra of the mitigated motions. The Tokachi-Oki spectra contains a wide spread of frequencies, while the Northridge spectra contains mainly higher frequency motion. As shown in Figure 13, the broad-band motion of a real earthquake becomes a modal response in the soft caisson system. Natural modes are stimulated at the expense of other vibration, and as the model does not include damping these peak responses are overestimated. A structure sharing a modal frequency with this system would undergo significant accelerations, but other structures are seen to experience much lower input motions. For this reason, it would be prudent to ensure the vibrational modes of the soft caisson system are different to structural modes.

FUTURE RESEARCH DIRECTIONS

The technique of soft layers has therefore shown the potential for positive as well as negative outcomes, and each case must therefore be assessed in its own right without recourse to a one-size-fits all strategy. The simplified mass-spring model has shown some promise in this when applied to the soft caisson structure tested, and would by portable to similar schemes, but alternative uses of soft materials, such as walls only or the V shaped layers investigated by Flora et al. (2018) would need some verification that comparable mass-spring models are valid for these cases. This may be achieved through use of numerical modelling, and both the work in this chapter and also the wider literature have indicated how that might be achieved.

The major challenge is therefore to transfer this potential to site usage, and that involves broaching two major challenges. The first relates to static stability. The dynamic behaviour has proved beneficial and predictable for the cases shown here and in the literature, but this rather assumes that the soft material can exist in equilibrium with the structure without compromising its integrity or producing

Figure 13. Mitigated and unmitigated 5% damping response spectra of strong motions; a) Urahoro station, E-W component, Japan earthquake 26 Sep 03, b) Century City station, E-W component, Northridge earthquake 17 Jan 94

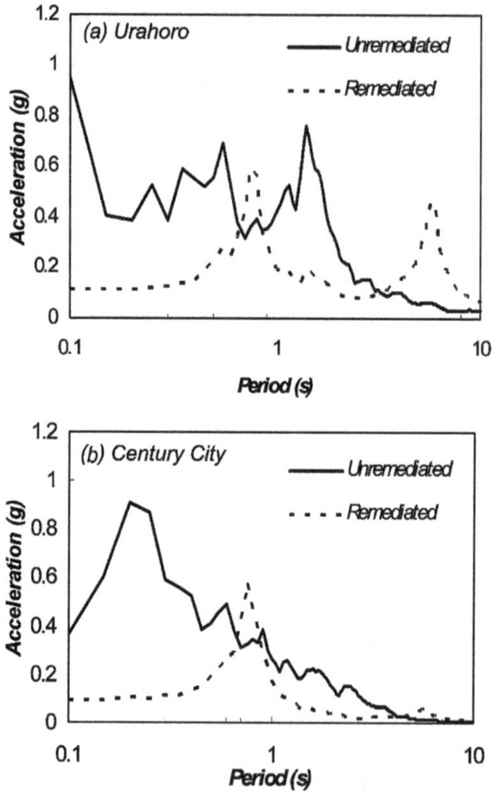

excessive settlements. This must not just apply to the final as-built condition, but also to the transient phases during which the material is installed and occupies only a fraction of the finished volume, or possesses a transient strength or stiffness due to curing times. Further, this must be demonstrated to be adequate through physical experience, particularly large trials, in order to convince stakeholders that this is a safe and reliable engineering activity.

The second challenge is in producing the material itself. Various prototype laboratory substances have been mooted, and feature as the target material, but the ability to mass produce a suitable material reliably and possessing predictable material properties is the next, crucial, step to method roll-out.

CONCLUSION

A number of observations may be made based on these models.

A soft caisson was built beneath a model structure's foundation and maintained static stability during a dynamic centrifuge test. Both the physical and numerical models showed a decrease in peak accelerations at surface level, suggesting that a well-designed soft caisson may significantly reduce earthquake motions reaching structures. Both physical and numerical models showed significant improvement of the response of an example structure, in terms of relative accelerations.

Two modes were identified, an in-phase response and an antiphase response. Rocking in antiphase to applied loading was produced in the models, and although such rigid-body rotation does not directly affect structural joint loading, this acted to further reduce accelerations felt at the structure base. For a truly soft caisson, this response should always be dominant, as the lower fundamental period should be much closer to the dominant periods in the earthquake load.

A simple mass-spring model produced a passable representation of the behaviour demonstrated by the more sophisticated models. This is important, as it indicates that different soft materials and different geometries may be easily explored by designers and material developers in order to quickly assess the usefulness of a given design.

For the system of a soft caisson, as shown in this chapter, the mass-spring model showed that attenuation could be achieved providing the fundamental periods of the soft caisson differed from the earthquake loading by more than 10%. Real strong motions applied to the simplified model showed that broad-band input motion triggers almost exclusively modal response. This should make the response of surface structures easier to predict.

ACKNOWLEDGMENT

This research was supported by the EU funded GROWTH project (GRD1-40457) "NEMISREF". It also required the excellent assistance of technical staff at the Schofield Centre, Cambridge, which is acknowledged with thanks.

REFERENCES

Ahmed, S., & Al Hussaini, T. M. (1991). Simplified design for vibration screening by open and in-filled trenches. *Journal of Geotechnical Engineering*, *117*(1), 67–88. doi:10.1061/(ASCE)0733-9410(1991)117:1(67)

Beskos, D. E., Dasgupta, B., & Vardoulakis, I. G. (1986). Vibration isolation using open or filled trenches Part 1: 2-D homogeneous soil. *Computational Mechanics*, *1*(1), 43–63. doi:10.1007/BF00298637

Brennan, A. J., Klar, A., & Madabhushi, S. P. G. (2013). Mitigation of seismic accelerations by soft caissons. *International Journal of Geotechnical Earthquake Engineering*, *4*(2), 1–17. doi:10.4018/ijgee.2013070101

Brennan, A. J., Madabhushi, S. P. G., & Houghton, N. E. (2006) Comparing laminar and equivalent shear beam (ESB) containers for dynamic centrifuge modelling. In *Proceedings of 2nd Int. Conf. on Physical Modelling in Geotechnics* (pp.171-176). Hong Kong. Taylor and Francis/Balkema. 10.1201/NOE0415415866.ch17

Brennan, A. J., Thusyanthan, N. I., & Madabhushi, S. P. G. (2005). Evaluation of shear modulus and damping in dynamic centrifuge tests. *Journal of Geotechnical and Geoenvironmental Engineering*, *131*(12), 1488–1497. doi:10.1061/(ASCE)1090-0241(2005)131:12(1488)

Celebi, E., Firat, S., Beyhan, G., Cankaya, I., Vural, I., & Kirtel, O. (2009). Field experiments on wave propagation and vibration isolation by using barriers. *Soil Dynamics and Earthquake Engineering*, *29*(5), 824–833. doi:10.1016/j.soildyn.2008.08.007

Flora, A., Lombardi, D., Nappa, V., & Bilotta, E. (2018). Numerical Analyses of the Effectiveness of Soft Barriers into the Soil for the Mitigation of Seismic Risk. *Journal of Earthquake Engineering*, *22*(1), 63–93. doi:10.1080/13632469.2016.1217802

Gazetas, G., & Mylonakis, G. (1998) Seismic soil-structure interaction: new evidence and emerging issues. In Proceedings of Geotechnical Earthquake Engineering and Soil Dynamics III (pp. 1119-1174). ASCE Geotechnical Special Publication 75.

Ghosh, B., & Madabhushi, S. P. G. (2005). Comparison of free-field ground response of layered soil under liquefiable conditions. *Indian Geotechnical Journal*, *35*(2), 199–219.

Ishibashi, I., & Zhang, X. (1993). Unified dynamic shear moduli and damping ratios of sand and clay. *Soil and Foundation*, *33*(1), 182–191. doi:10.3208andf1972.33.182

Itasca. (2002). *FLAC Ver. 4.0 user manual*. Itasca Group.

Iwasaki, T., Tatsuoka, F., Tokida, K., & Yasuda, S. (1978) A practical method for assessing soil liquefaction potential based on case studies at various sites in Japan. *Proceedings of 2nd Int. Conf. on Microzonation for Safer Construction – Research and Application*, 885-896.

Kirtas, E., & Pitilakis, K. (2009). Subsoil Interventions Effect on Structural Seismic Response. Part II: Parametric Investigation. *Journal of Earthquake Engineering*, *13*(3), 328–344. doi:10.1080/13632460802347471

Le, T., Lee, V. W., & Luo, H. (2016). Out of plane (SH) soil-structure interaction: A shear wall with rigid and flexible ring foundation. *Earth Science*, *29*(1), 44–55. doi:10.100711589-016-0139-2

Lombardi, D., Flora, A., Lirer, S., & Silvestri, F. (2014). Soil grouting for the seismic protection of existing buildings. *Ground Improvement*, *167*(3GI3), 206–218. doi:10.1680/grim.13.00010

Madabhushi, S. P. G., Schofield, A. N., & Lesley, S. (1998). A new stored angular momentum (SAM) based actuator. *Proceedings of Centrifuge*, *98*, 111–116.

May, T. W., & Bolt, B. A. (1982). Effectiveness of trenches in reducing seismic motion. *Earthquake Engineering & Structural Dynamics*, *10*(2), 195–210. doi:10.1002/eqe.4290100203

Mitrani, H., & Madabhushi, S. P. G. (2008). Centrifuge modelling of inclined micro-piles for liquefaction remediation of existing buildings. *Geomechanics and Geoengineering*, *3*(4), 245–256. doi:10.1080/17486020802483730

Mylonakis, G., & Gazetas, G. (2000). Seismic soil-structure interaction: Detrimental or beneficial. *Journal of Earthquake Engineering*, *4*(3), 277–301. doi:10.1080/13632460009350372

Schnabel, P. B., Lysmer, J., & Seed, H. B. (1972) *SHAKE: A computer program for earthquake response analysis of horizontally layered sites* (Report No. UCB/EERC-72/12). Earthquake Engineering Research Center, University of California, Berkeley, CA.

Schofield, A. N. (1980). Cambridge geotechnical centrifuge operations. *Geotechnique*, *30*(3), 227–268. doi:10.1680/geot.1980.30.3.227

Seed, H. B., Wong, R. T., Idriss, I. M., & Tokimatsu, K. (1986). Moduli and damping factors for dynamic analyses of cohesionless soils. *Journal of Geotechnical Engineering*, *112*(11), 1016–1032. doi:10.1061/(ASCE)0733-9410(1986)112:11(1016)

Tokimatsu, K., Mizuno, H., & Kakurai, M. (1996) Building damage associated with geotechnical problems. In Soils and Foundations special edition January 1996, on geotechnical aspects of the January 17th 1995 Hyogoken-Nambu earthquake (pp. 219-234). Academic Press. doi:10.3208andf.36.Special_219

Wang, J. G., Sun, W., & Anand, S. (2009). Numerical investigation on active isolation of ground shock by soft porous layers. *Journal of Sound and Vibration*, *321*(3-5), 492–509. doi:10.1016/j.jsv.2008.09.047

Wiegel, L. R. (1970). *Earthquake Engineering*. Englewood Cliffs, NJ: Prentice-Hall Inc.

Yang, Y. B., & Hung, H. H. (1997). Parametric study of wave barriers for reduction of train-induced vibrations. *International Journal for Numerical Methods in Engineering*, *40*(20), 3729–3747. doi:10.1002/(SICI)1097-0207(19971030)40:20<3729::AID-NME236>3.0.CO;2-8

Ye, L., & Tan, Y. (2011). Mitigation of building responses to DDC impacts by soft and stiff barriers. *Journal of Vibration and Control*, *17*(2), 259–277. doi:10.1177/1077546309355297

Yegian, M. K., & Catan, U. (2004). S*oil isolation for seismic protection using a smooth synthetic liner. Journal of Geotechnical and Geoenvironmental Engineering*, *130*(11), 1131–1139. doi:10.1061/(ASCE)1090-0241(2004)130:11(1131)

Yegian, M. K., & Kadakal, U. (2004). Foundation isolation for seismic protection using a smooth synthetic liner. *Journal of Geotechnical and Geoenvironmental Engineering*, *130*(11), 1121–1130. doi:10.1061/(ASCE)1090-0241(2004)130:11(1121)

Chapter 6
Numerical Simulation of Seismic Interaction of Pile Foundation and Structure

Pavan Kumar Emani
Graphic Era (Deemed University), India

Janaki Rama Raju Patchamatla
Graphic Era (Deemed University), India

Shashank Kothari
Graphic Era (Deemed University), India

Ritesh Kumar
Tokyo Tech, Japan

Vedula Srinivasa Phanikanth
Bhabha Atomic Research Centre, India

ABSTRACT

Modeling a semi-infinite soil domain has always been a challenge, especially in problems where the interaction between the soil domain and structure domain is a dynamically responsive one, as in the case of seismic wave loading. Under such conditions, researchers used various strategies to simulate, both physically and numerically, the behavior of pile foundations under seismic loading. While describing the challenges and hitherto applied strategies, two comprehensive numerical studies are described: one at behavioral state in terms of inelastic response spectrum and the other at the limiting state of collapse.

DOI: 10.4018/978-1-5225-6948-0.ch006

INTRODUCTION

Seismic soil-foundation-structure interaction (SFSI) is considerably simpler in shallow foundations due to negligible interaction among various layers or depths of soil, and also due to the absence of interactions between the foundation components. In contrast, for a pile foundation, the coupling among the various soil strata and the pile group effects do not allow much simplification, especially when combined with non-linear pile-soil separation, soil- plastification, pore-pressure or liquefaction-related effects, and wave-propagation effects. Among laboratory experimental works, centrifuge modeling is the closest to field conditions, as it can model all the above aspects. Analytical or numerical studies have isolated or uncoupled one or more aspects. Continuum Finite Element Modeling (FEM) of the soil skeleton with pore-pressure coupled dynamic formulation, contact-element-based pile-interface with consistent boundaries based on Boundary Element Method (BEM), is so far the closest simulation model to the actual behavior. The researchers studied various aspects that are prominent for the particular situation of concern. The present work provides a clear perspective on the state-of-art in numerical modeling, and discusses two such numerical studies in detail.

BACKGROUND

The numerical modeling of seismic soil-pile interaction essentially consists of strategies for modeling the seismic wave propagation through soil, inelasticity and pore-pressure development in soil, nonlinearity at soil-pile interface, interconnecting soil and pile, and avoiding unrealistic reflections from truncated interfaces. In the following paragraphs, an attempt is made to review the existing strategies for modeling these phenomena.

Continuum vs. Discrete Modeling of Spatial Dimension

Continuum modeling of soil-pile system essentially consists of four components (Figure 1): pile or pile group with pile cap, interface between pile and soil, near-field soil that can undergo plasticity, and far-field soil that accounts for the stiffness and damping effects of truncated semi-infinite soil. This continuum modeling can be computationally expensive, but it naturally simulates the a) wave propagation effects, b) soil plasticity, c) soil-pile separation, and d) pile group effects (Figure 2).

Figure 1. Finite Element Mesh of a continuum model showing pile elements, Near-field soil elements and far-field soil elements: a) plan and b) isometric view

(a)

(b)

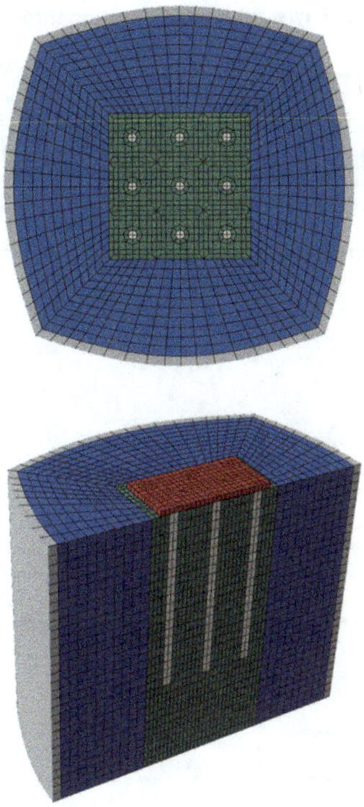

For practical or design purposes that neither need modeling the seismic wave propagation effects nor require pile group effects, the simple alternative of discrete modeling of soil using Winkler springs exists. Again, Winkler modeling can use static or dynamic p-y curves with or without dampers. For modeling radiation damping, more complex series of springs and dampers can be used (Nogami, Otani, Konagai, & Chen, 1992). In all discrete models, pile is modeled as a beam with rotational degrees of freedom. Figure 3 shows the typical details of such models, with the pile properties (modulus of elasticity E and linear density m of pile material, radius r_0, cross-sectional area A, length L) and the near-field soil properties (soil density ρ, shear modulus G_s, radius of near-field r_1 stiffness proportional damping ξ_k, mass proportional damping ξ_m). In fact, such models are derived on the basis of experimental results or continuum modeling.

Figure 2. Abilities of continuum models to simulate a) seismic wave propagation, b) soil plasticity, c) soil-pile slippage and gapping, and d) pile-group effects

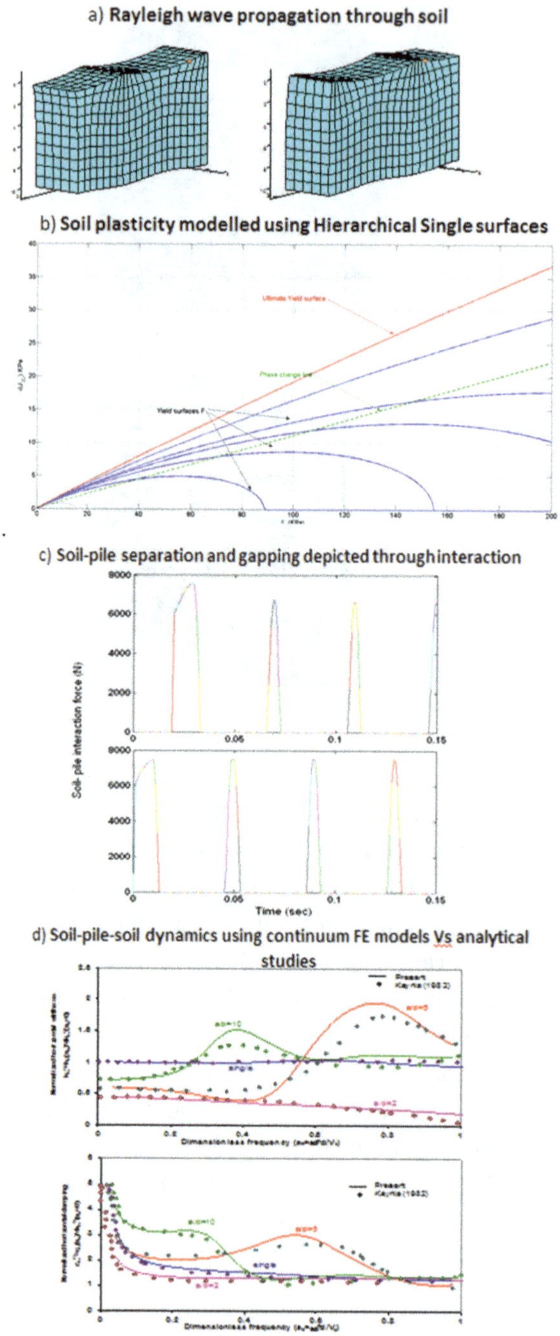

Figure 3. Discrete modeling of soil-pile systems using complex soil springs

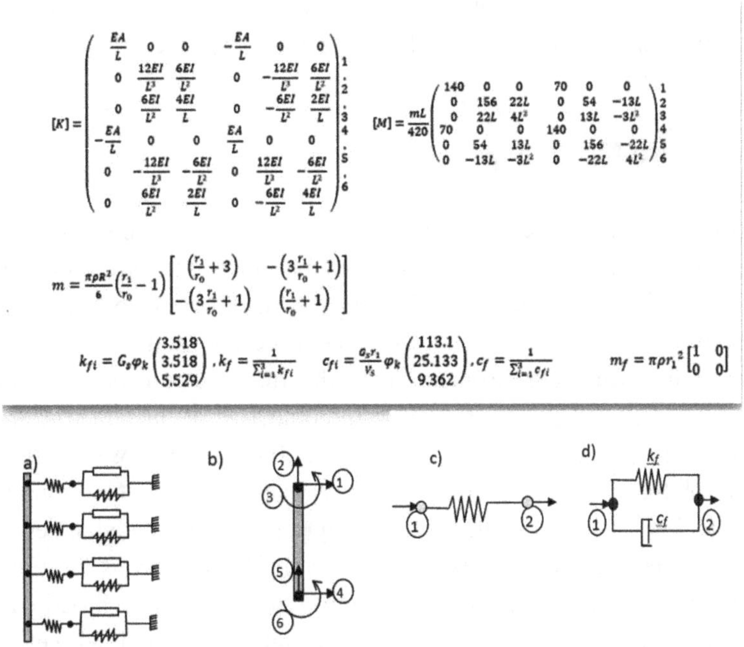

Frequency Domain vs. Time Domain Modeling of Time Dimension

Seismic/Dynamic analysis for both continuum and discrete modeling can be implemented conveniently through frequency domain procedures when only the steady-state condition is of interest. In such procedures, the frequency-dependent aspects of modeling, like the radiation boundary conditions, become considerably accurate. To obtain transient response using frequency domain procedures is computationally very expensive and must implement Fast Fourier Transforms (FFT) recursively. Further, the nonlinearities are very difficult to account for using frequency domain procedures. In the time domain, time stepping or time marching is done, which naturally accounts for the gradual development of nonlinearity and inelasticity.

Seismic Modeling

The seismic loading is quite complex, in that the seismic motion has complex distribution spatially, temporally, and frequency-wise. It varies with geographic location; topographic location; depth and stiffness of foundation; and presence of super-structure. It demands Ground Response Analysis (GRA), kinematic interaction analysis, and inertial interaction analysis. These analyses are often done in an uncoupled manner, which can be justified based on the phase lag in their incidence (Wang et al., 1998).

Pore-Pressure Modeling

Usually pore-pressure-related phenomena dominate in loose sands/soils, especially those used for land reclamation. Such phenomena can lead to partial liquefaction of the soil surrounding the piles, further leading to loss of lateral soil support to piles. In the case of sloping ground with a free face, excessive soil pressures caused by laterally spreading soils can take place. Again, the study can be made using continuum modeling or discrete modeling, based on the objective of the study (Cubrinovski et al., 2008; Genes & Kocak, 2005; Lehmann, 2005; Rahmani & Pak, 2012; Tang, Maula, Ling, & Su, 2014; Yazdchi, Khalili, & Valliappan, 1999). Continuum models rely on the ability of the constitutive modeling of soil to simulate the generation of pore-pressures and volume changes in soil-skeleton. Many rigorous and approximate models are available for this, and all of them fall within three broad categories: u-p-W formulation, u-U formulation, and u-p formulation (Zienkiewicz et al., 1999).

Stochastic Modeling Parametric uncertainty is inherent in geotechnical engineering. Traditionally the variability is dealt with using the global factor of safety approach, which cannot explicitly reflect the uncertainty of the individual underlying parameters and their correlation structure (Chan & Low, 2012). Geological processes and human-made influences cause fluctuations of properties within the soil. Even within an apparently homogeneous soil layer, variability in geotechnical properties is a well-established fact. Several researchers have investigated this soil variability in their work. Phoon and Kulhawy (1999) categorized the uncertainty of soil properties in their work to model soil variability. Considering the spatial variability of soil properties, the random fields can be considered an appropriate tool for reliability analysis in geotechnical engineering. Modeling the stochastic character of soil profiles is crucial in geotechnical performance prediction. Even within nominally homogeneous soil layers, engineering soil properties may exhibit considerable spatial variation. Therefore, a model is necessary for characterizing spatial variability, considering variability at a point and spatial dependence (1D, along with the depth of pile). Variability at a point determines what could be the possible values the soil parameter can acquire at a particular location, given a probability density function, over which spatial dependence imposes a conditional probability, depending on the values of surrounding points and the corresponding distances. For instance, uncertainty could be assumed for geometric soil stiffness. Generally, Moving Average (MA), Direct Fourier Transform (DFT), Fast Fourier

Transform (FFT), Turning Bands (TB), Generalized Stochastic Subdivision (GSS), Local Average Subdivision (LAS), and Covariance Matrix Decomposition (CMD) are some of the commonly used methods for generation of the random field. Figure 4 shows the typical variation of a random variable (Geometric soil stiffness, *Es*) along the depth of pile.

Analysis Methods

When the objective of the study is to determine the collapse load or ultimate load or safe load, then nonlinear effects and inelastic effects are to be incorporated, and usually a static nonlinear analysis, in the form of limit analysis or quasistatic pushover analysis, is performed. Such analyses involve studying the collapse mechanism and the possible modes of failure, involving a combination of phenomena like bearing failure of soil, bending failure of pile, buckling failure of piles, shear failure of piles, or a combination of any of them (Bhattacharya, 2003; Bhattacharya & Madabhushi, 2008; Dash, Bhattacharya, & Blakeborough, 2008). For example, Finn and Fujita (2002) and Wu and Finn (1997) studied the transient response of socketed single-pile in terms of pile head displacement or pile bending moments during seismic loading, and compared them with the measured responses during centrifuge tests or field tests. Discrete modeling of soil using p-y, t-z springs, and Kelvin elements essentially uncouples the interaction between the layers and requires complicated empirical rules to simulate the real behavior of pile groups, especially considering the pore-pressure development and liquefaction. Boulanger, Curras, Kutter, Wilson, and Abghari (1999) established the applicability of dynamic p-y curves for seismic soil-pile interaction using comparisons with centrifuge test results.

In the present chapter, two important cases of seismic soil-pile modeling are presented: generation of inelastic seismic response spectrum for designing structures

Figure 4. Typical model for random soil variable

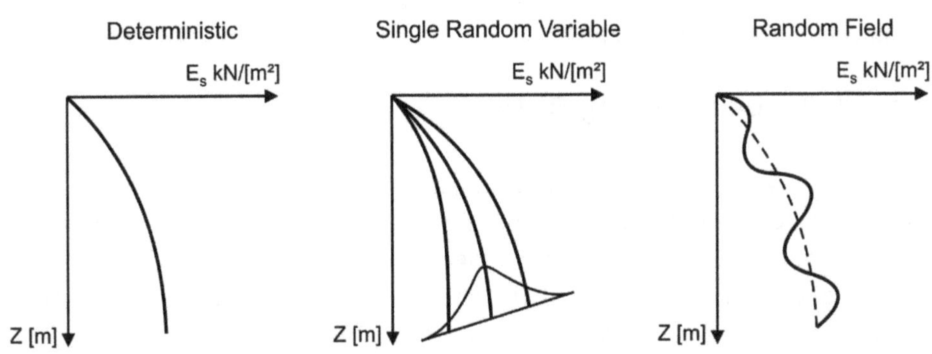

resting on a single pile; and collapse load analysis of a two-pile group under design axial load with monotonously increasing lateral load.

INELASTIC RESPONSE SPECTRUM FOR SINGLE PILE STRUCTURE

Design of pile foundations in soils in general and in liquefying soils in particular has received great attention in the recent past, due to the discovery of new failure modes. Analysis of pile-soil dynamics is especially complicated because of the presence of semi-infinite soil and a limited zone of nonlinearity. This is further complicated by liquefaction and consequent loss of lateral soil support, passive earth pressures from moving soil mass, and the destabilizing effect of axial loads on the pile (Madabhushi, Knappett, & Haigh, 2010). Both continuum modeling (Maheshwari & Emani, 2014; Emani & Maheshwari, 2009) and discrete or Winkler spring modeling of soils (Nogami et al., 1992; Wang et al., 1998) are carried out by researchers for nonlinear seismic analysis of soil-pile systems, even under liquefaction conditions (PhaniKanth, Choudhury, & Reddy, 2013; Maheshwari, Truman, Gould, & El Naggar, 2005). For design practices, the importance of response spectra is identified quite early. Housner (1959) and Newmark (1967) contributed to early developments of design spectra. Newmark and Hall (1973) presented the detailed inelastic behavior of elastic-perfectly plastic structures, while Riddell and Newmark (1979) presented guidelines for deriving inelastic response spectra based on statistical methods. Tarquis & Roesset (1992) studied the applicability of previously derived design spectra for narrow-banded earthquakes.

The present work is undertaken with a motivation to simulate the behavior of piles and pile groups in liquefying soils under seismic vibrations using continuum modeling, and to generate inelastic response spectra for a structural single degree of freedom (SDOF) system. In the present work, only the approximate solution is presented using discrete modeling of the soil support. This approximation involves the use of one-dimensional soil springs and dampers to simulate the stiffness and damping of a semi-infinite domain.

Soil-Pile Modeling Using BNW

The soil-pile system consists of a single 10-m-long circular pile with a diameter of 0.4 m, embedded completely in a nonlinear soil. The soil-pile system is modeled as a beam supported by nonlinear Winkler soil (BNW) springs in the near field and a series of dampers in the far field. The spring stiffness values, as well as those for the damping of the far-field dampers, are taken from Nogami et al. (1992). The inertia of the pile is consistently modeled, while the inertia of the soil system is also suitably modeled by point masses lumped at the appropriate height of the pile beam (Maheshwari et al., 2005). The various aspects of the numerical (Finite Element) model of the soil-pile system are described in detail in Table 1 below.

Loading

The present work considers the following loads. Seismic load is applied at the base of the pile in the form of displacement time history, which is a synthetic time history compatible with the design spectrum of Indian Standards (2009) IS 1893:2002 (Part-I). Seismic inertial load at the top of the pile is induced by a point/lumped mass at a height of 4 m above the pile top. The mass forms the SDOF system representative of the mass of the structure supported by the pile. The stiffness of the structure is represented by the stiffness of the beam element connecting the mass to the pile

Table 1. Single-pile interaction with nonlinear soil under external dynamic load

Feature	Description (Figure 5)
Pile	
Geometry	Circular pile of length 10 m and 0.4m diameter
Material	E=25E9 Pa, $\rho = 2500$ kg/m^3
Meshing	Linear flexural beam element with consistent mass, and no damping. Ten elements are used in 10m length
Boundary conditions	A harmonically horizontal displacement load is applied on the top and the base is kept fixed
Near-Field Soil	
Geometry and meshing	Two noded nonlinear springs are used to model the exponential degradation of soil stiffness due to plasticity. Also, lumped mass elements model the inertia of NF soil. Lumped mass on pile side is 188 kg and on far-field side is 110 kg. No damping in near-field zone
Plasticity (Figure 6)	Exponential degradation of strain hardening is modeled as $F = F_y + Q_\infty \left(1 - e^{-bup}\right)$ where F_y=6000 N, Q_∞=600000N, b=0.3 are used
Far-Field Soil	
Geometry and meshing	The stiffness and damping of the far-field soil is modeled by spring-dashpot assembly in series. The equivalent stiffness and damping are 1.9066e+007 and 4.9968e+005. Mass lumped on pile-side is 35 kg.
Soil-Pile Interface	
Elements	Gap elements are used to model the separation between soil and pile

Figure 5. FE model of the single-pile soil system

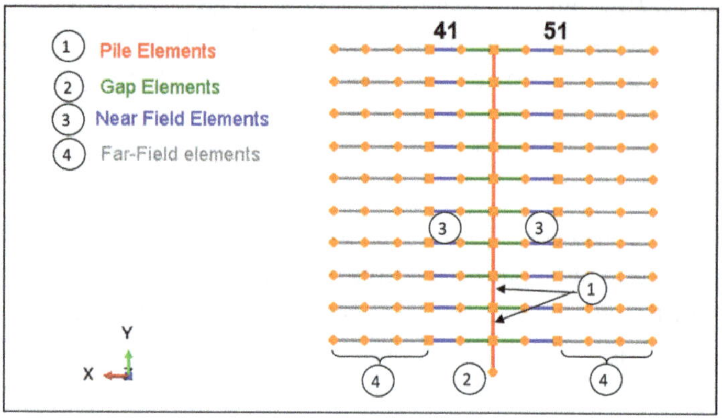

Figure 6. Force-displacement relation of nonlinear near-field soil springs (after)

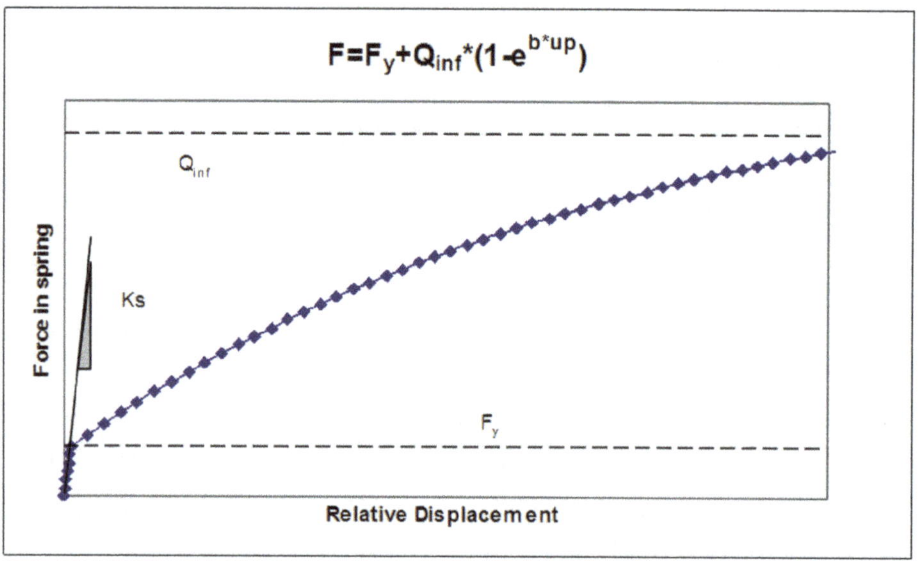

top. This SDOF gives both translational and rotational effects of the superstructure (Figure 7).

For verification problems, harmonic (displacement) load applied laterally at the top of the pile is used. Researchers use harmonic loading to study the frequency response of structures and systems (e.g., Cheung, Tharn, & Lie, 1995). The input displacement and its characteristics are described in the results.

Figure 7. Seismic loading on the soil-pile-structure system

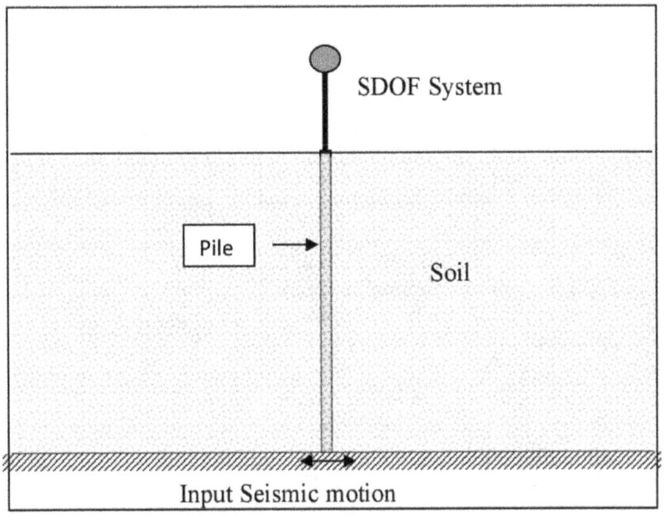

Generation of Inelastic Response Spectrum

The flowchart in Figure 8 shows the methodology adopted to generate the inelastic response spectra of structural SDOF system.

The generation of seismic response spectra for inelastic structures (Chopra, 2001; Humar, 1990) provides a useful summary of the seismic response. Synthetic strong ground motions (Gasparini & Vanmarcke, 1976) were applied at the base of the pile to generate inelastic response spectra. A total of six earthquake signatures were taken and modified to get input response spectra (as per IS 1893(Part 1):2002) using the MATLAB program and SeismoMatch software

Synthetic time histories were obtained for all three response spectra (i.e., hard soil, medium soil, and soft soil) as specified in IS 1893(Part 1):2002. A total of six strong ground motions (earthquake signatures) were selected (out of 10 strong ground motions with strong fit to the input response spectra) for generation of synthetic acceleration time histories.

Results and Discussion

To investigate the effect of seismic SSI, synthetic time histories were applied at the base of the pile. An inelastic response spectrum for structural SDOF system came out with significant magnification in magnitude, compared to specified response spectra in IS 1893(Part 1):2002. This is because response spectrum of IS 1893(Part 1):2002 is generated by applying the time histories at the base of the SDOF system,

Figure 8. Flow chart for generation of inelastic response spectra of structural SDOF system.

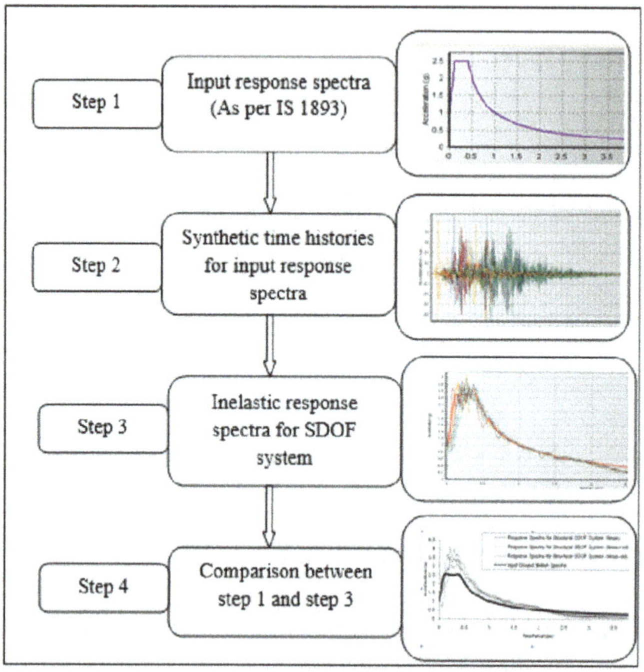

and thus does not consider the effect of ground response analysis (GRA) along with other phenomena of seismic soil structure interaction (i.e., kinematics and inertial interactions), as well as the inelasticity of the soil support. This study applied strong ground motion (synthetic earthquake time history) at the base of the pile. The motion is then modified with the help of GRA (using the transfer function in MATLAB code) for each level of the soil-pile system. The GRA, kinematics interaction, and the soil inelasticity will modify the ground motion (excitation) experienced by the SDOF at its base (Kramer, 1996; Wolf & Deeks, 2004). In fact, even the inertial interaction modifies the motion at the base of the structure (or the top of the pile).

While investigating the inelastic response spectra for the SDOF system, it was found that for soft soil the magnification factor was between 1.16 and 1.5 for all six synthetic time histories. Similarly, the factor was between 1.12 and 1.38 for medium soil, and 1.1 to 1.34 for rocky/hard soil (Figure 9, Figure 10, and Figure 11 respectively).

Figure 9. Inelastic response spectra for structural SDOF system for soft soil

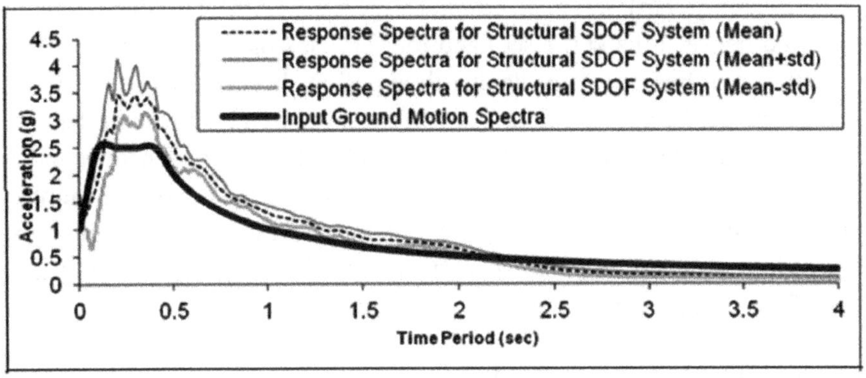

Figure 10. Inelastic response spectra for structural SDOF system for medium soil

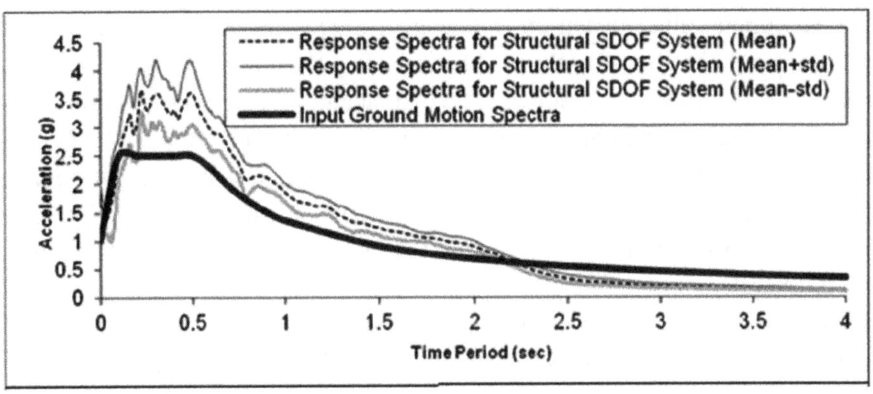

Figure 11. Inelastic response spectra for structural SDOF system for hard soil

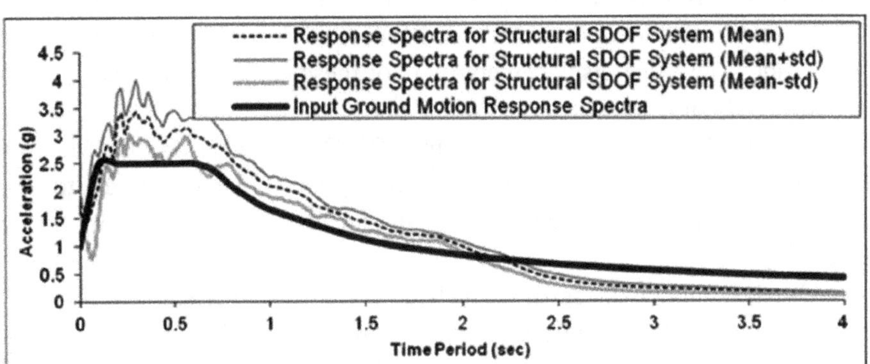

Figure 9 makes clear that the magnitude of inelastic response spectra of the structural SDOF system is less than the input response spectra for very rigid structures (time period between 0 and 0.1 sec) and for very flexible structures (time period more than 2 sec). For structures having time periods between 0.1 sec and 2 secs, the magnitude has been magnified. The same kind of trend is found in the case of medium soil and soft soil (Figure 10 and Figure 11). The trend of generated inelastic response spectra for a structural SDOF system also depends upon the number of synthetic time histories considered for the investigation and their compatibility with the particular site. Hence, a band is generated for the inelastic response spectra with a range of $\pm\sigma$ (standard deviation) about the mean.

Verification of MATLAB Code

The MATLAB program and all its features (i.e., geometry modeling, stiffness modeling, mass modeling, damping modeling, loads and boundary conditions, along with the dynamic solver algorithm and the post-processing algorithms) are verified by comparing the results with those obtained using the commercial Finite Element Analysis (FEA) software ABAQUS. The displacement responses at various levels of the pile are shown in Figure 12. The comparison between the MATLAB results and those from ABAQUS is excellent. The input motion (displacement boundary condition) at the top of the pile can be seen as a harmonic motion of amplitude 0.01 m. The effect of inertia can be seen from the phase lag in the response at various levels of the pile.

Displacement and force history in the nonlinear near-field soil spring can also be verified using the ABAQUS results. In Figure 13, the displacement history of near-field node 45, node 95, and pile node 6 is shown. Figure 12 shows the force history of the near-field elements at the top of the pile (Element No. 41 and Element No. 51 on either side of the pile top as labeled in Figure 13). Again, the comparison between MATLAB response history (Figure 14a) and that from ABAQUS (Figure 14b) is excellent. The force histories clearly show the formation of gap/separation on either side. The force-displacement plots of these elements are also shown on the left side of Figure 14a.

Similarly, the Bending Moment Diagrams (BMD) obtained from both MATLAB and ABAQUS (Figure 15a and Figure 15b) are in good agreement, except for small changes in the location of inflexion points and locations of peak bending moment values.

A numerical model for seismic analysis of a single vertical pile supported by inelastic soil is developed and verified. All aspects of the results—namely, the displacement histories, bending moment histories, soil-pile interaction force histories—are verified and found to be in close agreement with the solutions given

Figure 12. Displacement responses at various levels of the pile in MATLAB and ABAQUS

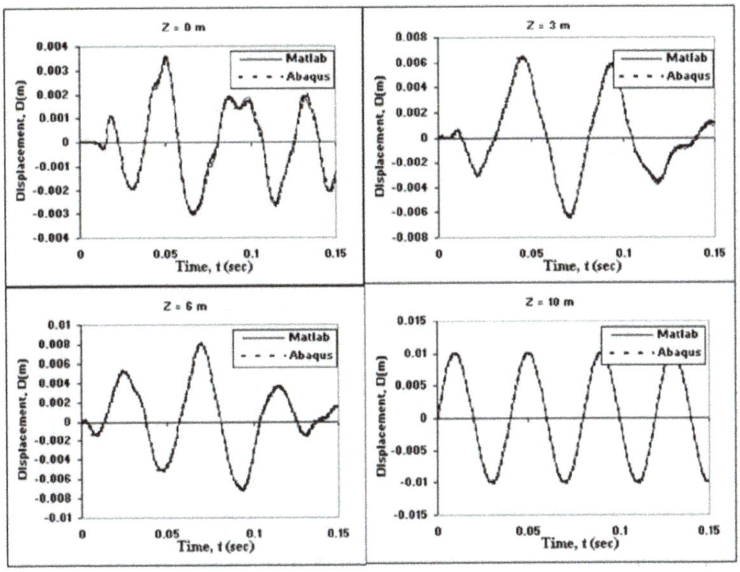

Figure 13. (a) Displacement history of near field nodes (b) Location of nodes in model

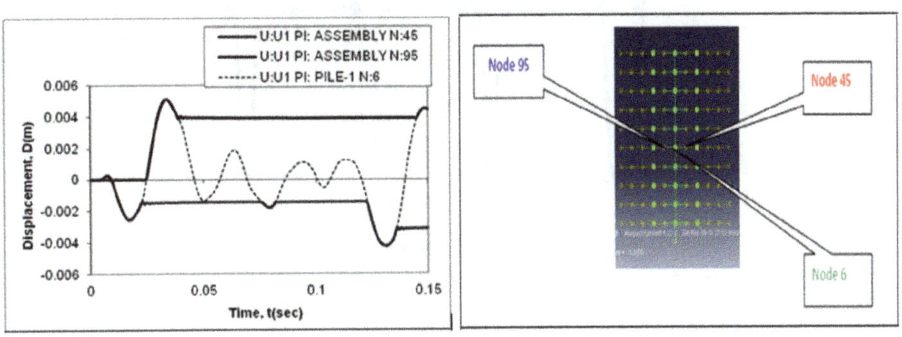

by the well-tested commercial FEA package (ABAQUS). Synthetic time histories compatible with the seismic design spectrum of IS 1893 are generated and modified to account for 1D ground response at different levels of the stratum. The modified ground motions are applied at the base of the Finite Element model of the inelastic soil-pile system to simulate the effects of kinematics and inertial interactions. Thus, all aspects of seismic soil-pile interactions are accounted for. The inelastic responses are postprocessed to present in the form of response spectra. The results show that

Figure 14. (a) Force history of the near-field elements at the top of the pile using MATLAB (b) Force history of the near-field elements at the top of the pile using ABAQUS

(a)

(b)

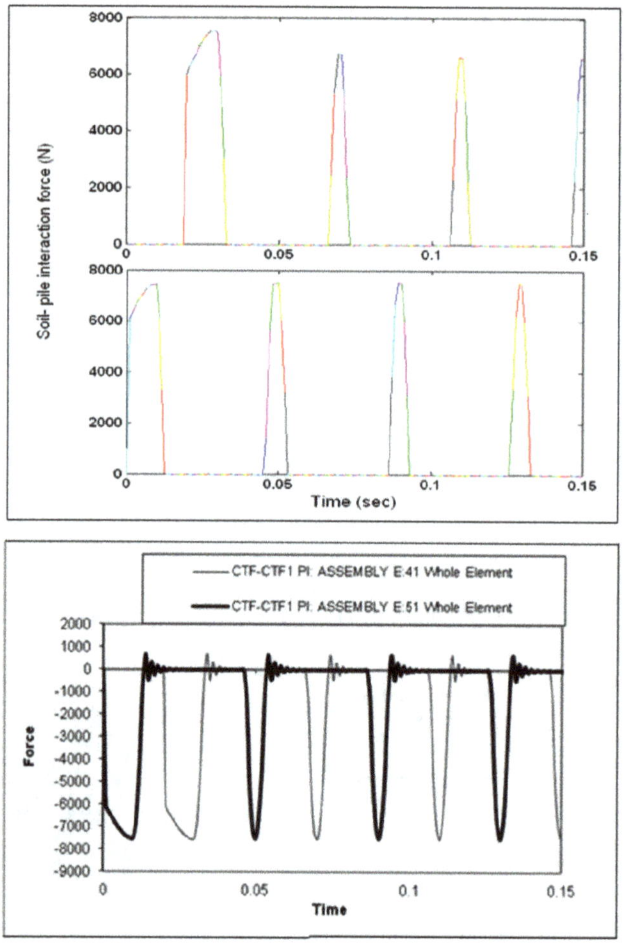

a structure supported on piles attracts accelerations up to 1.5 times higher than a fixed-base structure.

COLLAPSE LOAD ANALYSIS OF RC PILE GROUP UNDER LATERAL LOADING

Seismic analyses carried out for design purposes are rather simple compared to those analyses performed for research and verification studies. Typically, the seismic

Figure 15. (a) BMD of pile at t=0.1 sec b) BMD of pile at t=0.15 sec

(a)

(b)

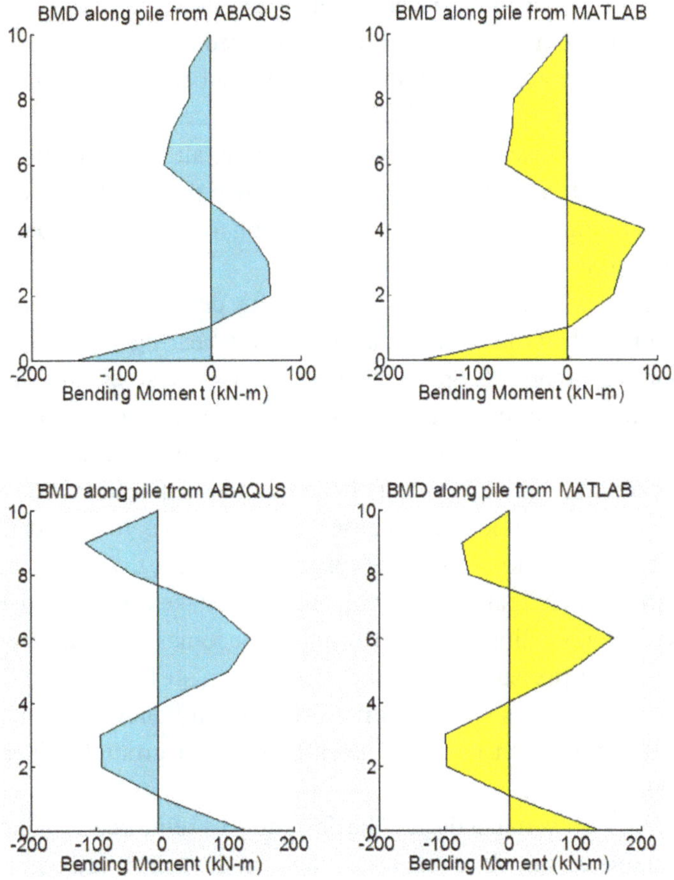

design codes recommend static nonlinear analyses for most structures. The present section means to demonstrate methods for design engineers. The ultimate load analysis of RC pile groups has assumed much significance under liquefying soil conditions, especially due to post earthquake studies following the 1964 Niigata, 1995 Kobe, and 2001 Bhuj earthquakes. The present study reports the results of numerical simulations on pile groups subjected to monotonically increasing lateral loads under design amounts of pile axial loading. The soil liquefaction has been considered through nonlinear p-y relationship of the soil springs, which can vary along the depth/length of the pile. Again, this variation is related to the liquefaction potential of the site and the magnitude of the seismic shaking. As the piles in the group can reach their extreme deflections and rotations during increased amounts of lateral loading, a precise modeling of the inelastic behavior of the pile cross-section is done, considering the complete stress-strain behavior of concrete, with and without confinement, and reinforcing steel, including the strain-hardening portion. The possibility of the inelastic buckling of the individual piles is considered in the overall collapse modes. The model is analyzed using Riks's analysis (1972) in finite element software to check the postbuckling behavior and plastic collapse of piles. The results confirm the kinds of failure modes predicted by centrifuge test results reported by researchers on the pile group, although the pile material used is significantly different from that of the simulation model. The extension of the present work promises an important contribution to the design codes for pile groups in liquefying soils.

Piles are considered to be most promising deep foundations to support the superstructures when the underlying soil has low bearing capacity. Along with the axial compressive loads of superstructure, piles also must resist the lateral loads due to seismic activities in earthquake-prone areas, leading to liquefaction of soil. In such cases, the collapse load analysis of piles under liquefaction for lateral loading is quite important. One of the most challenging issues in the field of geotechnical engineering is liquefaction that causes great damage to the structures during earthquakes. Liquefaction was a chief cause of damage to the pile foundations during studies of the 1964 Niigata and 1995 Kobe earthquakes (Kramer, 1996).

Under seismic loading, piles performance becomes a complex problem due to excess pore-water pressures, as well as decrease in stiffness of the saturated soils. Past earthquakes studies prove that most damage in liquefiable soils has occurred due to inertial forces on piles and lateral forces on ground (Liyanapathirana & Poulos, 2005).

Studies have proved that during the designing of reinforced concrete piles, if the length-to-diameter ratio for a pile is less than 30, the stress on the pile is always less than Euler's stress. Therefore, there is no chance of buckling failure, and the pile section, designed to resist the expected bending moments, will also be safe against buckling and formation of plastic hinge (Mokhtar,Abdel-Motaal & Wahidy, 2014).

The Beam on Nonlinear Winkler Foundation (BNWF) model, the most versatile soil-pile interaction model, assumes soil to be a set of springs attached to the pile along the whole length of piles. For seismic loading, the lateral soil springs (referred as p-y springs) play a vital role in determining the response of piles. These p-y relationship curves vary with excess pore-water pressure, which in turn influences the lateral pile response subjected to displacement loading (Dash et al., 2008)

A centrifuge model test found that at low strain level, the liquefied soil shows negligible stiffness and strength, but strain-hardening behavior with increase in stiffness at large strain. This behavior also must be included in exact modeling of p-y curves for soil springs (Lombardi et al., 2017).

During elastic modeling of piles, generally beam elements are modeled with a linear force-displacement relationship, assuming zero tensile strength of concrete. However, with constant axial forces and increasing bending moment, in the initial stage the moment-curvature relationship may be linear, but with variation of the compressed part of a section in later stages, moment-curvature relationship becomes nonlinear and results in inconsistent effective moment of inertia. This effect of tension shift provides a more realistic force-displacement relationship (Bosco, Ghersi, & Leanza, 2008).

Description of the Model

The results reported in this study are of a socketed two-pile group connected by a relatively rigid pile cap embedded in a liquefiable soil stratum and subjected to monotonously increasing static lateral load, under service axial load conditions.

Piles

The piles are RC piles in which M25-grade concrete and Fe 415-grade steel are used. Each of the piles has an outer diameter of 0.5 m and is 10 m long (Figure 16). The concrete cover to the confining reinforcement is 50 mm. The main longitudinal reinforcement is 0.82%. The confining reinforcement and its effect on the stress-strain behavior of confined concrete is described in Figure 17 (Park & Pauley, 1975). A value of Z = 52.6 describes the confinement of concrete in the RC pile. Figure

Figure 16. Model layout of socketed two-pile group connected by rigid pile cap embedded in liquefiable soil strata

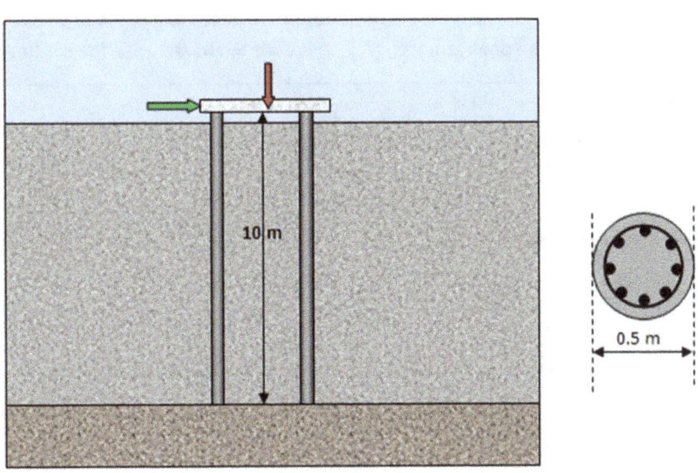

18 shows the moment-curvature relationship for such circularly confined concrete reinforced with strain-hardening steel reinforcement. The onset of strain hardening of compression and tension steel can be seen from the plot. The moment-curvature behavior is also a function of the axial load, which can be assumed not to vary during seismic loading. The results reported in the present work are for High Yield Strength Deformed (HYSD) Bars, so the strain-hardening effects are not present.

Soil Conditions

The soil surrounding the pile is assumed to be nonlinearly elastic. The nonlinear p-y curves for the soil springs are obtained as per the American Petroleum Institute (API) guidelines (1991), but modified to consider the change in behavior under liquefaction. The ultimate soil resistance (P_u) is obtained using the API formula for sand having internal friction of 30° and mass density of 1800 kg/m^3, and subjected to noncyclic loading. The nonlinear p-y curves for the soil stiffness also vary based on the extent of liquefaction. The zone of liquefaction (Z_L) governs the r_u-profile, and based on the r_u-value, the p-y curves are as given in Figure 19 (Dash et al., 2008). For the results presented herein, a depth of z_L =1 m, 3 m, and 7 m is assumed to be completely liquefied.

Figure 17. Stress-strain behavior of confined concrete(

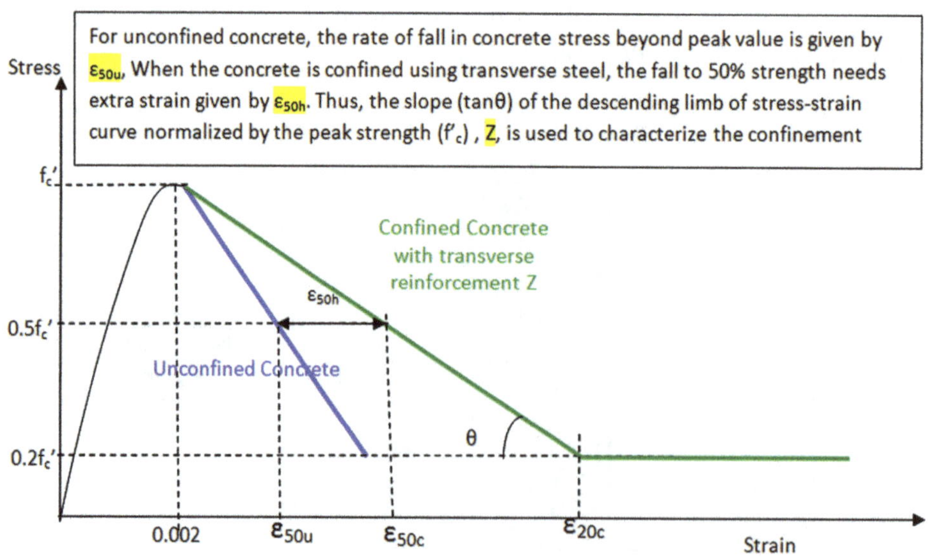

Figure 18. Moment-Curvature relationship for the pile section under various confinements

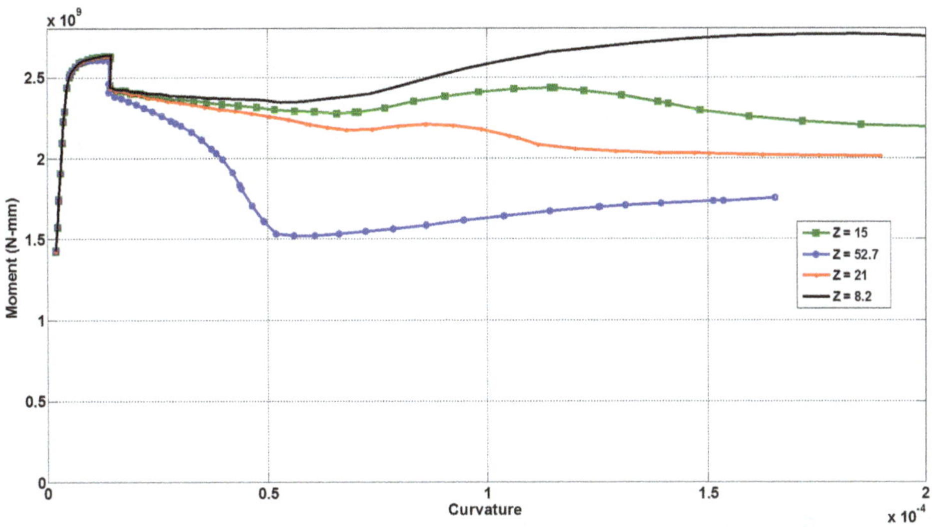

Loading

The pile group is subjected to static nonlinear loads, so that the analysis results are of some relevance to seismic conditions and liquefaction. Only lateral load is varied, while the axial load is kept constant at $0.3f_{ck}D^2$. Basically, a push-over analysis is performed on the pile group while considering the possibility of formation of failure mechanism. The failure is identified as the case for which convergence could not be reached.

Numerical (Finite Element) Modeling of Pile-Soil System

The pile-soil system is modeled in ABAQUS 6.13 software, as Beam on Nonlinear Winkler Foundation (BNWF), where both the piles of the group are modeled as two-noded cubic-beam elements. The cap is also modeled as a rigid-beam element. For ensuring plastic section behavior of the pile cross-section, rotational springs with desired moment-rotation properties are used at 0.5 m spacing between the elastic beam elements. Also, the soil springs are modeled as axial connector elements with nonlinear elastic behavior. For a monotonic loading, the nonlinear elastic behavior simulates the inelastic behavior, as there will be no unloading.

Figure 19. Nonlinear p-y curves at different depths of liquefied strata

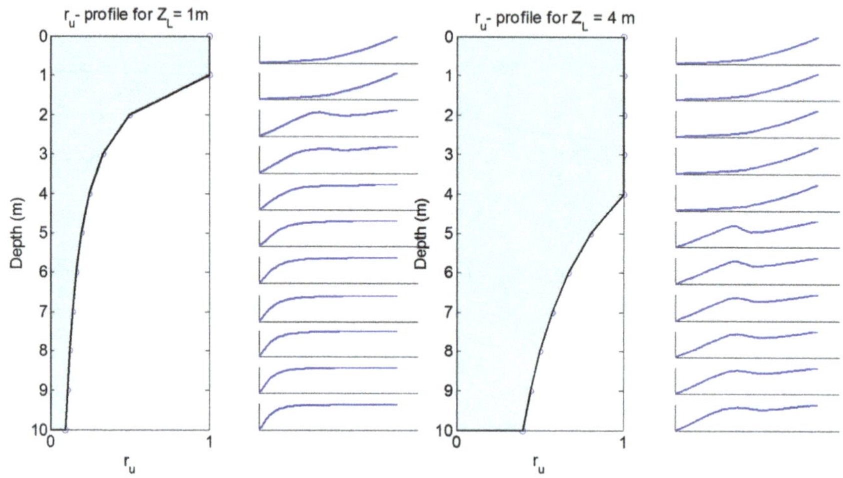

RESULTS AND DISCUSSION

The collapse load is identified during Riks's analysis (1972, 1979) as an unstable state wherein numerical convergence is not achieved. This can happen due to formation of enough plastic hinges, leading to development of mechanism; buckling of both the piles; and a combination of plastic-hinge formations and buckling of individual pile(s). The collapse loads for the three degrees of liquefaction are shown in the Figure 20. Figure 21 shows bending moment at critical locations in piles and location of plastic hinges.

Figure 20. Deformed shapes just before collapse for various degrees of liquefaction

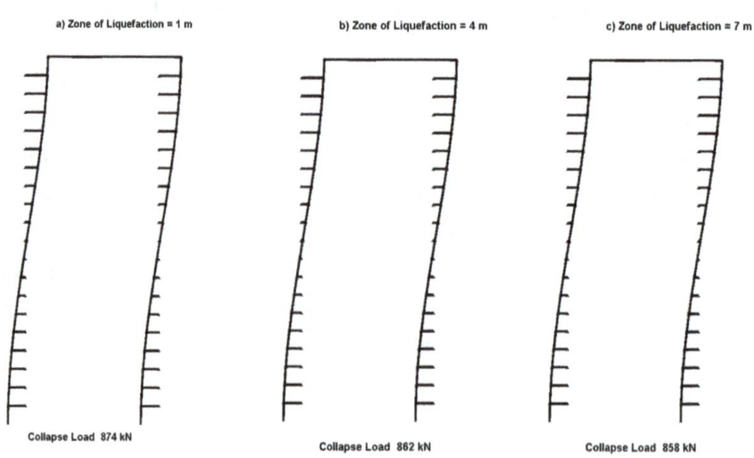

Figure 21. Bending moments at critical locations in piles and location of plastic hinges

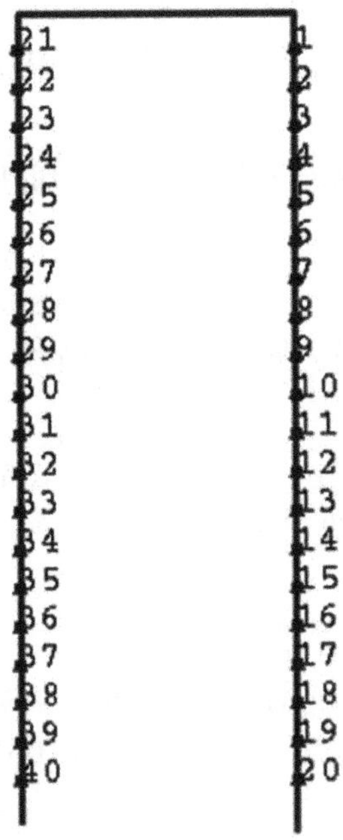

Location (See ←)	Moment (kN-m)		
	Z_L= 1m	Z_L= 4m	Z_L= 7m
1	**2256.487**	**2255.831**	**2254.917**
2	2055.156	2058.086	2058.526
3	1844.072	1850.578	1852.367
4	1620.804	1630.916	1634.065
5	1385.994	1400.514	1405.022
16	-1338.06	-1349.14	-1352.48
17	-1556.49	-1575.66	-1582.67
18	-1763.97	-1791.83	-1802.62
19	-1959.47	-1996.31	-2010.9
20	**-2142.01**	**-2187.83**	**-2206.17**
21	**2324.003**	**2323.643**	**2323.983**
22	2067.617	2071.838	2073.311
23	1815.705	1822.741	1824.819
24	1566.012	1575.945	1578.655
25	1317.395	1331.107	1334.471
36	-1276.22	-1286.57	-1289.55
37	-1507.49	-1525.51	-1531.87
38	-1739.92	-1766.36	-1776.28
39	-1974.12	-2009.52	-2023.1
40	**-2210.71**	**-2255.37**	**-2272.67**
Note: **M_p= 2324 kN-m**			

CONCLUSION

The research reports the strategies for seismic soil-pile-structure interactions. It also demonstrates two important studies for seismic analysis of structures on pile foundations. The first study reports the inelastic response spectrum for a given pile foundation, and the second reports the lateral collapse load under liquefying soil conditions. Both studies consider the inertial interactions from structure; but dynamic interaction appears only in the former case, while in the latter case, a static nonlinear analysis was needed.

REFERENCES

American Petroleum Institute. (1991). *RP2A: Recommended practice for planning, designing and constructing fixed offshore platforms* (19th ed.). Washington, DC: American Petroleum Institute.

Bhattacharya, S. (2003). *Pile instability during earthquake* (Unpublished Ph.D. thesis). University of Cambridge, UK.

Bhattacharya, S., & Madabhushi, S. P. G. (2008). A critical review of methods for pile design in seismically liquefiable soils. *Bulletin of Earthquake Engineering, 6*(3), 407–446. doi:10.100710518-008-9068-3

Bosco, M., Ghersi, A., & Leanza, S. (2008). Force-displacement relationships for R/C members in seismic design. *Proceedings of the 14th World Conference on Earthquake Engineering.*

Boulanger, R. W., Curras, C. J., Kutter, B. L., Wilson, D. W., & Abghari, A. (1999). Seismic soil-pile-structure interaction experiments and analyses. *Journal of Geotechnical and Geoenvironmental Engineering, 125*(9), 750–759. doi:10.1061/(ASCE)1090-0241(1999)125:9(750)

Chan, C. L., & Low, B. K. (2012). Probabilistic analysis of laterally loaded piles using response surface and neural network approaches. *Computers and Geotechnics, 43*, 101–110. doi:10.1016/j.compgeo.2012.03.001

Cheung, Y. K., Tharn, L. G., & Lie, Z. X. (1995). Transient response of single pile under harmonic excitation. *Earthquake Engineering & Structural Dynamics, 24*(7), 1017–1038. doi:10.1002/eqe.4290240707

Chopra, A. K. (2001). *Dynamics of structures*. Delhi: Pearson Education.

Cubrinovski, M., Uzuoka, R., Sugita, H., Tokimatsu, K., Sato, M., Ishihara, K., & Kamata, T. (2008). Prediction of pile response to lateral spreading by 3-D soil-water coupled dynamic analysis: Shaking in the direction of ground flow. *Soil Dynamics and Earthquake Engineering, 28*(6), 421–435. doi:10.1016/j.soildyn.2007.10.015

Dash, S. R., Bhattacharya, S., & Blakeborough, A. (2008). *Bending-buckling interaction: Investigation of this failure mechanism of pile foundations in seismically liquefiable deposits. Technical Report-2302/08*. Department of Engineering Science, University of Oxford.

Emani, P. K., & Maheshwari, B. K. (2009). Dynamic impedances of pile groups with embedded caps in homogeneous elastic soils using CIFECM. *Soil Dynamics and Earthquake Engineering, 29*(6), 963–973. doi:10.1016/j.soildyn.2008.11.003

Finn, W. D. L., & Fujita, N. (2002). Piles in liquefiable soils: Seismic analysis and design issues. *Soil Dynamics and Earthquake Engineering*, *22*(9-12), 731–742. doi:10.1016/S0267-7261(02)00094-5

Gasparini, D. A., & Vanmarcke, E. H. (1976). *Simulated earthquakes compatible with prescribed response spectra. Publication No. R76-4*. Cambridge, MA: Massachusetts Institute of Technology.

Genes, M. C., & Kocak, S. (2005). Dynamic soil-structure interaction analysis of layered unbounded media via a coupled finite element/boundary element/scaled boundary finite element model. *International Journal for Numerical Methods in Engineering*, *62*(6), 798–823. doi:10.1002/nme.1212

Housner, G. W. (1959). Behavior of structures during earthquakes. *Journal of the Engineering Mechanics Division*, *85*(4), 109-130.

Humar, J. L. (1990). *Dynamics of structures*. Englewood Cliffs, NJ: Prentice Hall.

Indian Standards. (2009). *IS 1893(Part 1):2002. Criteria for Earthquake Resistant Design of Structures*. General Provisions and Buildings.

Kramer, S. L. (1996). *Geotechnical earthquake engineering*. Prentice Hall.

Lehmann, L. (2005). An effective finite element approach for soil-structure analysis in the time-domain. *Structural Engineering and Mechanics*, *21*(4), 1–14. doi:10.12989em.2005.21.4.437

Liyanapathirana D. S., & Poulos H. G. (2005). Seismic lateral response of piles in liquefying soil. *Journal of Geotechnical and Geo-Environmental Engineering*, *131*(12), 1466e79.

Lombardi, D., Dash, S. R., Bhattacharya, S., Ibraim, E., Wood, D. M., & Taylor, C. A. (2017). Construction of simplified design p-y curves for liquefied soils. *Geotechnique*, *67*(3), 216–227. doi:10.1680/jgeot.15.P.116

Madabhushi, G., Knappett, J., & Haigh, S. (2010). *Design of pile foundations in liquefiable soils*. London: Imperial College Press.

Maheshwari, B. K., & Emani, P. K. (2014). Three-dimensional nonlinear seismic analysis of pile groups using FE-CIFECM coupling in a hybrid domain and HiSS plasticity model. *International Journal of Geomechanics*, *15*(3), 04014055-1-12.

Maheshwari, B. K., Truman, K. Z., Gould, P. L., & El Naggar, M. H. (2005). Three-dimensional nonlinear seismic analysis of single piles using finite element model: Effects of plasticity of soil. *International Journal of Geomechanics*, *5*(1), 35–44.

Moghtar, A. S., Abdel-Motaal, M. A., & Wahidy, M. (2014). Lateral displacement and pile instability due to soil liquefaction using numerical model. *Ain Shams Engineering Journal*, *5*(4), 1019–1032. doi:10.1016/j.asej.2014.05.002

Newmark, N. M. (1967). Design criteria for nuclear reactors subjected to earthquake hazards. *Proc., IAEA Panel on a Seismic Design and Testing of Nuclear Facilities.*

Newmark, N. M., & Hall, W. J. (1973). *Procedures and criteria for earthquake resistant design. building practice for disaster mitigation*. Washington, DC: National Bureau of Standards.

Newmark, N. M., & Riddell, R. (1980, September). Inelastic spectra for seismic design. *Proc. 7th Worm Conf. on Earthquake Engineering*, 4, 129-136.

Nogami, T., Otani, J., Konagai, K., & Chen, H. L. (1992). Nonlinear soil-pile interaction model for dynamic lateral motion. *Journal of Geotechnical Engineering*, *118*(1), 89–106. doi:10.1061/(ASCE)0733-9410(1992)118:1(89)

Park, R., & Paulay, T. (1975). *Reinforced concrete structures*. John Wiley & Sons, Inc. doi:10.1002/9780470172834

Phanikanth, V. S., Choudhury, D., & Reddy, G. R. (2013). Behavior of single pile in liquefied soil deposits during earthquakes. *International Journal of Geomechanics*, *13*(4), 454–462. doi:10.1061/(ASCE)GM.1943-5622.0000224

Phoon, K. K., & Kulhawy, F. H. (1999). Evaluation of geotechnical property variability. *Canadian Geotechnical Journal*, *36*(4), 625–639. doi:10.1139/t99-039

Rahmani, A., & Pak, A. (2012). Dynamic behavior of pile foundations under cyclic loading in liquefiable soils. *Computers and Geotechnics*, *40*, 114–126. doi:10.1016/j.compgeo.2011.09.002

Riddell, R., & Newmark, N. M. (1979). *Statistical analysis of the response of nonlinear systems subjected to earthquakes. University of Illinois Engineering Experiment Station. College of Engineering*. University of Illinois at Urbana-Champaign.

Riks, E. (1972). The application of Newton's method to the problem of elastic stability. *Journal of Applied Mechanics*, *39*(4), 1060–1065. doi:10.1115/1.3422829

Riks, E. (1979). An incremental approach to the solution of snapping and buckling problems. *International Journal of Solids and Structures*, *15*(7), 529–551. doi:10.1016/0020-7683(79)90081-7

Tang, L., Maula, B. H., Ling, X., & Su, L. (2014). Numerical simulations of shake-table experiment for dynamic soil-pile-structure interaction in liquefiable soils. *Earthquake Engineering and Engineering Vibration, 13*(1), 171–180. doi:10.100711803-014-0221-5

Tarquis, F., & Roesset, J. M. (1992). Inelastic response spectra for narrow-banded earthquakes. *Proceedings of 10th World Conference on Earthquake Engineering,* 2, 645-650.

Wang, S., Kutter, B. L., Chacko, M. J., Wilson, D. W., Boulanger, R. W., & Abghari, A. (1998). Nonlinear seismic soil-pile structure interaction. *Earthquake Spectra, 14*(2), 377–396. doi:10.1193/1.1586006

Wolf, J. P., & Deeks, A. (2004). *Foundation vibration analysis: A strength-of-materials approach.* Oxford, UK: Elsevier.

Wu, G., & Finn, W. D. L. (1997). Dynamic nonlinear analysis of pile foundations using finite element method in the time domain. *Canadian Geotechnical Journal, 34*(1), 44–52. doi:10.1139/t96-088

Yazdchi, M., Khalili, N., & Valliappan, S. (1999). Dynamic soil-structure interaction analysis via coupled finite-element-boundary-element method. *Soil Dynamics and Earthquake Engineering, 18*(7), 499–517. doi:10.1016/S0267-7261(99)00019-6

Zienkiewicz, O. C., Chan, A. H. C., Pastor, M., Schrefler, B. A., & Shiomi, T. (1999). *Computational geomechanics.* Chichester, UK: John Wiley & Sons.

Chapter 7
Seismic Behavior and Dynamic Site Response of Municipal Solid Waste Landfill in India

Naveen B. P.
Amity University Gurgaon, India

Sitharam T. G.
Indian Institute of Science, India

Sivapullaiah P. V.
GITAM University Bangalore, India

ABSTRACT

Seismic behavior of landfills need for a better understanding of the dynamic properties of municipal solid waste (MSW) and site response of MSW landfill during seismic events. This chapter presents unit weight, shear wave velocity, strain-dependent normalized shear modulus reduction, and material damping ratio relationships for Mavallipura landfill based on field testing, laboratory measurements, and validated using semi-empirical methods. In addition, one-dimensional seismic response analysis by an equivalent linear method for Mavallipura landfill, Bangalore is done using software like SHAKE2000 and DEEPSOIL. Results indicated that the MSW landfill has less shear stiffness and more amplification due to the loose filling and damping, which need to be accounted for by seismically safe MSW landfill design in India. Also, results of seismic response analyses performed by the authors and other researchers are examined to assess the impact of stiffness and height of the landfill refuse fill on the overall response.

DOI: 10.4018/978-1-5225-6948-0.ch007

INTRODUCTION

Disposal of municipal solid waste in engineered landfills is one of the most widely accepted waste management practices in the India and worldwide. During its design life, landfilled MSW undergoes several complex mechanisms controlled by physical, chemical, thermal, and biological processes and their interrelated behaviors. A thorough understanding of mechanisms of landfilled MSW interactions is critical in designing a stable, effective and well-operational landfill in seismic active regions in Japan. However, to date, the recent practices associated with mathematical modeling as well as long-term monitoring of landfill performance are mostly empirical and limited to site-specific conditions only. Moreover, they fail to substantially quantify the changes in the geotechnical properties of MSW that result from coupled processes, especially the highly uncertain biological processes that result in MSW degradation in landfills. Furthermore, the spatially and temporally varied waste composition, heterogeneous and anisotropic nature of field MSW together with leachate and landfill gas production due to biodegradation results in a typical differential MSW settlement and, therefore, can adversely impact the long-term performance of landfills (Reddy et al., 2017).

MSW landfills, are an integral part of waste management and disastrous consequences happen if seismic vulnerability of these landfills is ignored. Therefore, seismic response analysis of MSW landfills is receiving considerable attention these days. The Federal Resource Conservation and Recovery Act (RCRA, 1993) of the United States Environment Protection Agency (USEPA) was the first regulatory legislations that have addressed the importance of seismic loading on MSW landfills. RCRA stated that new MSW landfills or lateral expansions of the landfill located in the seismic zone must be designed to resist the maximum horizontal acceleration. The dynamic force is generated due to earthquakes and produces relative movements within the MSW landfill mass, cover and bottom liner system, foundation and their interfaces. This relative movement in landfill results in cover cracking, geomembrane tears and damage to the appurtenant systems for leachate and gas collection. Such damages were noticed in US landfills during the 1994 Northridge earthquake and was documented by Augello et al., (1995).

However, the seismic response of MSW landfill is very important, especially in seismically active zones (Krinitzsky et al., 1997). Due to seismic event, dynamic loading induces relative movements take place within the waste mass and the foundation, which results increased tensions in the landfill liner material may lead to tearing due to excessive stretching and also cracks developed at the top of the landfill thereby disrupts the function of leachate and gas collection system, ultimately leading to failures of these landfills.

Seismic behaviour of MSW landfills gained major importance only after 1987 Whittier Narrows earthquake (Mw=6.0). For observation of the MSW landfills during earthquakes, USEPA installed two strong-motion instruments, one at the bottom of the landfill and one at an outcrop adjacent to the Operating Industries Incorporated (OII) landfill site was operated since 1989. These OII landfill instruments have captured ground motion records from five earthquakes of magnitude greater than 5.0 with PHGA from 0.05g to 0.26g at the crest and 0.03g to 0.25g at the outcrop (Hushmand Associates, 1994). Similarly, the observation database of landfill performance during 1989 Loma Prieta earthquake, 1992 Landers earthquakes and 1994 Northridge earthquake has provided importance of the seismic performance of MSW landfills. Based on the database that the engineering landfill has proven a good ability to withstand earthquakes of at least 0.2g without landfill failures/ damages to human health and the environment.

Site response analysis of MSW landfill was performed by various researchers (Idriss, 1990; Kramer, 1996; Hashash & Park, 2001) in order to account for the effects of ground motion propagation during an earthquake. The estimation of dynamic properties of solid waste is vital importance for performing reliable ground response analysis of landfills. The most essential MSW dynamic parameters used for the seismic response analysis are unit weight, shear wave velocity and the strain-dependent normalized shear modulus reduction (G/Gmax) relationships and the material damping ratio function. The shear modulus (G), which relates shear stresses to shear strains, is an main material property in the estimation of seismic response of MSW. Though many researchers (Augello et al., 1998; Matasovic & Kavazanjian, 1998; Zekkos et al., 2008) have conducted studies on the ground response of landfills by performing field tests or by back-calculations in several parts of the world, the current research is the first of its kind to be performed on India. The lack of data availability of the dynamic properties of MSW and site response of MSW landfills in Indian has led to the implementation of this work.

In this paper, the unit weight (γwaste), shear wave velocity (Vs), strain-dependent normalized shear modulus reduction (G/Gmax) and material damping ratio (ζ) relationships for Mavallipura landfill is developed based on field testing, laboratory measurements and also validated using semi-empirical methods. Finally, seismic response analysis by an equivalent linear method for Mavallipura landfill using the software like SHAKE2000 and DEEPSOIL. Shear wave velocity is essential input for carrying seismic response analysis, have been obtained from multichannel analysis of surface waves (MASW) test. The response of layered soil profiles is calculated from the recorded acceleration time history at underlying rock half space considering vertically propagating waves. Peak ground acceleration, response spectra, and amplification ratio were determined, and the results of SHAKE2000 and DEEPSOIL have been compared.

PROJECT SITE DETAILS

Mavallipura landfill site is about 40.48 hectares located in Mavallipura and about 35 acres is used for landfill. The landfill was operated by M/s Ramky Environmental Engineers opened in 2007 can sustain about 600 tonnes of waste. However, the Bruhat Bengaluru Mahanagara Palike (BBMP) has been sending almost 1,000 tonnes of garbage from Bengaluru city every day. Citizens around Mavallipura village demand that the landfill site must be stopped immediately as it is illegal and unscientifically managed, and thus it is now closed for land filling. Figure 1 shows the Mavallipura site location in Bengaluru. Waste was placed in the landfill from 2007 to 2013 at the rate of approximately 4,500 tons per day.

The waste consisted of approximately 60% biodegradable waste and 40% non-biodegradable waste. The typical height of solid waste dumping is 6m, and the deposit consisted of uncompacted waste. The GPS coordinate of the landfill is Latitude 13°50'North and Longitude 77°36' East. Auger drilling operation using a 150mm diameter is carried out in landfill site. The purpose of the auger drilling operation was to characterize the municipal solid waste visually, to retrieve bulk samples of waste from different depths, of different degree of degradation and different age as shown in Figure 2.

Figure 1. Borehole location on Mavallipura landfill site
Source: Google earth images, https://maps.google.com/map

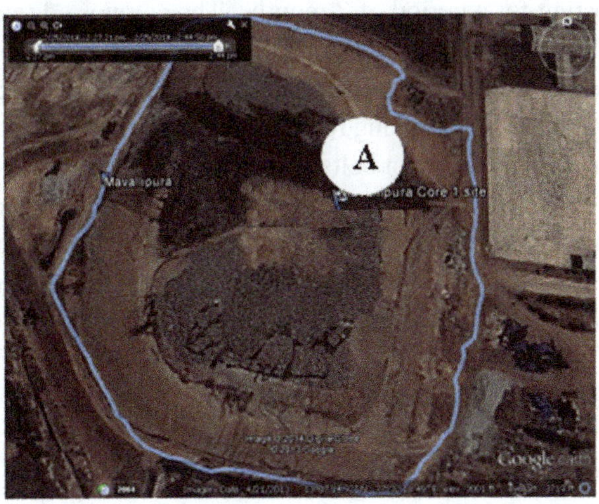

Figure 2. Typical bore log data and auger drilling for collection of waste samples

MSW SAMPLE COLLECTION

The MSW sample used in this study is the composted MSW reject collected from the Mavallipura landfill site, Bengaluru. Composting has been adopted as a potential pre-treatment method. Hand sorting of recoverable waste was followed by aerobic windrow composting for a period of 2 months. The composted reject had particle sizes 35mm to 4mm. Composted MSW is then shifted to coarse segregation area having mechanical arrangements to segregate the rejects like plastic, cloth, metals, a glass of the size above 14mm using rotary screen (trommel) of 35mm and 15mm sizes. Feeder, hopper and conveyor belts are provided with trommel for taking away rejects and semi-finished compost separately. Above semi-finished compost is then further shifted to curing section for further digestion. After curing, compost is shifted to refinement section that consists of vibratory screen or trommel of 4-6mm size. Disposal of reject material (Reject material like gravels, sand, stones, glass, metal and other inert materials above 4 mm are separated in this section). This reject material goes to landfill. Figure 3 shows the dry waste at the landfill site. From this heap, sufficient quantity of dry waste is collected and kept in the closed bags and carried to the laboratory for testing.

Figure 3. MSW landfill

MSW SAMPLE PREPARTION

The particle size of land filled MSW components is too large to conduct the geotechnical properties in a laboratory test, hence the component of the land filled MSW is cut with scissors/ crushed to a certain volume before the preparation of MSW samples.

A representative sample of the municipal solid waste is one that contains a subset of all the constituents of a municipal solid waste in a landfill in exactly the same proportion as they are present in the target solid waste landfill.

- The sample is placed on the floor and thoroughly mixed by mechanical shovel.
- The sample is then placed in a uniform pile of approximately 0.8 m high in a square box.
- The uniform pile is divided into four quarters using straight lines perpendicular to each other.
- A portion of the sample is collected from all the quadrants to have a representative sample.
- The process is repeated until the desired representative sample size i.e.10-20kg is obtained.

The conventional approach is used to establishing the geotechnical properties of in-situ sampling and laboratory testing is of limited value in evaluating the MSW properties. Due to limitations of this approach include constraints on MSW sampling mainly related to health and safety issues associated with sampling and testing waste material, sample disturbance, and the size of the waste constituents. Considering these limitations, in-situ measurements and observations of field performance provide best means of evaluating the MSW properties for seismic analysis.

ENGINEERING PROPERTIES OF MSW

Moisture Content

Moisture content is defined as dry gravimetric moisture content: the ratio of the mass of water in a waste sample to the mass of solids in the waste sample expressed as a percentage. During the moisture content determinations, the temperature was maintained at 60° C to avoid combustion of volatile materials. The average moisture content of MSW is about 36%, due to an operation carried out in a landfill. All the experiments were performed as per the standard procedure established by the American Society for Testing and Materials (ASTM,2006).

Grain Size Distribution

The particle size distribution test was carried out as per ASTM D 422. A particle size distribution curve describes the percentage by mass of particles of the different size ranges. The horizontal axis represents the particle size on a logarithmic scale. The vertical axis represents the percentage by weight of particles that are finer than a specific size on the horizontal axis. Figure 4 shows the typical gradation of MSW, and it can be seen that the MSW sample consisted of coarser particles with sizes ranging from 0.08 to 5mm, but approximately 85% of weight basis of waste particles with sizes less than 10mm.

In Situ Unit Weight

The unit weight is a basic material property in landfill engineering. Knowledge of the unit weight of Municipal Solid-Waste (MSW) is required for many engineering analyses of landfill systems including static and dynamic slope stability, landfill capacity evaluation and geomembrane puncture. However, evaluating the unit weight of MSW plays a significant source of uncertainty in landfill engineering analyses due to various factors, including the heterogeneous nature of solid waste and lack of systematic analysis of the available data. Previous researchers reported values in the literature shows the significant scatter exists in the total unit weight of MSW. Hence, it is very challenging for an engineer to estimate with confidence the MSW unit weight profile for use in engineering landfill analyses.

In Mavallipura landfill, using the in-situ method of measuring the unit weight by replacing the waste with calibrated gravel. The weight of waste removed from a 0.5-1.0m length of the borehole was measured, and the volume of the material was evaluated by backfilling the 0.5-1.0 m interval with gravel of known unit weight. The unit weight of MSW is expressed as a kN/m3. The unit weight of waste with

Figure 4. Particle size distribution curve

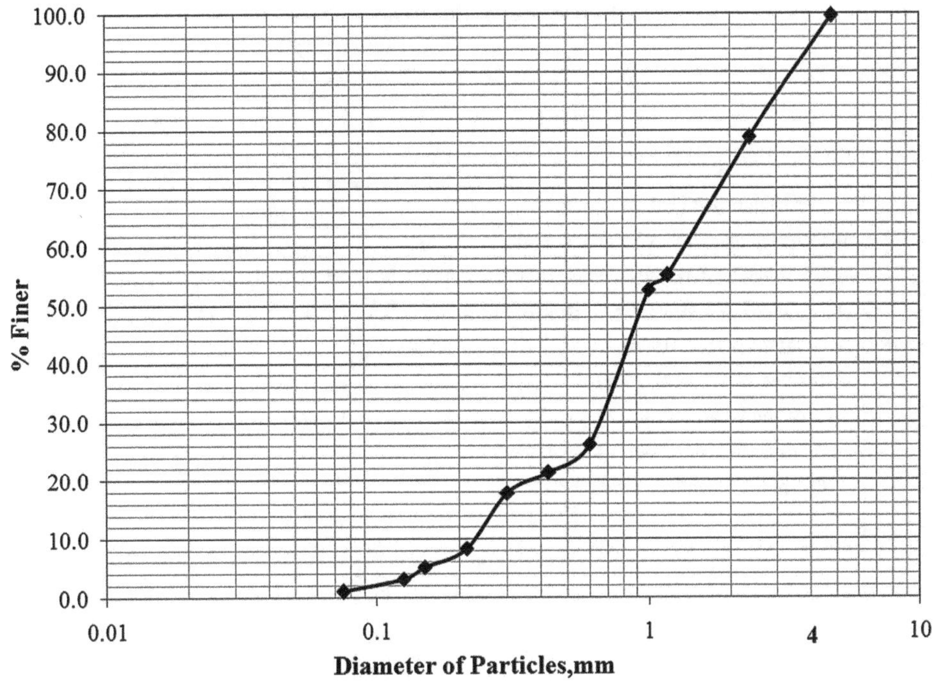

depth results obtained from field measurements are plotted. The range of typical unit weight with depth for various landfills available from the literature is plotted. Figure 5 shows a consistent trend of unit weight increasing with increasing depth in response to the increase in overburden stress.

The values obtained by field measurements lie close to the profile suggested by Kavazanjian et al. (1995) and Oweis and Khera (1998). The value obtained by field measurement found to be the lower end of the range of data, due to uncompacted or poorly compacted waste immediately after placement and the OII landfill values found to be higher because of high overburden pressures or good standards of practice for waste compaction and daily cover application. Based on the results of previous research's studies, the unit weight of MSW was used to capture the effects of waste composition as well as compactness. The unit weight with depth profile was developed for use in engineering analysis of the seismic response of engineering MSW landfills.

Figure 5. Plot of variation of unit weight of MSW landfills with depth

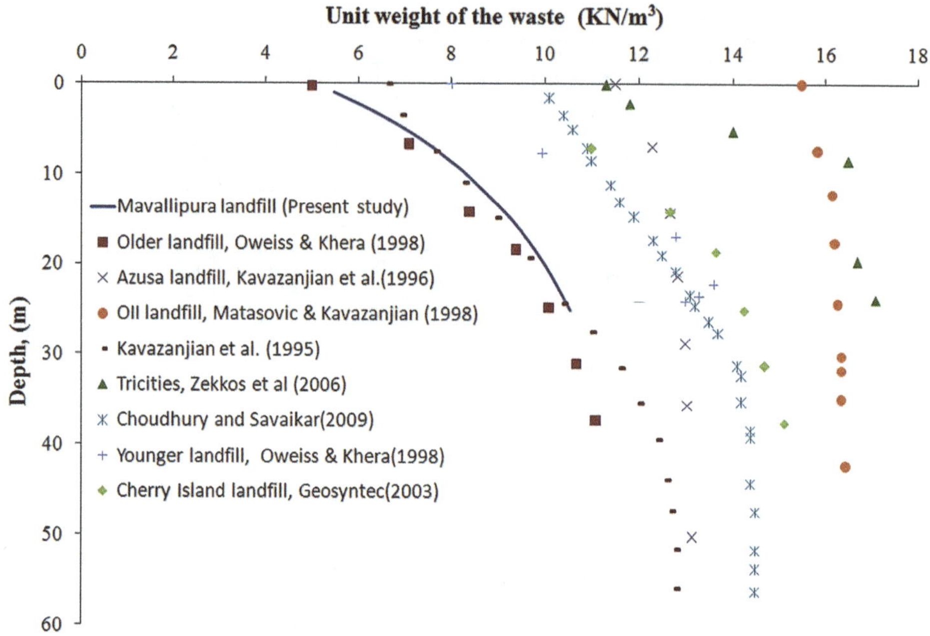

MULTICHANNEL ANALYSIS OF SURFACE WAVES (MASW)

Shear wave velocity of MSW landfills is important property that relates to the stiffness of the landfill and is useful in evaluating the dynamic properties of landfills. The various investigators have reported in situ measurements of MSW shear wave velocity from geophysical tests like seismic cross-hole test, suspension logging test, seismic up hole and downhole test, SASW (Spectral Analysis of Surface Wave) test and MASW (multichannel analysis of surface waves) test are available to estimate the variation of shear wave velocity with depth in the field. Kavazanjian et al. (1996) reported an average shear wave velocity profiles from eight California MSW landfills and the values ranged from 80 m/s near the surface to 300 m/s at a depth of 30 m. Sharma et al. (1990) measured the shear wave velocity ranging from 198.3 m/s at a depth of 15.3 m for San Pablo Bay landfill, Richmond, California. Carey et al. (1993) measured a shear wave velocity ranging between 185 and 478 m/s. In the present study, shear wave velocity were evaluated by using the MASW testing. The typical MASW test setup used for the present study consists of 24-channel geode seismograph and 24 geophones of 4.5 Hz capacity. Seismic data were recorded using Geode seismograph with sledgehammer source on one side of the landfill and geophones on the other side with different spacing of geophones (1m). The landfill was surveyed up to a length of about 25m at the top level. The deposit consisted of uncompacted waste up to a depth 6m. Eight Borehole drilling was performed at the top level of the landfill. Landfill area is divided into three grids. Grid 1 consists of two bore holes, grid 2 and grid 3consists of three boreholes each is shown in Figure 6. Eight 1-D columns generated from the data are shown in Figure 7. It can be noted from Figure 7 that the MSW at the site has a thickness varying from 1.5 to 13.5m, followed by silt soil having a thickness of 13.5-22.5m, below which is a soft weather rock having a shear wave velocity of about 395m/sec.

The data on the variation of shear wave velocity of MSW landfills with depth for various landfills available from the literature are plotted in Figure 8. The values of shear wave velocity of 55m/s at the top surface of landfills. The values obtained by MASW testing lie close to the profile suggested by Bray and Rathje (1998) and Kavazanjian et al. (1995). Kavazanjian et al. (1995) report on shear wave velocities as low as 80m/s near the surface and to over 256 m/s at a depth of 75m from SASW measurements. Bray and Rathje (1998) reported a shear wave velocities as low as 50m/s near the surface and to over 212 m/s at a depth of 80m. Woodward Clyde Consultants (1987) report an average shear wave velocity of between 206 m/s and 244 m/s from seismic refraction surveys for OII landfill. For the same landfill, Hushmand Associates (1994) reported that average shear wave velocities of between 244 m/s and 366 m/s from ambient vibration measurements and back calculation from small magnitude earthquakes. Choudhury and savaikar (2009) reported a shear wave velocity of between 108 m/s near surface and 391m/s at a depth of 85m by using the curve-fitting technique. Carey et al. (1993) reported a shear wave velocity

Figure 6. (a) Typical borehole locations in landfill. (b) MASW carried out in landfill site

(a): Typical borehole locations in landfill *(b): MASW carried out in landfill site*

Figure 7. Municipal Solid Waste (MSW) columns generated for site response analysis

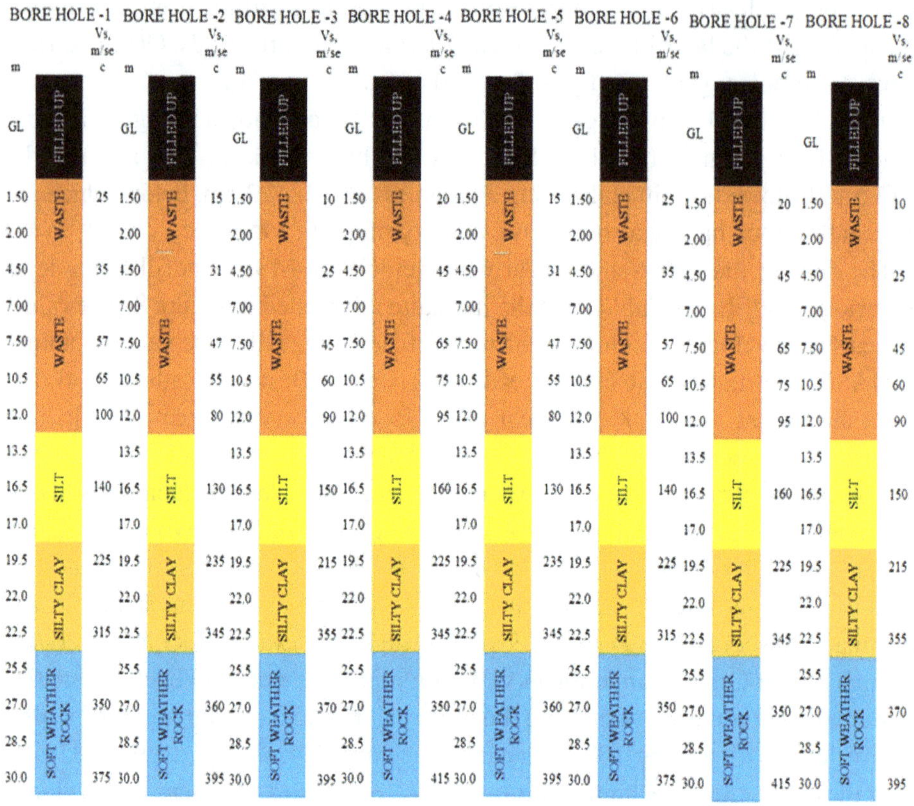

of between 185 to 478 m/s for Brookhaven landfill, New York from Cross-hole shear wave velocity. The shear wave velocity with depth profile was developed for use in seismic response.

LABORATORY TESTING

Small-Strain Shear Modulus (Gmax)

In this analysis, the small-strain shear modulus (Gmax) is obtained from ultrasonic technique and shear wave velocity (Vs) is determined from field measurements such as multichannel analysis of surface wave (MASW). Field measurements are considered more reliable than laboratory measurement of shear wave velocity or small strain modulus. However, in field technique for measurement of the dynamic modulus of large strains and material damping ratio are not available.

Figure 8. Shear wave velocity vs. depth

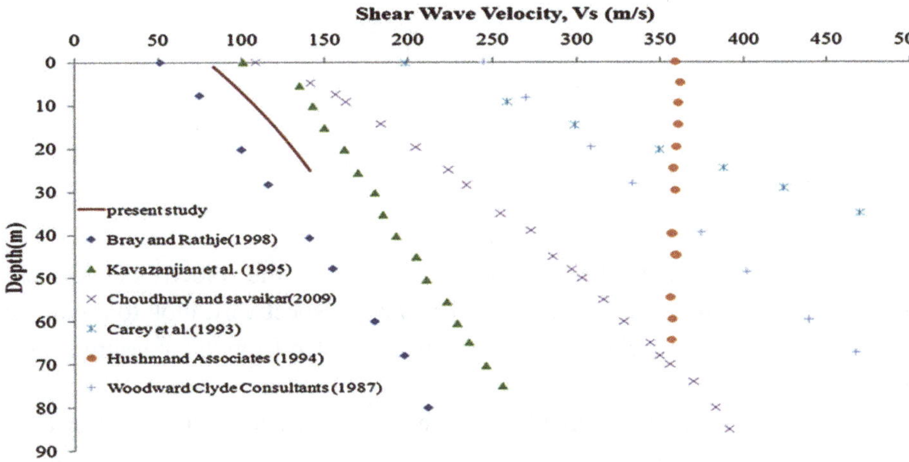

The small-strain shear modulus (Gmax) is obtained from ultrasonic pulse technique, and testing were conducted as per ASTM D2845. The receiver is positioned on a plane opposite to the plane to which the transmitter is pressed. The velocities of either P- or S-waves (Vp, Vs) are calculated from the measured travel time and the distance between transmitter and receiver. Figure 9 shows the material used for sample preparation.

Velocities are calculated from travel times measured and the distance, d, between transmitter and receiver by using the equations:

Figure 9. Material used for sample preparation: (a) ultrasonic wave velocity meter (b) positions of transmitter and receiver on the specimen

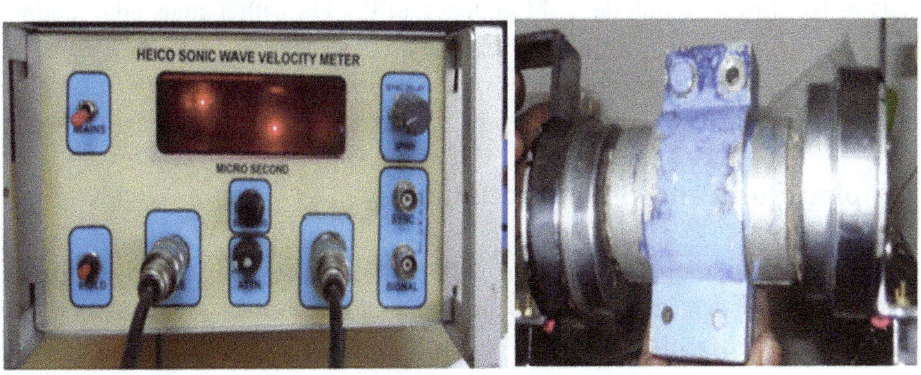

$$V_p = \frac{d}{t_p}. \tag{1}$$

$$V_s = \frac{d}{t_s}. \tag{2}$$

where Vp is the velocity of the longitudinal wave, Vs is the velocity of the shear wave, tp and ts are the times that the P and S-wave, respectively, took to travel the distance d. The ultrasonic elastic constants are calculated from the measured wave velocities.

The small-strain shear modulus (Gmax) related to the shear wave velocity (Vs) as given by Equation (3.0)

$$G_{\max} = \rho V_s^{2}. \tag{3}$$

where Vs is the shear wave velocity of the MSW, ρ is the mass density of MSW

Cyclic Triaxial Test

The municipal solid waste sample with a maximum dry density of 7.0 kN/m3 at a different confining pressure of 50, 100 and 150 kPa were prepared into a cylindrical specimen of height 10cm and a diameter of 5cm, an aspect ratio of 2. All the municipal solid waste samples tested under cyclic loading with a void ratio of 0.7 and specific gravity is 1.15 and porosity are 40% respectively. After the sample was prepared sample enclosed in an acrylic cell (Figure 10).

The loading ram and load cell are aligned in coincidence with their respective vertical axis. In cyclic shearing phase, the specimens were subjected to repetitive loads of fixed load amplitude and frequency of 0.1 Hz with a sinusoidal loading waveform. In monotonic loading, the samples were sheared under a single constant strain rate of 0.6 mm/min. The values of shear modulus and damping ratios were calculated from ASTM D 3991-91.

Damping Ratio and Shear Modulus

The cyclic stress-strain behaviour of solid wastes were analyzed to determine shear strain dependent shear modulus and damping ratio values. Solid waste under repetitive or cyclic loading undergoes degradation of shear modulus and increase in

Figure 10. Materials used for sample preparation: (a) rubber membrane (0.3 mm thick), (b) split mould, (c) solid waste sample enclosed in membrane, (d) sample inside the tri-axial cell with applied confining pressure

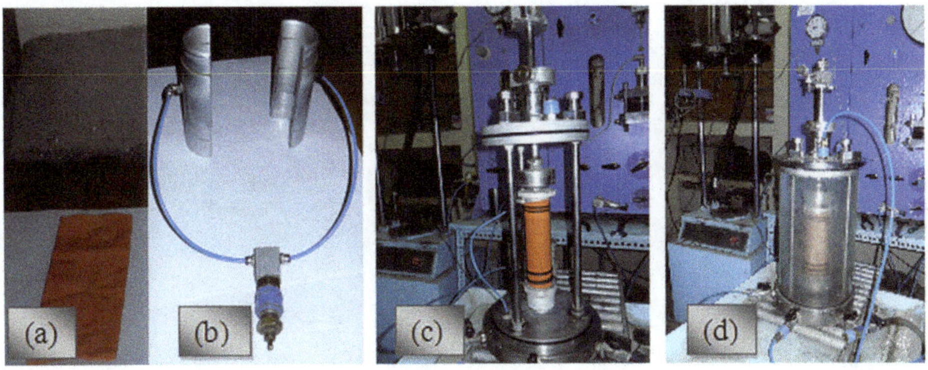

damping ratio with shear strain. In a cyclic triaxial testing system, after considering the effects of system compliance and applicable corrections ASTM-D5311, a typical hysteresis loop obtained on solid waste sample is shown in the Figure 10. values of shear strain (%), shear modulus (MPa) and damping ratio (%) were estimated for each cycle of loading from the hysteresis loops obtained in stress controlled cyclic loading. The methods followed to determine the shear modulus (G), and damping ratio from cyclic loading are depicted Figure 11.

Figure 11. Hysteresis curve for deviator stress and axial strain

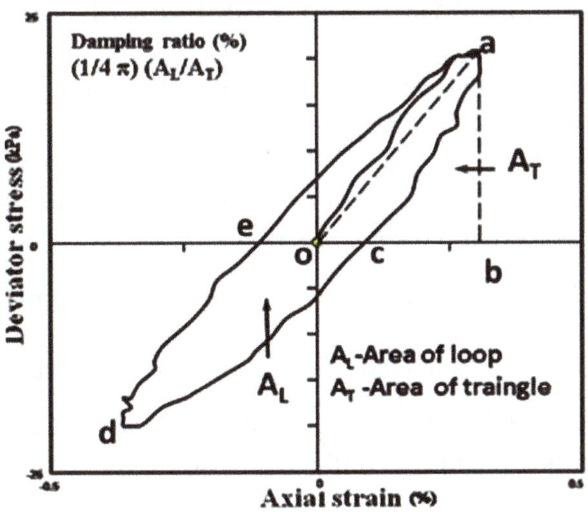

From the plot of deviator stress v/s axial strain. The dynamic young's modulus (E) and hence shear modulus (G) of the solid waste sample were estimated using the Equations (4), (5) and (6).

$$E = \frac{\sigma_d}{\varepsilon} \tag{4}$$

$$\gamma = (1+\nu)\varepsilon \tag{5}$$

$$G = \frac{E}{2(1+\nu)} \tag{6}$$

where, E is the dynamic Young's modulus, σ_d is the deviator stress, ε is the axial strain, γ is the shear strain(%), ν is the Poisson's ratio, and G is the shear modulus.

$$D = \frac{A_L}{4\pi A_T} \tag{7}$$

AL is the area enclosed by hysteresis loop (total energy supplied); AT is the area of the triangle (percentage of energy lost). The estimation of damping ratio was done using the Equation (7) for different axial strain amplitudes.

Normalized Shear Modulus and Damping Ratio Relationship

The current practice for evaluating shear modulus and damping ratio for seismic response analysis involves: (a) determining or measuring the Vs in the field and (b) estimating or measuring the variation of damping and shear modulus with shear strain primarily in the laboratory. It is common practice to the normalized shear modulus (G) by dividing by Gmax.

A variation plot of $\frac{G}{G\max}$. with shear strain is called a normalized modulus reduction curve. However, in the present analysis, normalized shear modulus $\frac{G}{G\max}$. profiles suggested by Zekkos (2005) from laboratory work on three different waste compositions for the Tri-Cities landfill in California have been used.

Zekkos (2005) proposed a hyperbolic model expressed by the following Equation (8.0).

$$\frac{G}{G\max} = \frac{1}{1+\left(\frac{\gamma}{m}\right)^{n}}. \tag{8}$$

where γ is the percentage shear strain, G_{max} is the small-strain shear modulus, G is the secant shear modulus at the shear stain, and m and n are two hyperbolic parameters given in Table 1. These two hyperbolic parameters depends primarily on the solid waste composition and confining stress ranges from 25 to 90 kPa, for three different waste composition: solid waste containing 8-25% particles less than 20mm size, 62-76% particles less than 20mm size and 100% particles less than 20mm size.

Variation of Normalized Shear Modulus

Based on the observations of the effect of confining stress on the $\frac{G}{G\max}$ and damping curves, that data collected for 100kPa are considered applicable. As confining stresses increases, the $\frac{G}{G\max}$ curve shifts to the right, i.e. lateral strain increases. The shear modulus reduction and material damping data for MSW from this study at confining stress 100kPa, which were shown in Figures 12 and 13 are compared to relationships previously recommended by different researchers.

The shear modulus can be computed using proposed Equation (8.0). The values obtained by Equations (8.0) lie close to the profile suggested by Bray et al. (1998)

Table 1. Hyperbolic parameters (m and n) to be used for computation of the normalized shear modulus $\frac{G}{G\max}$ for landfills with confining stress of 25-90kPa reporting by Zekkos (2005)

Hyperbolic Parameters	Percentage of Waste Specimen Containing Less Than 20mm		
	100	62-76	8-25
m	0.118	0.265	0.697
n	0.886	0.819	0.925
R^2	0.972	0.978	0.981

and also close to the profile suggested by Idriss et al.(1995) for landfills composed of waste materials with 100% particles sized less than 20mm. Hence, Equations (8.0) suggested above can be used for the design of a landfill in the absence of landfill-specified field data on normalized shear modulus. Thus, it clear shows that proposed equations for material damping ratio and shear modulus are validated with the profiles suggested by various researchers and it has been used extensively in seismic response analysis and design. For the preliminary design of landfill facilities, above mentioned equations and the range of values can be used confidently, and they can serve as a guideline for seismic response analysis and the design of landfills in the absence of landfill specific data.

Variation of Material Damping Ratio

In the present analysis, strain dependent material damping ratio ζ is calculated by using the curve-fitting technique, a generalized mathematical Equation (9.0) as given

Figure 12. Plot of variation of normalized shear modulus of MSW landfill with the percentage shear strain

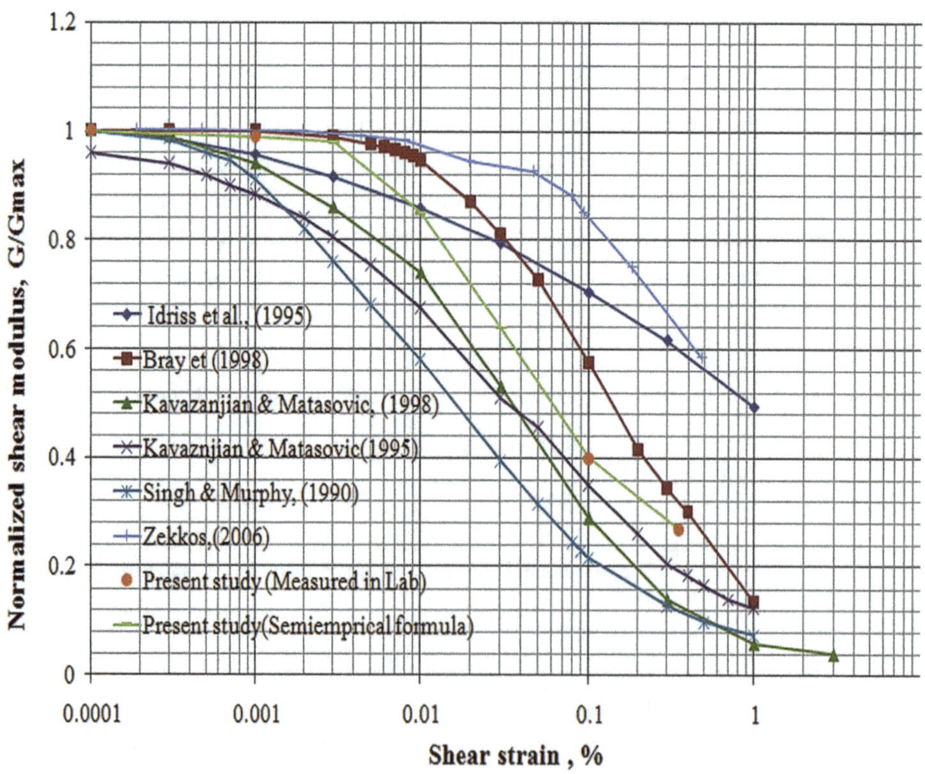

Figure 13. Plot of variation material damping ratio of MSW landfill with the percentage shear strain

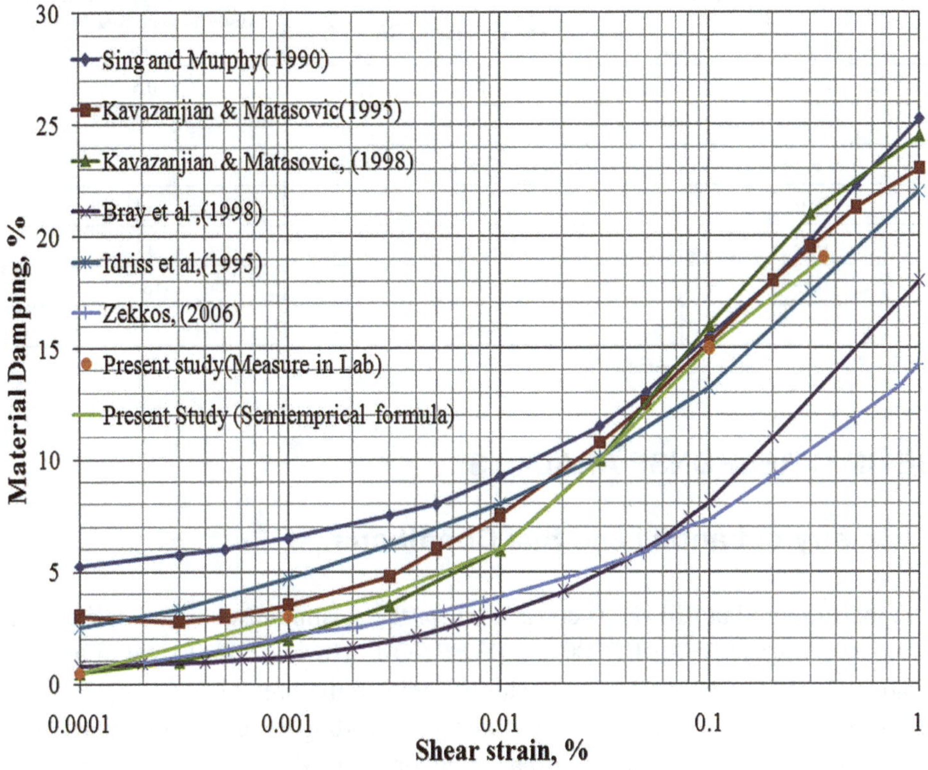

below, which yields $R^2 = 0.998$. Strain-dependent material damping ratio results are shown in Figure 13. Strain-dependent normalized shear modulus reduction and material damping ratio for Mavallipura landfills with confining stress of 100kPa as shown in Table 2.

$$\zeta = 30\left(1 - e^{-1.4\gamma^{0.36}}\right) \tag{9}$$

Where γ is the percentage shear strain.

From the variation of material damping ratio with shear strain shown in Figure 13. The material damping ratio can be easily computed using Equation (9.0). The values yielded by Equation (9.0) lie between those suggested by Kavazanjian and Matasovic (1995) and Kavazanjian and Matasovic (1998) for the OII landfill. Equation (9.0) Suggested above can be used in the absence of landfill-specific field data on material damping.

Table 2. Measured data on strain dependent material damping ratio for mavallipura landfills with confining stress of 100kPa

Shear Strain γ (%)	Normalized Shear Modulus, G/G_{max} (%)	Material Damping Ratio ζ (%)
0.0001	1	0.5
0.001*	0.99	3
0.003*	0.98	6
0.01	0.85	14
0.03	0.64	17
0.1*	0.40	18.5
0.35*	0.27	19.0

Note: * Measured values in laboratory

SEISMIC RESPONSE ANALYSIS

Geometry of Landfill and Site Conditions

The mavallipura landfill site is considered for the seismic response analysis using computer program SHAKE 2000 and DEEP SOIL. However, the height of landfill is about 10m. The borehole investigations indicate the presence of a top layer of waste. The waste is loose and unconsolidated, extending down to about 10m depth below existing ground level. After that, strata of sand, silty clay, stiff clay and hard clay are encountered. These lower strata appear to be highly disintegrated rock, which extend down to refusal strata as shown in Figure 14.

Figure 15 shows the foundation of MSW landfill. The site responses analysis performed for the couple response of the foundation soil and waste mass. Basic input to SHAKE and DEEPSOIL includes the soil stratification, waste, soil and rock properties, and the input time history, soil properties include the shear wave velocity and unit weight for each layer plus curves relating the reduction in modulus and damping ratio to shear strain for each layer. The normalized shear modulus reduction and material damping profiles recommended by Kavazanjian and Matasovic, (1995) for waste material and Seed and Idriss (1970) for sand and seed & sun (1989) for clay and Schnabel (1973) used for rock and 5% damping are assumed for the bedrock.

Seismic Ground Motion

The selection of appropriate ground motion is important in the seismic response analysis. Literature indicates that the several seismic response analysis was carried

Figure 14. A typical display of bore log sheet

BORE HOLE -1		
DEPTH m	DESCRIPTION	N VALUES
GL	FILLED UP	
1.50	WASTE	8
2.00		
4.50	WASTE	9
7.00		
7.50		11
10.50		12
12.00		
13.50	SILTY SAND	14
16.50		18
17.00		
19.50	SILTY CLAY	20
22.00		
22.50		35
25.50	SOFT WEATHER ROCK	45
27.00		
28.50		48
30.00		50

Figure 15. Foundation for MSW landfills as used in present 1-D analysis

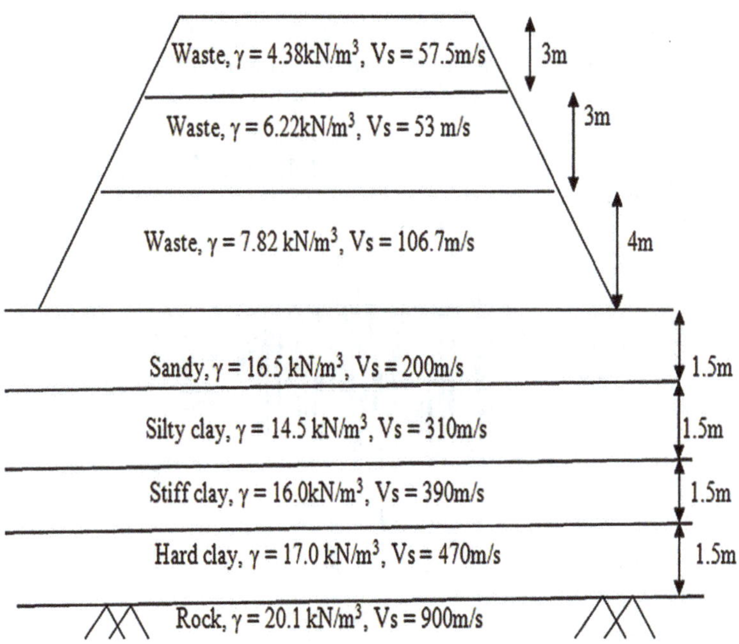

out in MSW landfill in the inter-plate region. Hence, in this study, 1-D equivalent linear analysis, intraplate ground motion records having acceleration peaks of 0.15g and 0.05 second of time interval were used to study the influence of variation of seismic accelerations and time period on seismic response of Mavallipura landfill. Seismic ground motion is shown in Figure 16.

The response spectrum is used extensively in earthquake engineering practice to indicate the frequency content of an earthquake motion. The response spectrum describes the maximum response of a damped single degree of freedom (SDOF) system to a particular input motion as a function of the natural period/frequency and damping ratio of the SDOF system.

Figure 17 shows that top of the landfill with 5% damping ratios obtained by SHAKE 2000 and DEEPSOIL. It is observed that peak spectral acceleration at 5% damping value is 0.7g for 0.07 sec in SHAKE 2000 and peak spectral acceleration at 5% damping value is 0.63g for 0.04 sec in DEEPSOIL. The study shows that amplification and spectral signature of intraplate MSW landfill are different from interplate MSW sites, which need to be accounted in seismic design of intraplate MSW landfill sites. When input motion were applied at the base of the rock profile significant reductions in the peak ground acceleration and spectral acceleration values are observed even for low amplitudes such as 0.15g. Figure 18 shows a variation of peak ground acceleration (PGA) at the surface; it is observed PGA has decreased from 0.3g to 0.15g in DEEPSOIL and PGA has decreased from 0.33g to 0.15g in SHAKE2000.

Figure 16. Synthetic seismic ground motion generated for Bangalore

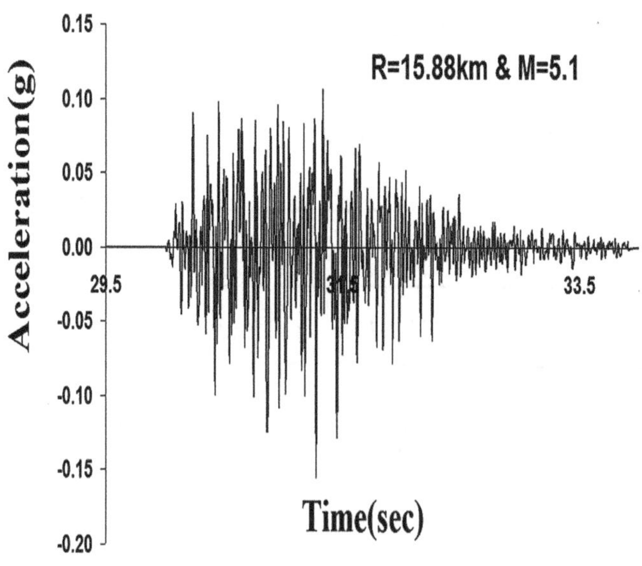

Figure 17. Variation of spectral acceleration for 0.15g amplitude

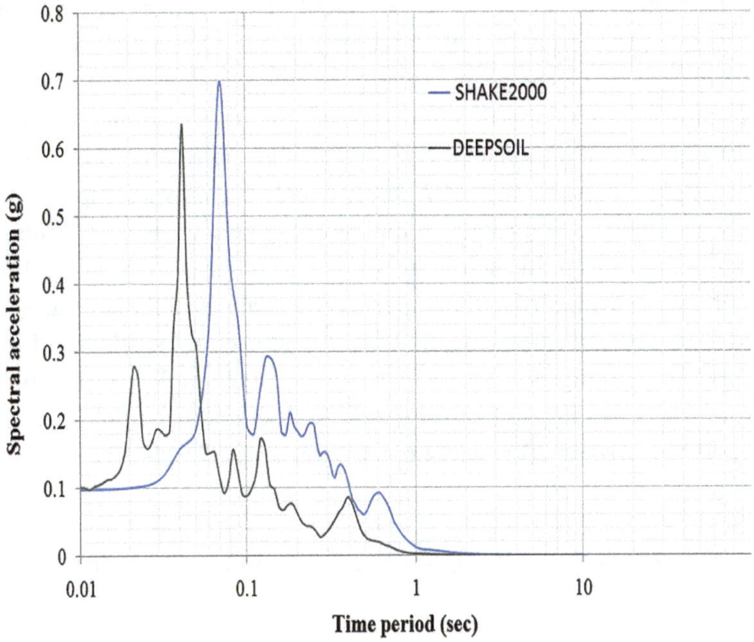

Figure 18. Variation of PGA for 0.15g input acceleration

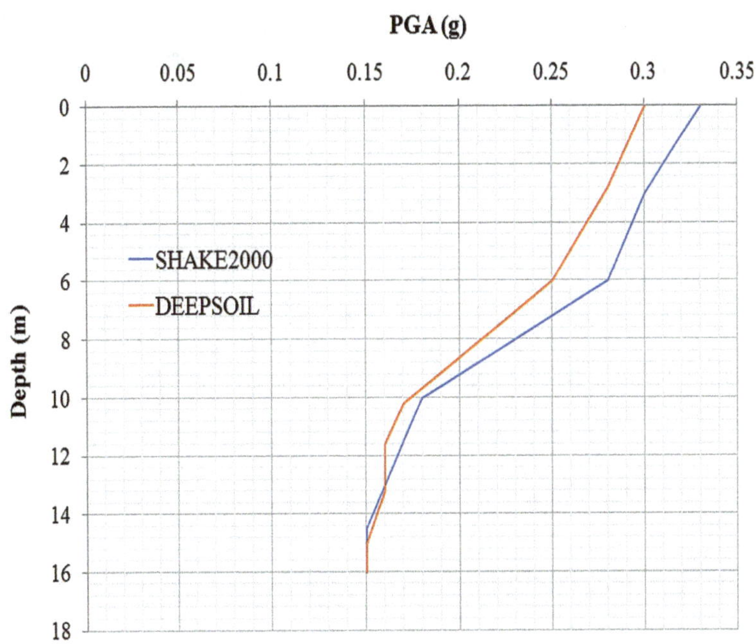

The plot of acceleration on top of each layer shows that input acceleration is lightly amplified due to composite layer at the bottom of MSW. The presence of composite layer i.e. medium to the dense soil above bedrock is enhancing amplification of MSW, which was also observed by Choudhury and Savoikar (2009). Site response study of intraplate MSW landfill site response also shows similar conclusion, but the spectral signature of intraplate MSW landfill site is different from Repetto et al. (1993). The study shows that the MSW landfill underlined by a composite layer and rock can cause amplification for lower and intermediate levels of shaking. Though the results obtained from this study are similar to other seismic ground response studies with respect to amplification, the response analysis of the several MSW sites is required to arrive at the amplification factor and spectral signature of intraplate MSW landfill sites. Figure 19 illustrates the variation of amplification ratio (A) and predominant frequency for the landfill site. It is observed that amplification ratio is 6.11 at 1.11 Hz in SHAKE2000 and 4.65 at 2.67Hz in DEEPSOIL.

Figure 19. Variation of amplification ratio for 0.15g amplitude between bottoms to top layer

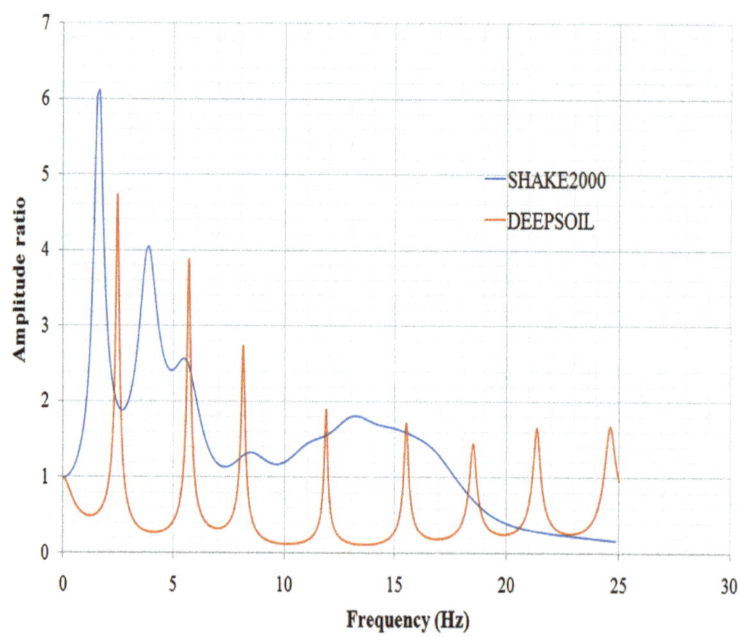

SIMPLIFIED METHOD OF LANDFILL RESPONSE EVALUATION

The seismic response of municipal solid waste landfill (MSWL) are typically estimated using procedures developed for earth dams. Generally, evaluation of MSWL stability under seismic conditions are carried out using deformation analysis. Typical deformation simulation employ a combination of 1-D seismic response analyses. For an accurate assessment of the material properties, especially the dynamic properties are essential for use in the above analyses. The landfill refuse is a highly heterogeneous material and its dynamic properties are yet to established through insitu test data. Another aspect of the MSWL which mainly differ from a well compacted homogeneous earth dam. All above issues raised concerns regarding the relevance of the conventional geotechnical earthquake engineering procedures to municipal solid waste landfill refuse slope. Therefore, the simplified procedures for preliminary assessment of the amplification effects of the landfill refuse fill to base motions.

The factors that could influence earthquake loading in landfills are: height of landfill refuse, shear wave velocity of refuse, dynamic properties of refuse, presence of clay liner, refuse geometry and characteristics of base motions. In initial estimation of landfill response, two major factors that should be considered namely shear wave velocity (stiffness) and height of landfill refuse influence the basic vibrating characteristics of the solid waste landfills. Conventional methods of cross hole and down hole measurements have been successfully implemented in solid waste landfill sites. However, this methods require drilling or puncturing a probe into the refuse, most recently, a new method of using multi analysis of surface wave (MASW) measurements was successfully used to estimate the shear wave velocity of refuse.

Measurements made of landfills typically illustrates the average shear wave velocity ranges between 150m/sec to 300m/sec. Although shear wave velocity in excess of 300m/sec have also been reported. The shear wave velocity profiles at some landfills sites show that the shear wave velocity increases with depths. Also, it should be noted that unlike conventional soil fills, the landfill refuse fill compacts under its own weight with time, thus the shear wave velocity increase with time.

Here author's present simplified charts using the landfill refuse height and the average shear wave velocity to estimate the response of landfills. The relationships shown in Figure 20 are derived from the authors' experience, published results, and field instrumentation records on landfill response during earthquakes. Results of seismic response analyses for different levels of base motion. The horizontal axis represents the anticipated ground motion at the base of the landfill. The vertical axis represents the anticipated ground acceleration at the top of the landfill. By using these charts for preliminary design purposes, this can be determined either from probabilistic seismic hazard or conventional deterministic evaluation. Figure 20 shows the response of 60m high landfill.

The authors developed curve; they correspond to average refuse shear wave velocity of 121m/sec and 213 m/sec. It should be noted that due to the limited instrumentation on landfills, Figure 20 also incorporate some of the authors' analytical work as well as simulations presented by others. Similarly, Harder (1991) reported a earth dams chart, where he compared the acceleration recorded at the base of the earth dam with those recorded at the top of the dam. The authors presume that he suggested upper bound correlation by Harder can be also viewed as an upper bound for response of solid waste landfills. The assessments are made on the basis of the (a) 2-D geometry of a typical earth dam will respond more than that of a landfill which responds more in a one-dimensional manner, (b) the solid waste landfill refuse will have a higher damping than that of an earth fill dam.

CONCLUSION

The lack of data availability of the dynamic properties of MSW and site response of MSW landfills in Indian has led to the implementation of this work. The four

Figure 20. Approximate relationship between maximum accelerations at the base and crest

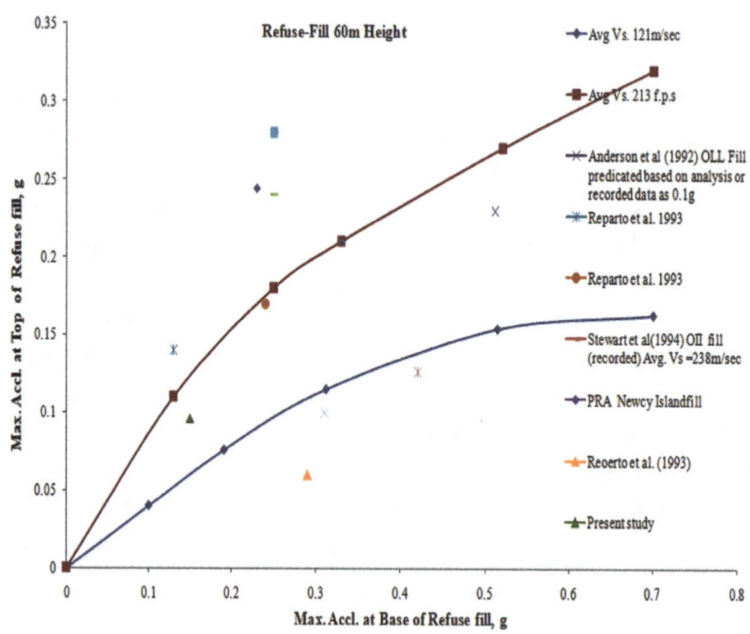

important properties of MSW unit weight, shear wave velocity, normalized shear modulus and material damping are extensively used in seismic response of MSW landfills. Though, evaluation of these four properties of MSW in the field, laboratory poses several problems. Hence, these properties can be determined by profiles suggested by different researchers or back-analysis of available seismic data.

Dynamic properties of the waste from Mavallipura landfill, Bangalore have determined. Based on the studies conducted the following observations are made:

- Based on the results of previous research's studies, the unit weight of MSW was used to capture the effects of waste composition as well as compaction effort. Figure 5 shows a consistent trend of unit weight increasing with increasing depth in response to the increase in overburden stress. The insitu unit weight results lie close to the profile suggested by Kavazanjian et al. (1995) and Older landfill (1998).
- Laboratory data on waste modulus and damping at small strain modulus is determined from ultrasonic technique and high strain modulus, and damping is determined from the cyclic triaxial test. Normalized shear modulus and damping ratio relationship for mavallipura landfill site are developed by using hyperbolic model expression proposed by Zekkos (2005). The proposed material damping ratio curve lie between those suggested by Kavazanjian and Matasovic (1995) and Kavazanjian and Matasovic (1998) for the OII landfill. Similarly, the proposed normalized shear modulus reduction curve lie close to the profile suggested by Bray et al. (1998) and also close to the profile suggested by Sing and Murphy (1990) for landfills composed of waste materials with 100% particles sized less than 20mm.
- Peak spectral acceleration at 5% damping value is 0.7g for 0.07 sec in SHAKE 2000 and peak spectral acceleration at 5% damping value is 0.63g for 0.04 sec in DEEPSOIL. Amplification ratio is 6.11 at 1.11 Hz in SHAKE2000 and 4.65 at 2.67Hz in DEEPSOIL. Peak ground acceleration (PGA) for the landfill site, it is observed PGA has decreased from 0.3g to 0.15g in DEEPSOIL and PGA has decreased from 0.33g to 0.15g in SHAKE2000. High amplification at the landfill site is due to the presence of higher thickness of the soft layer, low SPT value that results in low shear wave velocities.
- Comparison of SHAKE2000 and DEEPSOIL indicates that SHAKE2000 software showing higher value compares with DEEPSOIL, but both software's values are in similar range.
- The height and shear wave velocity of the landfill refuse material in MSWL are the significant factors which influence the basic vibrating characteristics of the MSW landfills.

REFERENCES

Augello, A. J., Bray, J. D., Abrahamson, N. A., & Seed, R. B. (1998). Dynamic properties of solid waste based on back-analysis of OII landfill. *Journal of Geotechnical and Geoenvironmental Engineering*, *124*(3), 211–222. doi:10.1061/(ASCE)1090-0241(1998)124:3(211)

Augello, A. J., Matasovic, N., Bray, J. D., Kavazanjian, E., Jr., & Seed, R. B. (1995). Geotechnical Special Publication: Vol. 54. Evaluation of solid waste landfill performance during the Northridge earthquake. Academic Press.

Bray, J. D., & Rathje, E. M. (1998). Earthquake-induced displacements of solid-waste landfills. *Journal of Geotechnical and Geoenvironmental Engineering*, *124*(3), 242–253. doi:10.1061/(ASCE)1090-0241(1998)124:3(242)

Carey, P. J., Koragappa, N., & Gurda, J. J. (1993). *A case study of Brookhaven landfill. In WasteTech Proceedings.* Washington, DC: National Solid Waste Management Association.

Choudhury, D., & Savoikar, P. (2009). Simplified method to characterize municipal solid waste properties under seismic conditions. *Waste Management (New York, N.Y.)*, *29*(2), 924–933. doi:10.1016/j.wasman.2008.05.008 PMID:18682319

Fassett, J. B., Leonards, G. A., & Repetto, P. C. (1994). Geotechnical properties of municipal solid wastes and their use in landfill design. *Waste Tech*, *94*, 13–14.

Geosyntec Consultants, Inc. (1996). *Operating Industries, Inc. Landfill, Monterey Park, CA. Technical Reports SWP-1 through SWP-9.* Huntington Beach, CA: GeoSyntec Consultants, Inc.

Harder, L. S. Jr. (1991). Performance of Earth Dams During the Loma Prieta Earthquake. *Proceedings Second International Conference on Recent Advances in Geotechnical Earthquake Engineering and Soil Dynamics*, 1673.

Hashash, Y. M., & Park, D. (2001). Non-linear one-dimensional seismic ground motion propagation in the Mississippi embayment. *Engineering Geology*, *62*(1), 185–206. doi:10.1016/S0013-7952(01)00061-8

Henke, K. F. (1985). Standsicherheit von Deponien aus Mull-und Abfallstoffen. *Mull-und Abfall besitigung*. Kennzahl 4546.

Hushmand Associates. (1994). *Landfill response to seismic events, Report prepared for USEPA Region XI.* Laguna Niguel, CA: Hushmand Associates.

Idriss, I. M. (1990). Response of soft soil sites during earthquakes. *Proceedings Symposium to Honor H. B. Seed*, 273–289.

Kavazanjian, E., Matasovic, N., Stokoe, K. H., & Bray, J. D. (1996). In situ shear wave velocity of solid waste from surface wave measurements. In M. Kamon (Ed.), *Proceedings of Environmental Geotechnics,* 97–102.

Kavazanjian, J. E., & Matasovic, N. (1995). Seismic analysis of solid waste landfills. Geoenvironment 200 Proceedings, 1066-1080.

Kramer, S. L. (1996). *Geotechnical Earthquake Engineering*. Englewood Cliffs, NJ: Prentice-Hall.

Landva, A. O., & Clark, J. I. (1986, August). Geotechnical testing of waste fill. *Proceedings, 39th Canadian Geotechnical Conference,* 3714385.

Matasovic, N., & Kavazanjian, E. Jr. (1998). Cyclic characterization of OII landfill solid waste. *Journal of Geotechnical and Geoenvironmental Engineering, 124*(3), 197–210. doi:10.1061/(ASCE)1090-0241(1998)124:3(197)

Matasovic, N., & Vucetic, M. (1995). Seismic response of soil deposits composed of fully-saturated clay and sand layers. *First International Conference on Earthquake Geotechnical Engineering*.

Naveen, B. P., Sitharam, T. G., & Sivapullaiah, P. V. (2014). Evaluating the Dynamic Characteristics of Municipal Solid Waste for Geotechnical Purpose. *Current Advances in Civil Engineering, 2*(1).

Naveen, B.P., Sitharam, T.G., & Sivapullaiah, P.V. (2015). Seismic Analysis of Municipal Solid Waste Landfill in India. *International Journal of Geotechnical Earthquake Engineering*, *6*(2), 35-55.

Oweis, I. S., & Khera, R. P. (1998). *Geotechnology of waste management.* PWS Pub.

Reddy, K. R., Kumar, G., & Giri, R. K. (2017). Modelling Coupled Processes in Municipal Solid Waste Landfills: An Overview with Key Engineering Challenges. *Int. J. of Geosynthetic. and Ground Engineering*, *3*(6), 1–21.

Sanchez-Alciturri, J. M., Palma, J., Sagaseta, C., & Canizal, J. (1993). Mechanical properties of wastes in a sanitary landfill. *Proceedings of the Green '93, Waste Disposal by Landfill,* 357-363.

Sharma, H. D., Dukes, M. T., & Olsen, D. M. (1990). *Field measurements of dynamic moduli and Poisson's ratio of refuse and underlying soils at a landfill site. Geotechnics of Waste Fills – Theory and Practice, STP 1070*. Philadelphia, PA: ASTM.

Sharma, H. D., Dukes, M. T., & Olsen, D. M. (1991). Field measurements of dynamic moduli and Poisson's ratios of refuse and underlying soils at a landfill site. *International Journal of Rock Mechanics and Mining Sciences & Geomechanics Abstracts*, *28*(6), A356. doi:10.1016/0148-9062(91)91335-O

Shumann, H. (1989). Standdestigkeitsberechnung fur Hausmulldeponien–Praktische Erfahrungen. *Fortschritte der Deponietechnik*, *89*, 153–179.

Watts, K. S., & Charles, J. A. (1999). Settlement characteristics of landfill wastes. *Proceedings of ICE. Geotechnical Engineering*, *137*(4), 225–233. doi:10.1680/gt.1999.370408

Wiemer, K. (1982). *Qualitative and quantitative Kriterien zur Bestimmung der Dichte von Abfällen in geordneten Deponien*. Academic Press.

Woodward Clyde Consultants. (1987). *Seismic stability evaluation operating industries Islandfill. A Report Prepared for Camp Dresser Mckee*. Santa Ana, CA: Woodward Clyde Consultants.

Zekkos, D., Bray, J. D., & Riemer, M. F. (2008). Shear modulus and material damping of municipal solid waste based on large-scale cyclic triaxial testing. *Canadian Geotechnical Journal*, *45*(1), 45–58. doi:10.1139/T07-069

Zekkos, D. P. (2005). Evaluation of static and dynamic properties of municipal solid waste, *University of California, Berkeley* (Ph.D. thesis). Academic Press.

Chapter 8
DEM Studies on the Liquefaction Behavior of Particles With Different Aspect Ratios

Anitha Kumari S. D.
M. S. Ramaiah University of Applied Sciences, India

Sitharam T. G.
Indian Institute of Science, India

ABSTRACT

Discrete element method (DEM) provides insight into the fundamental physical principles leading to a better understanding of the complex behavior of granular materials under different loading conditions. In the reported studies adopting DEM, real particle shape is ignored, and the particles are modeled as spheres. Spherical shapes are preferred as they provide the simplest case with least computational effort in calculations and efficient algorithms. Hence, in this study, a comprehensive and comparative study on the mechanical behavior of assemblies consisting of particles with different aspect ratios is reported. The results indicate that the assemblies with lower aspect ratio lose its strength completely at less number of cycles when subjected to lower confining pressures. At higher confining stresses, the particles are not quick enough to rearrange themselves resulting in the reduction of average coordination number. This causes a drop of shear strength leading to lower resistance to liquefaction for non-spherical assemblies at high confining stresses.

DOI: 10.4018/978-1-5225-6948-0.ch008

INTRODUCTION

Earthquake loadings simulate undrained conditions on sandy soils leading to a sudden surge in pore water pressure. This type of dynamic loading leads to degradation of stiffness which ultimately leads to a reduction in shear strength of the soils. An analysis of the effects produced by liquefaction indicates that it varies from the destruction of shallow buildings to the life lines of cities. Examples on the hazard impacts on the normal life due to liquefaction can be clearly observed from the earthquakes which occurred in a sequence during 2010-2011 in Christchurch New Zealand. Recent earthquakes which produced significant liquefaction related damage include Chile (2010), Christchurch New Zealand (2011) and Tohoku Earthquake, Japan (2011). The casualties and the hazards caused by liquefaction results in huge economic loss.

A large amount of research has been carried out to understand the behaviour of granular materials subjected to static and dynamic loading by analytical, physical and laboratory experiments. But it is well proven that these methods cannot shed light into the fundamental aspects of granular behaviour. In granular media, whether it is field observation or laboratory experiment or numerical modeling, it is understood that the macroscopic behaviour is dependent on the particle level mechanics. So understanding the internal structure and its evolution during loading is significant. Cundall & Strack (1979) proposed the Discrete element method (DEM) which tracks the movement of individual particles and further updates the contact forces and contact displacement of individual grains through a discrete element cycle. For numerical simulations, DEM is followed in this study.

BACKGROUND

The dynamic properties of the soil mass which include shear modulus and damping ratio are key parameters which control the behaviour of a soil deposit. The degradation of the strength parameters plays a very important role in the behaviour during cyclic loading. Laboratory cyclic triaxial tests and low strain tests have the limitations of providing the dynamic properties at the measured strain level. Also, in the case of laboratory tests, the difficulties associated with the preparation of samples having the same initial properties, the effect due to the rubber membrane and the non-uniform distribution of stresses and strain are very predominant. However, the numerical methods adopted for the modelling of the dynamic response of the soil subjected to earthquake loading effectively eliminates most of the problems faced by the laboratory tests. The initial arrangement of the particles in a sample or its fabric has significant effect on the liquefaction and post-liquefaction behaviour of sands. Recent studies reported by Wei & Wang (2017) indicates that an isotropic sample is having higher liquefaction resistance compared to samples with higher degrees of fabric anisotropy. The microscopic parameters obtained from the DEM studies can be related to the macroscopic behaviour. This aids in understanding the fundamental processes behind the mechanical behaviour of soils thereby helping to interpret the constitutive behaviour.

The behaviour of saturated soil during earthquake loading is highly influenced by the development of excess pore water pressure. As the pore pressure increases, the effective stress decreases leading to the loss of shear strength of the soil deposit. This underlines the necessity of understanding the progressive loss of strength during dynamic loading resulting in liquefaction. As a result of liquefaction, soil loses its strength and stiffness leading to failure of structures resting on this soil deposit. During cyclic loading, the soil response is directly related to the plastic contraction and dilation which are induced due to shear. These responses in terms of displacement is dependent on the effective stress at the considered level of the soil deposit. Among the various methods to model the gradual reduction of effective stress and the subsequent loss of shear strength, cyclic stress and cyclic strain approaches are widely adopted for liquefaction analysis. In this study, strain controlled numerical simulations are adopted as this approach directly relates the shear strain to the buildup of excess pore water in saturated sands during undrained cyclic loading. Moreover, in strain-controlled methods it is observed that the pore water pressure build up is mostly insensitive to a number of factors when compared to stress controlled methods. Researchers like Anderson and Stokoe (1978), Drnevich and Richart (1970), Seed and Idriss (1971) have reported that various factors in stress-controlled tests are affecting the cyclic shear strength and dynamic modulus of soils. Based on the above observations and underlining the fact that liquefaction is primarily related to the generation of excess pore pressure in any soil deposit, cyclic strain approach is found to be more appropriate compared to stress approach.

METHODS AND METHODOLOGY ADOPTED FOR THE STUDY

There are different methods adopted in practice for the evaluation of the liquefaction and the subsequent soil behaviour. Micromechanical theories and plasticity theories are the most common approaches adopted for developing the models for stress-strain behavior of soils. Many researchers (Ishihara et al., 1975; Mojtaba Rahimi et al. 2017; Seed et al. 1976; Vaid & Thomas, 1995; Yang et al. 2011) have reported the understanding of the liquefaction fundamentals, pore pressure changes, constitutive models for cyclic loading through analytical and laboratory experiments. Most of the analytical methods viz. Finn model, Disturbed State Concept (Desai et al. 1998) and Critical State Constitutive Model (Mojtaba Rahimi et al. 2017) are based on continuum approach.

The inherent discrete nature of the particulate matter has great influence on its behaviour and hence the modes of failure and the deformation may be affected by the assumption of continuum. In this context, DEM which takes into account the discrete nature of each particle provides a fundamental understanding of the deformation of the failure modes of the particulate mass. DEM pioneered by Cundall (1971) was developed to study the rock slope which was later extended to the study of application in granular media. The principle of DEM is based on Newton's second law of motion by monitoring the movement of each particle. Newton's laws of motion and force displacement law is alternatively used to get the updated positions of the particles. The basic unit/element of consideration in DEM is the contacts between the particles. The movement of each particle is monitored by solving the equations of motion. The relative displacement at the contacts gives rise to a contact force which is evaluated by force-displacement law. The equilibrium condition is maintained in the assembly by keeping the loading rates very low. When the loading rate is low, the inertial forces developed are very small. Compared to the average contact force in the assembly, inertial forces will form only a very small fraction of the average contact force.

Several researchers have adopted the technique of DEM for modelling liquefaction (Dinesh et al. (2004); Dobry & Ng (1992); Gong & Zha (2012); Nguyen et al. (2017); Sitharam (2003) and Wei & Wang (2017)) to understand the physics behind liquefaction. Most of the reported studies in literature consider the particles are spherical in shape. This is because the incorporation of various shapes increases the number of contacts and hence the solution becomes computationally expensive. Wensrich and Katterfeld (2012) have reported that rolling friction can be employed to simulate the effects of particle shape. Researchers have studied the granular soil behavior by modeling ellipse-shaped grains in 2D and ellipsoids in 3D simulations (Bagherzadeh-Khalkhali & Mirghasemi (2009), Lin & Ng (1997), Rothenburg & Bathurst (1992), Ting et al. (1993) and Wang et al. (2011)). Garcia et al. (2009) represented irregular particles as a group of overlapping spheres in DEM. According to Clayton and Heymann (2001) in low strain regions, particle shape can have significant impact on stiffness. Based on the above facts, in this study, assemblies consisting of spheres and a combination of spheres and ellipsoids are created. The comparative behaviour of these assemblies from a micromechanical point of view can be obtained when subjected to cyclic loading. Particle Flow Code (PFC^{3D}) which employs DEM is used for modeling the assembly in this study.

Laboratory undrained tests are generally performed on cylindrical samples. In this study, a numerical soil sample consisting of 3600 spherical particles as shown in Figure 1 is used for the study. The height to diameter ratio of this sample is kept at 2:1 conforming to the laboratory standards. To understand the effect of shape in terms of aspect ratio, a total of three samples with aspect ratio 1.0, 1.5 and 2.0 are selected for the study. In all the three samples, the total number of particles are restricted to 3600. The samples are denoted by Sample 1, Sample 2 and Sample 3. Sample 1 consists of 3600 spherical particles, whereas sample 2 consists of 30% ellipsoids with aspect ratio 1.5 and 70% spheres. In sample 3, 30% particles are ellipsoids with aspect ratio 2.0 and the remaining 70% consists of spherical particles with aspect ratio 1. The range of particles in each size is calculated in such a way that a particle size distribution as shown in Figure 2 is obtained for the study. The uniformity coefficient of the three samples are found to lie in the range of 2.4 whereas the coefficient of curvature of the three samples are 1.0, 1.1 and 1.0 respectively. In addition to the above 3 samples, Sample 4 was also generated consisting of 100% ellipsoidal particles having an aspect ratio of 1.5. Particles with different aspect ratios are generated by adopting the clump logic in PFC^{3D}. The advantage of adopting the clump logic is that the particles will act as a rigid body and will not break apart under the action of forces. Spherical particles can be joined together to get any shape and these particles remain at a fixed distance from each other. In addition, the constituent particles of a clump can be overlapped to any extent without any contact force being developed between them. A typical ellipsoidal particle is shown in Figure 3. While generating the clumps, the net volume of the clump is maintained the same as the sphere which it replaces. The packing density of the assembly can be controlled by the scaling option of clumped particles. This helps to maintain the void ratio of the various samples comparable for various tests. Table 1 shows the various parameters used for the particles in the simulation. These properties have been arrived after a series of parametric studies.

A series of numerical simulations of cyclic triaxial tests are performed on spherical and ellipsoidal assemblies to understand the effect of shape on the constitutive behaviour of granular materials. These simulations help to understand the micromechanical characteristics at the grain scale level along with the macroscopic response of monotonic and cyclic shear behaviour of granular media. The evolution of the microstructure of the assemblies under shearing is closely examined and related to the shape aspect of the particles constituting the assembly.

Figure 1. Cylindrical assembly

Figure 2. Particle size distribution

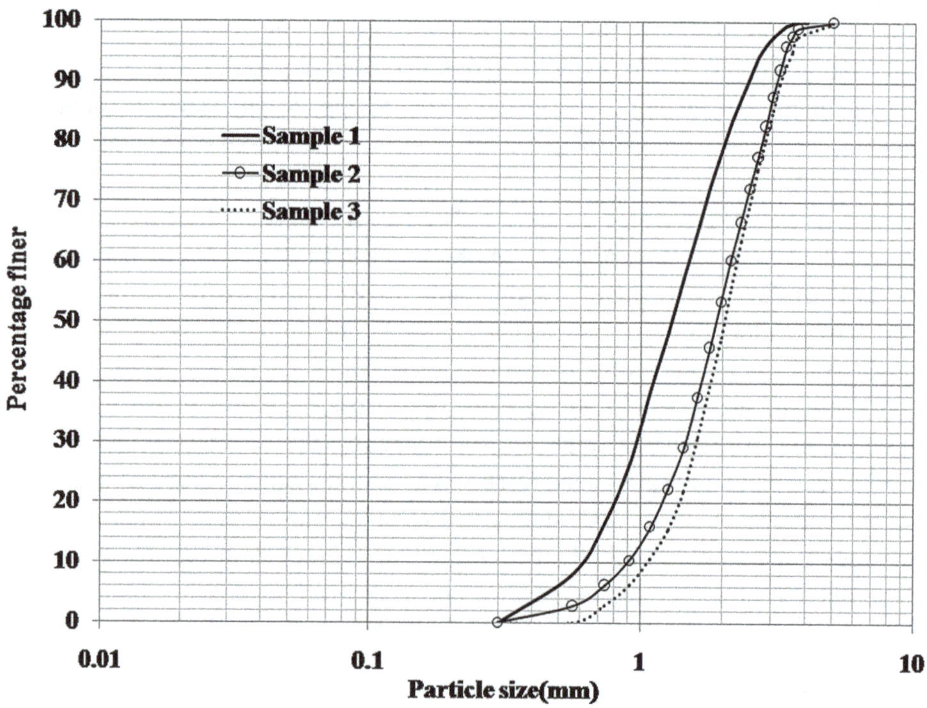

Figure 3. Typical ellipsoidal particle

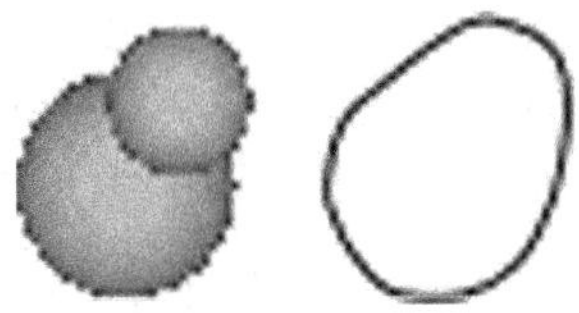

Table 1. Properties of the particles used for the simulation

Properties	Values
Normal contact stiffness of particle (K_n)	1e5N/m
Shear contact stiffness of particle(K_s)	1e5N/m
Wall stiffness (k)	1e6N/m
Particle density (ρ)	2650kg/m^3
Coefficient of inter-particle friction (μ)	0.5
Coefficient of friction between particle and wall(μ)	0.1
Damping coefficient	0.7
Acceleration due to gravity	-9.81m/s^2

In DEM simulations, undrained loading situations are generated without explicitly incorporating the pore pressure. This approach known as constant volume method is based on the assumption that during shearing, the effective stress path is independent of the total stress. This method takes into account the incompressibility of water. Here the effective stress is calculated by taking the difference between the total stress path and effective stress path (Dobry & Ng, 1992). In the simulations, the excess pore pressure is assumed to be zero at the start of shearing. Shearing is done by compressing the sample by applying load in the vertical direction. To keep the volume constant, the strains in the lateral directions are kept half that of the vertical direction. The variation of the axial, radial and volumetric strain are shown in Figure 4. Due to the conditions imposed, the magnitude of axial strains and the radial strains are equal, but opposite in direction resulting in a zero volumetric strain and thus enforcing the constant volume condition.

Cyclic loading under constant strain amplitude was done on isotropically consolidated samples at a confining pressure of 100kPa. The magnitude of cyclic strain amplitude was set as 0.006%. Considering the equilibrium of the system and the frequency adopted in the laboratory, a value of 1Hz for frequency was adopted for the study which was applied sinusoidally. Figure 5 shows the variation of strain amplitude for the simulations.

Figure 4. Constant volume condition: Variation of the axial, radial and volumetric strain

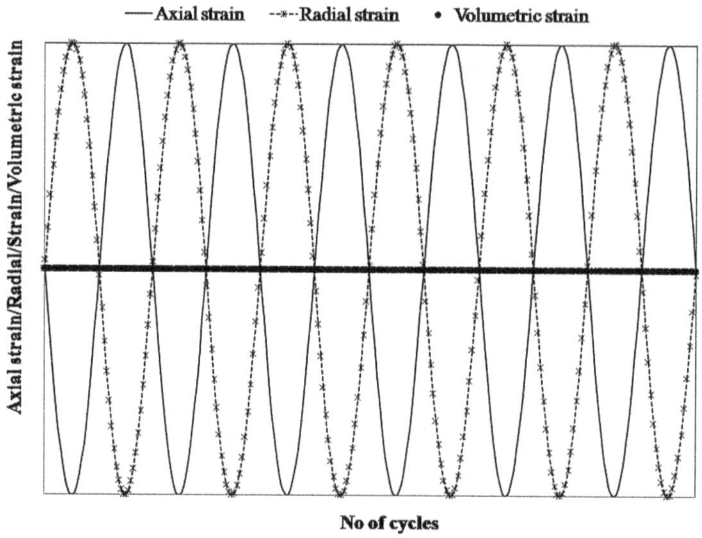

RESULTS AND DISCUSSIONS

When the sample is subjected to loading by applying the constant sinusoidal strain velocity, shearing progresses and there is a steady decrease in the deviatoric stress. As the loading progresses, there is a steady decrease in the deviatoric stress which is attributed to the development of excess pore pressure as shown in Figure 6 and Figure 7. A close examination of this stress strain curve indicates that the magnitude of deviatoric stress is higher in the compression side than extension side.

Figure 8 indicates the hysteresis loops of the samples for the first cycle of loading. Compared to the sample consisting of only spherical shaped particles, the strength of the sample consisting of 100% non-spherical particles is twice higher. This figure shows that the samples containing particles of higher aspect ratio is having higher shear strength at the same strain level resulting in a higher value of shear modulus. In addition, these steep loops are having lesser area compared to that of the assembly consisting of only spherical particles and thus results in lower damping value. As the cyclic loading progresses, it can be seen that the stress loops become more and more flat and finally the sample fails. The same trend is observed for all the four samples at different confining pressures. When the aspect ratio increases, more number of contacts per particle is possible due to the larger contact surface area compared to spherical particles. Moreover, rolling motion of these particles are also restricted to a greater extent due to the strong interlocking among them compared to that of the spherical particle assembly.

Figure 5. Magnitude of cyclic strain amplitude

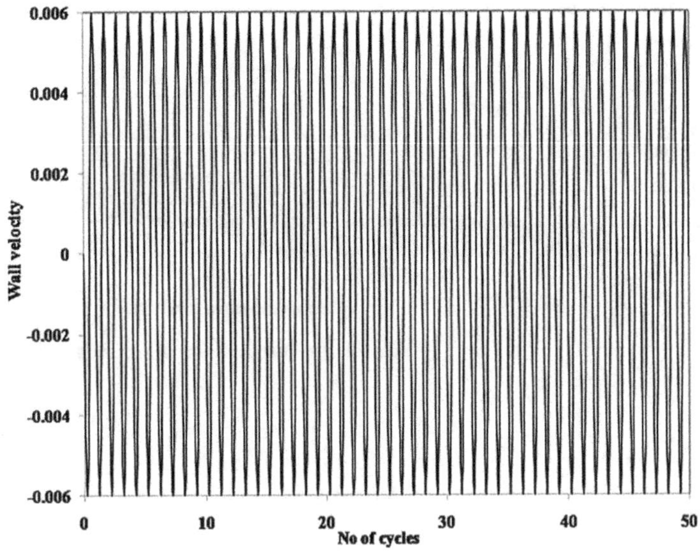

Figure 6. Deviatoric stress vs axial strain (Sample 1)

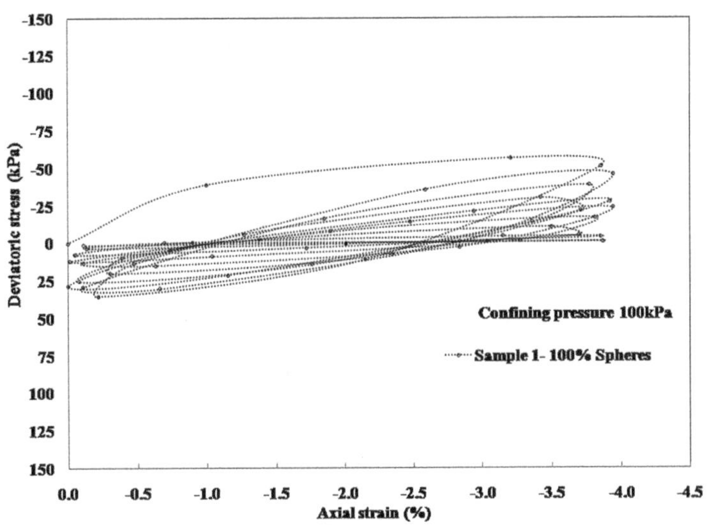

The rate of reduction in stiffness values is very high during the initial cycles of loading. But the rate of reduction in stiffness decreases as the loading cycle progresses. Idriss et al (1978) proposed a parameter known as degradation index (δ) to understand the reduction in stiffness during cyclic loading. This parameter can be expressed as:

Figure 7. Deviatoric stress vs axial strain (Sample 4)

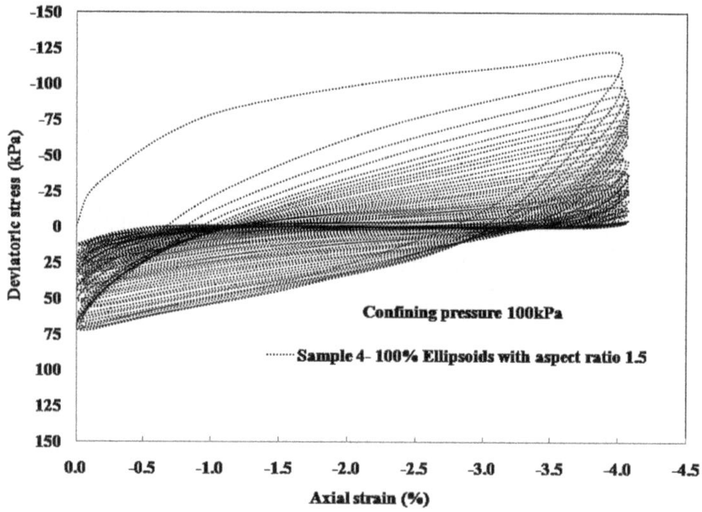

Figure 8. Hysteresis loops during the first cycle for the four samples

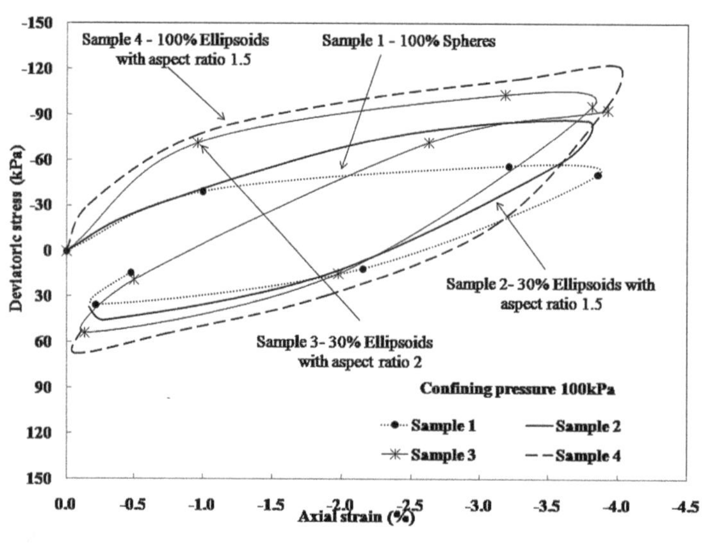

$$\delta = \frac{E_{N+\Delta N}}{E_N} = \frac{G_{N+\Delta N}}{G_N} = \frac{\sigma_{N+\Delta N}}{\sigma_N}$$

where E and G represents the cyclic deformation modulus and secant deformation modulus respectively. The cyclic shear stress at N+ΔN and Nth cycle can be represented as $\sigma_{N+\Delta N}$ and σ_N respectively. This index indicates the irreversible changes occurring in the structure of the soil due to cyclic loading. Figure 9 shows the variation of the degradation index with the number of cycles for the four samples. This plot reveals that the assembly consisting of only spherical particles is having a lower value of degradation index at any considered loading cycle. Moreover, it can be seen that in the case of sample 3, at intermediate cycles ranging from 10-12 and 17-20 there is almost nil degradation as the index remains constant.

When a sample is loaded cyclically in undrained condition, pore pressure gradually builds up in the sample leading to a reduction in the effective mean stress. This reduction in the mean stress will lead to a steady decrease in the average coordination number. As the effective stress reduces to a value closer to zero, the contact between the particles will be minimum leading to liquefaction. The cyclically loaded samples will liquefy at a much smaller stress level when compared to static undrained loading. Figure 10 shows the static undrained stress paths of the samples. The same samples when subjected to cyclic loading (Figure 11) indicates that the maximum deviatoric stress levels attained by all the samples are much lesser than the strength they have attained during monotonic undrained loading. The development of pore pressure in each of these samples is shown in Figure 12. As the loading progresses, the pore pressure ratio increases and has reached the maximum value of one at 11, 15, 25 and 40 cycles for the four different samples respectively indicating that the samples have liquefied. For samples with higher aspect ratio, the rate of pore pressure development is less during the initial stages compared to the sample

Figure 9. Variation of the degradation index for the four samples

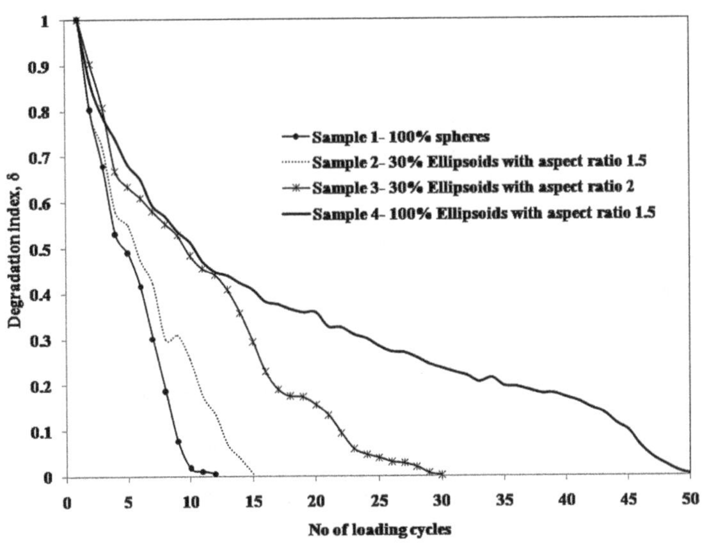

with 100% spherical particles. However, if the confining pressure is increased, it is observed that the rate of development of pore pressure, in samples with 100% spherical particles, is slower. Figure 13 shows the development of pore pressure ratio for samples at a confining pressure of 300kPa. Even after 300 cycles of loading, the pore pressure ratio developed is 0.6 and 0.8 for sample 1 and sample 2 respectively. This indicates that as the confining pressure increases, spherical particles are offering more resistance to liquefaction.

The evolution of the microstructure which ultimately is responsible for the shear strength of the assembly is dependent upon the contact force and its orientation. As the loading and unloading cycle progresses, the average coordination number decreases. Generally the average coordination number reduces suddenly during the first couple of cycles and then decreases gradually. However, when the average coordination number decreases to around 3, the assembly is completely collapsed and the mean effective stress is having a very negligible value. Figure 14 shows the variation of the contact force network in the sample at different stages of loading. The magnitude of the contact forces can be correlated to the thickness of the lines in the plots. The magnitude of maximum value of the contact force is 21.65N and is observed at the start of the shearing stage. As the loading progresses, the magnitude of average contact force decreases and is clearly seen from the reduction in the thickness of contact force diagrams plotted for 15, 30 and 50 cycles respectively. At 50 cycles, it is observed that the maximum contact force value is only 2.16N which is just 10% of the contact force at the start of the test. At the initial stages of loading,

Figure 10. Stress paths of the samples during monotonic undrained loading

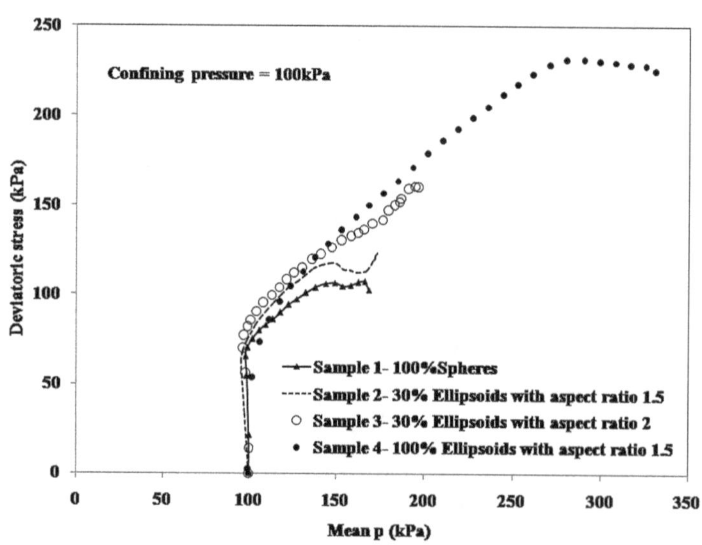

Figure 11. Variation of Deviatoric stress with no. of cycles during undrained cyclic loading

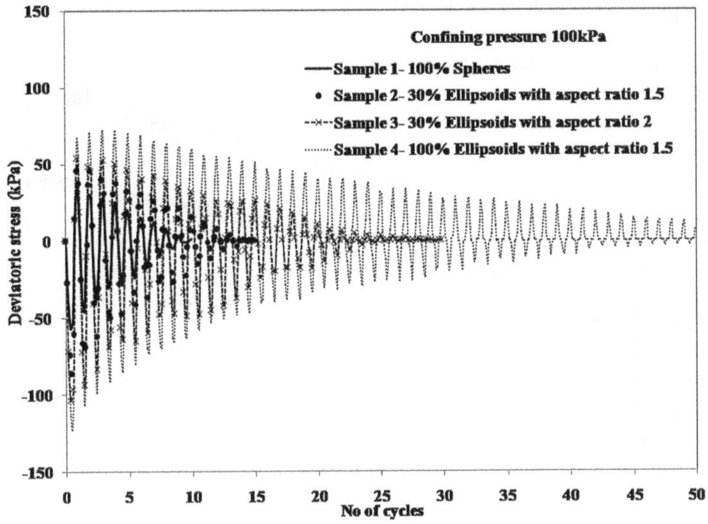

Figure 12. Development of pore pressure as the cyclic loading progresses

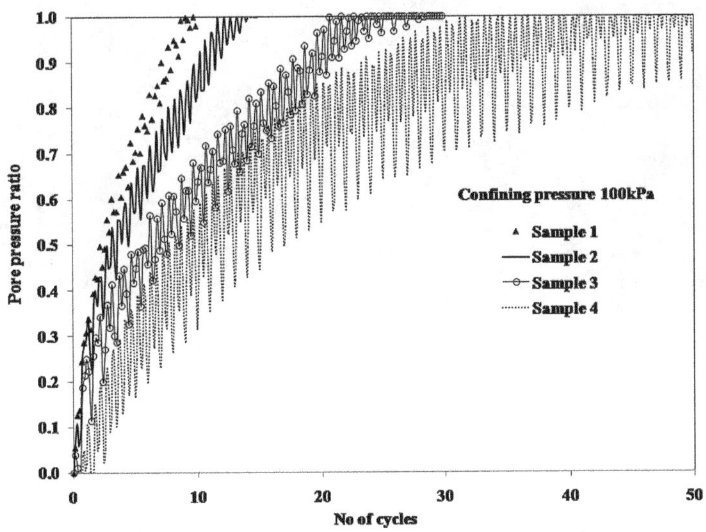

strong contacts are predominantly observed in the major principal stress direction. However as the loading progresses, there is considerable reduction of contact force magnitude in the axial direction and can be seen as the gradual decrease of average coordination number as shown in Figure 15.

Figure 13. Pore pressure vs No. of cycles at a confining pressure of 300kPa

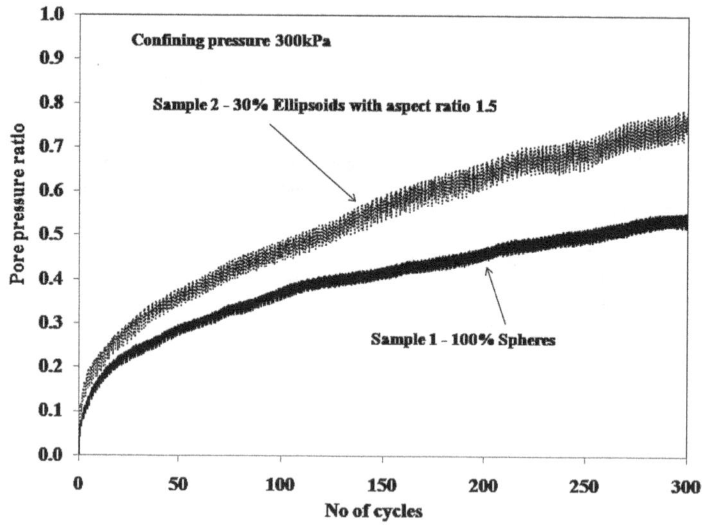

Figure 14. Variation in contact force as the loading cycle progresses

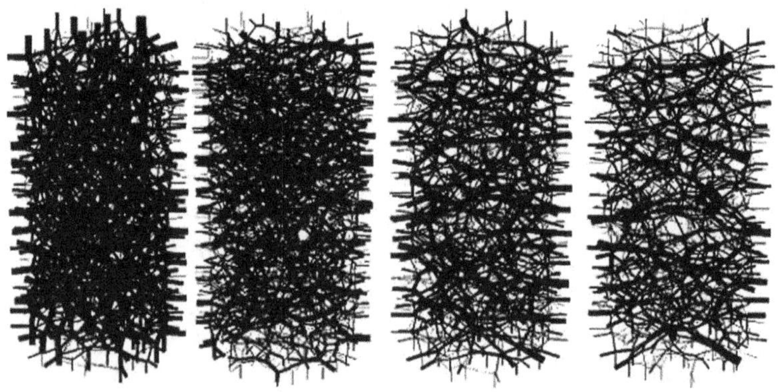

(i) Start – 21.65N (ii)15 cycles- 13.25N (iii) 30 cycles - 7.656N (iv) 50 cycles - 2.16N

CONCLUSION

An attempt has been made to understand the liquefaction behaviour of assemblies consisting of particles with different aspect ratios when subjected to undrained cyclic loading. The macroscopic and microscopic behaviour of the isotropically compressed samples were emulated by the simulations. The results indicate that the assemblies with lower aspect ratio lose its strength completely at less number of cycles at lower confining pressures. This can be attributed to the interlocking of

Figure 15. Average coordination number vs Mean stress

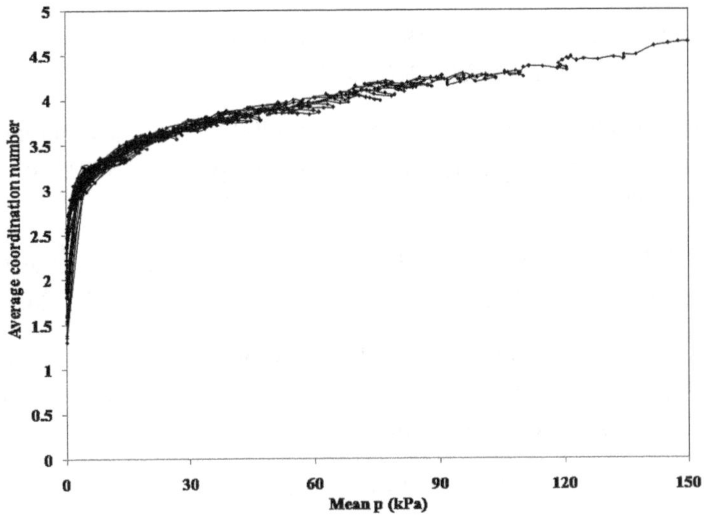

non-spherical particles resulting in higher contacts per particle and hence higher strength. When the loading direction changes, there is continuous reorientation of the contact normal which results in the loss of contacts. As the contacts gets lost, the average coordination number lowers. The loss of contacts is predominant in the minor principal stress direction. However at higher confining pressures, the rate of pore pressure generation of the assembly consisting of spherical particles is less compared to other samples. This is attributed to the increased resistance offered by the spherical particles for rotation. Also at higher confining stresses, when the loading changes from compression to extension, the particles are not quick enough to rearrange themselves to the new orientation. This results in the reduction of contacts and subsequent drop of shear strength leading to a lower resistance to liquefaction for non-spherical assemblies at high confining stresses. The ability of the assembly to resist the applied external stresses depends on the strong contact force chains. When the loading is static, the strong force chains are oriented along the major principal directions whereas when there is a reversal in the direction of major and minor principal stresses as in the case of cyclic loading, the strong force chains are disturbed. This leads to the gradual collapse of the assembly. The contact force diagrams at the liquefaction stage indicate that the assembly is highly unstable with the possibility of the particles floating in the pore water without any significant contacts. Further to this study, liquefaction resistance of assemblies with varying degree of anisotropy at the start of shear can be explored.

REFERENCES

Anderson, D. G., & Stokoe, K. H. (1978). Shear Modulus: A Time-Dependent Soil Property. *ASTM Special Technical Publication*, 66–90.

Bagherzadeh-Khalkhali, A., & Mirghasemi, A. A. (2009). Numerical and experimental direct shear tests for coarse-grained soils. *Particuology*, *7*(1), 83–91. doi:10.1016/j.partic.2008.11.006

Clayton, C. R. I., & Heymann, G. (2001). Stiffness of geomaterials at very small strains. *Geotechnique*, *51*(3), 245–255. doi:10.1680/geot.2001.51.3.245

Cundall, P. A. (1971). A computer model for simulating progressive large scale movements in Blocky Rock systems. *Proceedings of the Symposium of the International Society of Rock Mechanics*, 1, Paper No II-8.

Cundall, P. A., & Strack, O. D. L. (1979a). A Discrete Numerical Model for Granular Assemblies. *Geotechnique*, *29*(1), 47–65. doi:10.1680/geot.1979.29.1.47

Desai, C. S., Park, I. J., & Shao, C. (1998). Fundamental yet simplified model for liquefaction instability. *International Journal for Numerical and Analytical Methods in Geomechanics*, *22*(7), 721–748. doi:10.1002/(SICI)1096-9853(199809)22:9<721::AID-NAG942>3.0.CO;2-F

Dinesh, S.V., Sitharam, T. G., & Vinod, J.S. (2004). Dynamic properties and liquefaction behaviour of granular materials using discrete element method. *Current Science, Special Section: Geotechnics and Earthquake Hazards, 87*(10), 1379-1387.

Dobry, R., & Ng, T. T. (1992). Discrete modeling of stress-strain behaviour of granular media at small and large strains. *Engineering Computations*, *9*(2), 129–143. doi:10.1108/eb023853

Drnevich, V. P., & Richart, F. E. (1970). Dynamic pre straining of dry sand. *Journal of Soil Mechanics and Foundation*, *96*, 453–469.

Garcia, X., Latham, J. P., Xiang, J., & Harrison, J. P. (2009). A clustered overlapping sphere algorithm to represent real particles in distinct element modelling. *Geotechnique*, *59*(9), 1–6. doi:10.1680/geot.8.T.037

Gong, G., & Zha, X.-X. (2012). DEM simulation of liquefaction for cohesionless media at grain scale. *Journal of Central South University*, *19*(9), 2643–2649. doi:10.100711771-012-1322-9

Ishihara, K., Tatsuoka, F., & Yasuda, S. (1975). Undrained deformation and liquefaction of sand under cyclic stresses. *Soil and Foundation*, *15*(1), 29–44. doi:10.3208andf1972.15.29

Lin, X., & Ng, T. T. (1997). A three-dimensional discrete element model using arrays of ellipsoids. *Geotechnique*, *47*(2), 319–329. doi:10.1680/geot.1997.47.2.319

Martin, G. R. (1975). Fundamentals of Liquefaction Under Cyclic Loading. *Journal of Geotechnical Engineering*, *101*(5), 423–438.

Nguyen, H. B. K., Rahman, M. M., & Fourie, A. B. (2017). Undrained behaviour of granular material and the role of fabric in isotropic and K0 consolidations: DEM approach. *Geotechnique*, *67*(2), 153–167. doi:10.1680/jgeot.15.P.234

Rahimi, M., Chan, D., & Nouri, A. (2017). Constitutive model for cyclic behaviour of cohesionless sands. *Geomechanics and Geoengineering*, *12*(1), 36–47. doi:10.1080/17486025.2016.1156168

Rothenburg, L., & Bathurst, R. J. (1992). Micromechanical features of granular assemblies with planar elliptical particles. *Geotechnique*, *42*(4), 79–95. doi:10.1680/geot.1992.42.1.79

Seed, H. B., & Idriss, I. M. (1971). Simplified Procedure for Evaluating Soil Liquefaction Potential. *Journal of the Soil Mechanics and Foundations Division*, *97*, 1249–1273.

Seed, H. B., & Lee, K. L. (1966). Liquefaction of Saturated Sands during Cyclic Loading. *Journal of the Soil Mechanics and Foundations Division*, *92*, 105–134.

Seed, H. B., Martin, P. P., & Lysmer, J. (1976). Pore-water pressure changes during soil liquefaction. *Journal of Geotechnical Engineering*, *102*(4), 323–346.

Sitharam, T. G. (2003). Discrete element modeling of cyclic behaviour of granular materials. *Geotechnical and Geological Engineering*, *21*(4), 297–329. doi:10.1023/B:GEGE.0000006036.00597.0b

Ting, J. M., Khawaja, M., Meachum, L. R., & Rowell, J. D. (1993). An ellipse-based discrete element model for granular materials. *International Journal for Numerical and Analytical Methods in Geomechanics*, *17*(9), 603–623. doi:10.1002/nag.1610170902

Vaid, Y. P., & Thomas, J. (1995). Liquefaction and post liquefaction behaviour of sand. *Journal of Geotechnical Engineering*, *121*(2), 163–173. doi:10.1061/(ASCE)0733-9410(1995)121:2(163)

Wang, J., Yu, H. S., Langston, P., & Fraige, E. (2011). Particle shape effects in discrete element modelling of cohesive angular particles. *Granular Matter*, *13*(1), 1–12. doi:10.100710035-010-0217-4

Wei, J., & Wang, G. (2017). Discrete-element method analysis of initial fabric effects on pre- and post-liquefaction behavior of sands. *Géotechnique Letters*, *7*(2), 161–166. doi:10.1680/jgele.16.00147

Wensrich, C. M., & Katterfeld, A. (2012). Rolling friction as a technique for modelling particle shape in DEM. *Powder Technology*, *217*, 409–417. doi:10.1016/j.powtec.2011.10.057

Yang, C., Huang, M. S., & Cui, Y. J. (2011). Constitutive model of unsaturated structured soils under cyclic loading. In Alonso & Gens (Eds.), Unsaturated Soils (pp. 987-992). Taylor & Francis Group.

Chapter 9
Effect of Repeated Blast Vibrations on Rock Mass Damage in Tunnels

Ramulu More
Central Institute of Mining and Fuel Research, India

T. G. Sitharam
Indian Institute of Science Bangalore, India

ABSTRACT

Blasting is the most common method of rock excavation technique in mining and civil construction and infrastructure projects. General blast-induced damage was extensively studied globally, but the studies on repeated blast induced damage is not yet reported, quantitatively. This chapter deals with the research work carried on the effect of repeated dynamic loading imparted on the jointed rock mass from subsequent blasts in the vicinity, on the jointed rock mass at Lohari Nag Pala Hydroelectric Power Construction Project. The blast-induced damage was monitored by borehole extensometers, borehole camera inspection surveys and triaxial geophones installed at three test sites of different joint orientations at the Main Access Tunnel of power house. The study reveals that the repeated blast vibrations resulted in extra damage of 60%. The results of the study indicate that repeated vibrations, resulted in inducing damage even at 23-33% of the critical vibrations (Vc) levels.

DOI: 10.4018/978-1-5225-6948-0.ch009

INTRODUCTION

Blasting produces seismic waves similar to those produced by earthquakes, but with relatively high frequency and low amplitude which can create damage to surrounding structures. The degree of structural damage depends on the total energy of explosion, distance from the source, and the characteristics of the medium. Extensive experimental investigation in this regard was carried out by Thoenen and Windes (1942) of United States Bureau of Mines (USBM), Leet (1946) and Crandell (1949). Based on the experimental results the USBM recommended that no structural damage occurs if the acceleration of vibration is less than 0.1 times the gravity (g) and 0.1 to 1g is caution range; and the acceleration greater than 1g is danger zone. Leet (1946) prefers to limit the size of blasts by the displacements that they produce in the structure and the index of damage is a limiting displacement of 0.03 inches. Crandell (1949) proposes to limit the size of the explosion by limiting the kinetic energy delivered to the ground, which is proportional to the quantity of the explosive. A number of studies attempted to correlate ground-motion levels with observed damage to structures. It is generally agreed that the amount of blast damage correlates best to the peak particle velocity (V_{max}). The blasting criteria for residential structures is generally less than 5 cm/s and for massive concrete structures is generally less than 25 cm/s (Charlie, 1985). Oriard (1989) observed that V_{max} of 8-10 ips normally and not damage the structure, because of the very high frequencies and the rapid, localized attenuation. Tart (1980) observed that at high frequencies the vibration levels of 275 ips generate minor cracks in old concrete. Rock mass damage in underground openings occurs mainly due to blast induced forces, stress redistribution and weathering. As underground excavations are carried out, the in-situ stresses redistribute around the boundary of the openings, leading to high stresses on the backs and corners of the excavations and the blasting activity creates initiation and extension of fractures in the surrounding rock mass. Blast damage is defined as either creation or extension of new fracture surfaces or opening of pre-existing geological discontinuities or both in the rock mass (Law et al, 2001). Blast induced damage weakens a rock mass, potentially leading to stability problems in the underground excavations. The stability of the underground structure is very much dependent upon the integrity of rock immediately surrounding the excavation. The blast damage can easily extend few meters into the rock and the loosened rock can give rise to serious safety and stability problems to the surrounding rock mass of the underground openings. The blast damage problem is more severe and vulnerable for the jointed rock mass in underground excavations (Singh & Xavier, 2005). In spite of recognition of the importance of duration of ground motion on the dynamic response, current engineering practice correlates damage to peak ground motion during an episode of dynamic loading, since it can be related directly to peak transient stress in the ground wave, and the second power of velocity is related to dynamic strain energy (McGarr, 1983). Unfortunately, there are no specific safety guidelines availble for the blasted tunnels with regards to the threshold limits of vibrations caused by repeated blasting activity in the close proximity. Many efforts have been made to study blast induced cracking and framing of safety guidelines in residential structures (Langefors & Kihlstroam, 1963; Dowding, 1985; Scott, 1996; Anon, 1997), but less attention was found to study blast induced damage to rock mass in underground openings (Persson et al, 1994). Studies on blast induced damage on underground openings are well documented by many researchers globally (Langefors & Kihlstrom, 1963; Hendron, 1977; Holmberg, 1993; Singh, 1993; Paventi et al, 1996; Yu & Vongpaisal, 1996; Chakraborty *et al.*, 1998; Zhang & Chang, 1999). In a series of experiments the Swedish Detonic Foundation has investigated the extent of cracking emanating from blastholes in controlled conditions (Olsson & Bergqvist, 1996; Ouchterlony, 1993, 1997). Joint orientation with respect to perimeter line of underground opening is one of the influencing parameters of blast induced damage (Connigham & Goetzsche, 1996). Singh and Xavier (2005) observed largest overbreak for the joint orientations of 45^0 and minimum overbreak for parallel and perpendicular joint orientations. Similar results were also obtained by Lewandowski (1996) in his trial blasts on influence of discontinuities for effective pre-splitting. In view of the large amount of underground excavations like tunnels and caverns it is imperative to develop threshold limits of ground vibrations induced by blasting within or outside the underground opening. In this paper, it was aimed at prediction and assessment of blast induced damage and deterioration due to repeated dynamic loading produced by adjacent tunnel blasting.

REVIEW OF LITERATURE ON ROCK MASS DAMAGE DUE TO REPEATED BLASTING

The rock mass damage problem will be manifold if the blast loading is applied for repeated number of times, in contrast to the conventional single episode blast loading. Repeated blast loading causes progressive accumulation of damage in joints, which may lead to achievement of residual strength state in joints, with resultant large displacement at the joint surface (Brady, 1990). The effect of repeated blast loading on jointed rock mass was qualitatively studied by many researchers globally (Atchison & Pugliese, 1964; Oriard, 1989; Otuonye, 1997; Law et al, 2001; Villaescusa, 2004). Brady (1990) states that, substantial progress has not been attained in the study of repeated exposures of dynamic loading on jointed rock mass in comparison to conventional blasting with single episode of loading. Brown and Hudson (1974) states that rock mass damage by blast loading is predominantly due to joint motion, which is consistent with the experimental observation that joints decrease in shear strength under cyclic shear loading. Model studies of excavations in jointed rock under cyclic loading by Barton and Hansteen (1979) confirmed that excavation failure occurred by accumulation of shear displacements at joints. On the basis of these findings, St. John and Zahrah (1987) stated that, under dynamic loading, it is the number of excursions of joint motion into the plastic range that determines damage to an excavation. Wagner (1984) provided an indication of the general inadequacy of dynamic design based on V_{max} of single blast round. A possible conclusion with regards to dynamic behavior under a range of V_{max} is that repeated dynamic loading may amplify problems of dynamic instability in jointed rock mass in the underground openings like, multiple excavations, tunnels and caverns.

Although far-field damage is not a severe problem at the excavation sites where the blasting faces moves away and vibration gets attenuated substantially, the authors observed that it is an acute problem when the rock mass is subjected to repeated vibrations due to multiple excavations in the vicinity as in the case of multiple tunnels and caverns. In view of the above arguments it was decided to go for the investigation of both near field and far field damage assessments at the Lohari Nag Pala Hydroelectric Power Project (LNPHPP).

SITE DESCRIPTION

General

LNPHPP is located in the Central Crystalline of Higher Himalayan Zone and situated in Uttarkashi District of Uttaranchal State about 230 km from Rishikesh on National Highway No.108. The Project of 600MW capacity is located on the right bank of Bhagirathi River. The location of the site falls at a Longitude of 78.53^0E and Latitude of 30.25^0N and is shown in Figure 1. This infrastructural project comprises of construction of 73.40 m wide Barrage across river Bhagirathi. The major underground structures under excavation include Head Race Tunnel (HRT), Power House and Tail Race Tunnel (TRT). The HRT is of 14 km long and 6.0 m diameter with horse-shoe shape on the right bank of river meant for water transmission to the Underground Power House. The dimensions of underground Power House to be excavated are 155 m X 22 m X 47 m. Water from the Power House is discharged back in to the Bhagirathi River through a 350 m long Tail Race Tunnel (TRT) with 7.5 m diameter (horse-shoe shape) tunnel. The Powerhouse is proposed to house four units of 150 MW each totaling 600 MW. The construction of 14000 m Head Race Tunnel will be accessed from three numbers of intermediate construction adits, apart from inlet and outlet ends, which will facilitate the excavation of HRT from eight faces. The power house is accessed with TRT, Main Access Tunnel (MAT) of 8m diameter (D-shape) and a Cable &Ventilation Tunnel (CVT) of 6m diameter (D-shape). There is also another adit to access the TRT. All these tunnels around powerhouse are called power house complex, which is shown in Figure 2. As there was repeated number of blast loadings on the underground openings, the experimentation was designed to determine the effect of repeated vibrations on rock mass damage, on the basis of previous experience of the authors (Ramulu, 2009). The damage assessment was carried out at both the side walls of the tunnel as the joint orientation with respect to tunnel perimeter line was making acute angle at the Left-side and obtuse angle at the Right-side of the tunnel. The damage assessment was also carried out at Crown rock mass of the tunnel where the joint orientation with respect to tunnel perimeter line was making 180^0 angle. The objective of damage assessment at three different locations is to know the effect of joint orientation with respect to the perimeter line of the tunnel. The tunnel profile where the experimentation was carried out is shown in Figure 3.

Geological and Geotechnical Information of Experimental Site

The LNPHPP falls in the Uttrakhand Himalayas and is located on the River Bhagirathii upstream of Uttarkashi district. The Uttarkahsi region is made up of two main tectonic units namely the Himalayan Central Crystallines and the Lesser Himalayan Formation. The slab of the Central Crystallines has been thrusted southward along the Main Central Thrust (MCT) over the quartzite and volcanic of Berinag Formation of the Lesser Himalayas or rock belonging to Garhwal Group. The Garhwal Group towards north is followed by the Central Crystallines which have been divided into three zones i.e. Upper Crystallines, Middle Crystallines and Lower Crystallines. Garhwal Group towards north is followed by the Central Crystallines which have been divided into three zones i.e. Upper Crystallines, Middle Crystallines and Lower Crystallines. The main rock type of powerhouse complex is schistose gneiss and augen gneiss with abundance of mica and geotechnically the rock mass is negotiating in "Fair Category" and it's having three prominent joint sets. The main two joint sets intersecting at right angle which makes wedge continuously. Some weak zone/ clay filling, altered rock, sheared rock mass and excessive flow of water at places makes the rock poor. In maximum area it is found that the regional trend of foliation is perpendicular to the tunnel alignment, another joint which is intersecting the foliation at right angle and creates wedge on roof. The strike of the foliation is going through along the tunnel alignment which is geologically not favorable because of probabilities of plane failure and wedge failure in presence of heavy joint planes. The detailed geological information is given in the Table 1. Core samples were collected from both the monitoring locations by underground coring machine. Engineering properties like Rock Quality Designation (RQD) compressive strength, tensile strength, density and compressional wave velocity (Vp) were determined from the core samples. In-situ compressive strengths were also determined by using Schmidt hammer rebound testing. The average intact rock properties of the Schistose/Augan gneiss at two different experimental locations are given in Table 2.

DETAILS OF THE BLASTING AND INSTRUMENTATION

The tunnel was excavated with heading and benching simultaneously by using drilling and blasting method. Rocket boomer was used for drilling of blast holes of 45 mm diameter. Wedge cut blasting pattern was adopted with a maximum hole depth of 3.5m for the tunnels of $32m^2$ area of cross section. The blast pattern practiced for the full face tunnel blasting at powerhouse complex is shown in Figure 4. The specific charge was varied depending on the static and dynamic properties of rock. The explosive used was cartriged emulsion with 80% strength. The in-hole velocity of detonation (VOD) of the explosive was determined as 3000-3100 m/s by using VOD monitoring instrument called SuperTrap (M/s MREL Inc., Canada). Long delay detonators from 1 to 10 were used to fire various rows in different delays. The

Figure 1. Location of the LNPHPP site on Bhagirathi River, Uttaranchal, India

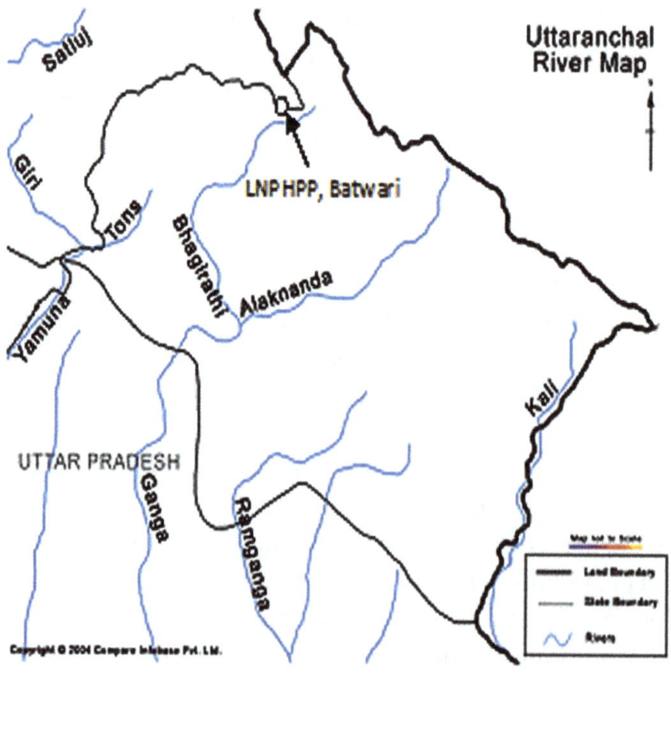

Figure 2. Plan view of various tunnels near the power house complex, LNPHPP

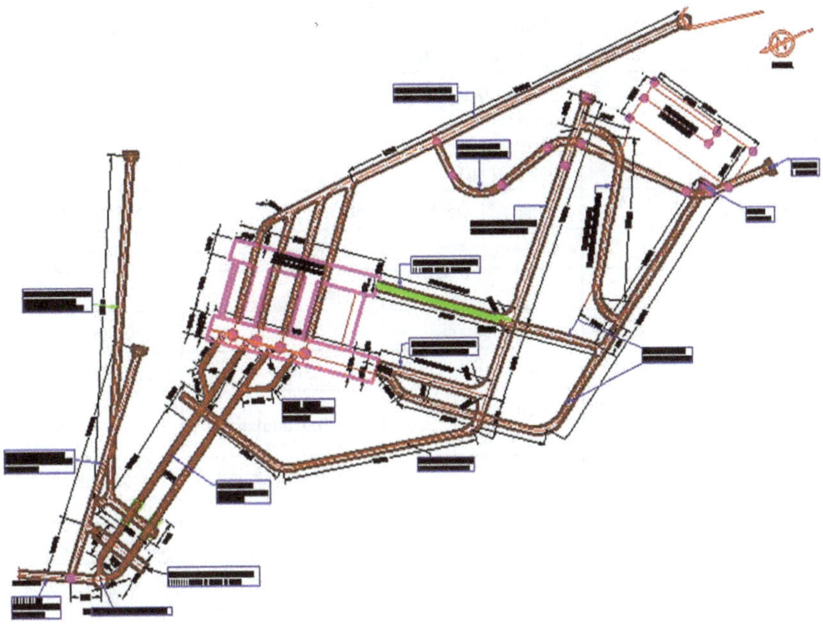

Figure 3. Experimentation tunnel with varying joint orientations at both left and right-sides

(a) Tunnel portal (b) Excavated tunnel profile

maximum charge per delay used in the blast round was 32 kg in the bottom holes and total charge per round was 169kg. Total number of holes used was 87 which include 13 dummy holes at the Crown periphery of the tunnel line. The specific charge used for Schistose/Augen gneiss was 1.75 kg/m^3. The progress per round with this blast pattern was observed as 2.5 to 2.75m out of 3m effective depth of blast round.

Table 1. General geological information

Location	Crown to Spring Level (Both Sides)	Below Spring Level (Left-Side)	Below Spring Level (Right-Side)
Rock type	Schistose/Augen gneiss with alternative bands of mica & quartz	Mica schist with alternative bands of quartz	Schistose/Augan gneiss with bands of mica and quartz
Critical joint angle	$50^0/52^0$, $210^0/50^0$	$095^0/55^0$, $175^0/45^0$	$095^0/55^0$, $175^0/45^0$
Seepage	Moderate	Continuous	Occasional
Wedge portion	Crown	None	Walls
Spacing	6-20, 20-60cm	6-20, 20-60cm	<6, 6-20cm
Opening	0.25-2.5mm	0.25-2.5mm	0.25-2.5mm
Joint Roughness	Rough, Planar& Undulating	Rough, Undulating	Smooth, Undulating
Joint Alteration	Altered joint, highly staining	Moderately altered walls	Moderately altered walls
Type of Filling	Mica, Quartz	Mica, Quartz	Quartz
Rock Strength	Weak-Medium strong (25-50MPa)	Weak-Medium strong (25-50MPa)	Weak (5-25 MPa)
Nos. of Joint Sets	Three joint sets + Random	Three joint sets + Random	Three joint sets + Random
Degree of Weathering	Slightly Weathered Rock	Moderately-Highly Weathered	Moderately Weathered
Water Inflow	Dripping (Low)	Damp	Seepage at places
Rock mass Quality (Q)	1.53-3.04 Poor rock	0.86-1.14, Poor rock	1.15-3.04, Poor rock
Rock mass Rating (RMR)	46-49, Class III, Fair	37-44, Class IV, Poor	52-57, Class III, Fair

Table 2. Average intact rock properties of the Schistose/Augan gneiss

Location	Rock Type	Mass Density, kg/m^3	Tensile Strength, MPa	Young's Modulus, GPa	P-Wave Velocity, m/s
Right-side wall	Schistose/Augan gneiss with acute joint orientation	2450	3.5	21.5	2958
Left-side wall	Schistose/Augan gneiss with obtuse joint orientation	2445	3.1	19.25	2610
Crown	Schistose/Augan gneiss with obtuse joint orientation	2420	3.3	20.18	2830

Instrumentation carried out in this study include, triaxial geophones for vibration monitoring, borehole camera survey for observing crack extension, loosening of joints and borehole extensometers for measurement of plastic deformations. It has become common practice, recently, to use peak particle velocity (V_{max}) as an indicator of the potential for rock mass damage, as the V_{max} is directly proportional to the dynamic strain (Jaeger & Cook, 1979). Numerous authors used V_{max} as criteria for blast damage in rock mass (Langefors and Kihlstrom, 1963; Kutter and Fairhurst, 1971; Holmberg and Persson, 1978; Holmberg and Persson, 1980; Oriard, 1982; Singh, 1993; Yu and Vongpaisal, 1996; Villaescusa et all, 2004). Application of borehole extensometers for blast damage inspections was reported by many authors globally (Niklasson, 1985; Stacey et al, 1990; Kim, et al, 1990; Villaescusa et al, 2004). Application of borehole camera for blast damage inspections was reported by many authors globally (Niklasson, 1985; Beyer and Jacobs, 1986; Stacey et al, 1990; Rocque et al, 1992; Singh, 1993; Andrieux et al, 1994; Doucet et al, 1996; Liu et al, 1998). A room of 0.5 m^3 inside the wall of MAT was made for installation of geophones to capture the blast vibrations from the surrounding excavations and for the installation of borehole extensometers. The damage monitoring stations were located at the Chainage of 45m inside the MAT, which was exposed to the vibrations from face blasting of MAT, CVT as well as TRT adit (Figure 6). A typical damage monitoring set-up at the tunnel wall is shown in Figure 5, where the geophones, borehole extensometers and borehole camera survey holes are shown. The location of borehole extensometers and geophones and blasting location is schematically shown in Figure 6. The details of installation of instrumentation are given in the following sections.

Installation of Geophones

It was required to measure the blast vibrations in the near-field as well as far-field zones with respect to blast site to assess the rock mass damage. Therefore a room of about 0.5 m^3 (1m x 1m x0.5m) was excavated inside the sidewall for installation of geophones at a height of 1m from the bottom. The geophone room excavation was carried out by controlled blasting by using mild explosive charges to avoid disturbance to surrounding rock mass. The geophone room was located at about 5m from the initial blasting face and the vibration monitoring was carried out continuously, while the blasting face receded away up to a distance of 45-50m. The geophone sensors of higher frequency and recording equipment with faster sampling rates were used for near-field monitoring and ordinary low frequency geophone sensors were used for far-field vibrations monitoring. Both the seismographs used for vibration monitoring are shown Figure 7.

Installation of Borehole Extensometers

Holes were drilled across the joint planes and foliations of the rock mass in the tunnel wall to install borehole extensometers. Conventional borehole extensometers available in the market were used for measurement of plastic deformations. Two borehole extensometers were installed at the Left-side and two more were installed at the Right-side of the tunnel wall, at a height of 1.5m from ground level. Each extensometer consists of two extension rods with varying lengths inserted into the borehole and installed as per the norms prescribed by ISRM (1992). Digital gauges were used to measure the positive or negative extension values of the extensometer at the free end in the mouth piece of extensometer. The lengths of borehole extension rods were 3 and 4m at the Left-side and 3.5 and 4.5m at the Right-side of the tunnel wall. The installation of borehole extensometer and geophones are shown in Figure 7.

Installation of Borehole Camera

Borehole camera inspection holes of 36mm diameter and 5m depth were drilled across the joint planes of the rock mass in the sidewall, approximately 0.5m away from the geophone location as well as extensometer (Figure 5). The location and direction of holes were made such that it can intersect maximum joint planes so that the joint displacements and possible new cracks can be detected by borehole cameras. The camera observation holes were made at the close proximity of geophone holes, so that the possible rock mass damage levels can be correlated with the measured peak particle velocity (V_{max}). The monitoring program consisted of surveying the inspection hole before and after each blasting event. The borehole camera used in this study was a robust unit with semi-rigid fiberglass signal cable. This camera contains a standard video output and can be connected with any TV or video-recording system with VCR input. Borehole surveys were made by using a front view lens attachments, which could capture images from all the sides of the hole. This gave a clearer picture of the borehole wall, before and after blasting. All surveys were recorded in a computer, attached to the camera and analysed on surface to determine the frequency of cracks and crack extensions before and after every blast.

Figure 4. Blast design at power house complex, LNPHPP

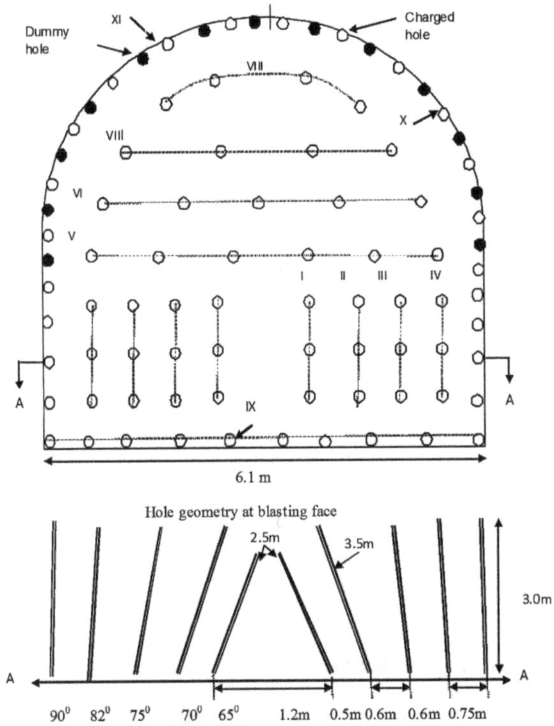

(The numbers in Roman letters denote delay numbers)

Figure 5. Schematic of installation locations of geophones, borehole extensometers and borehole camera inspection holes

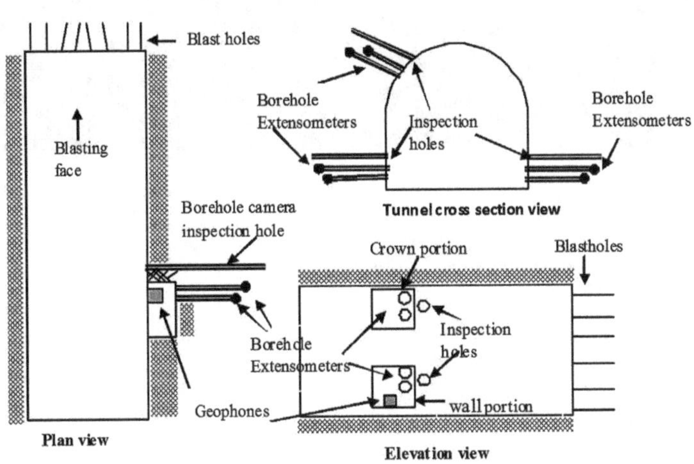

Figure 6. Schematic view of vibration and damage monitoring stations with respect to blasting location

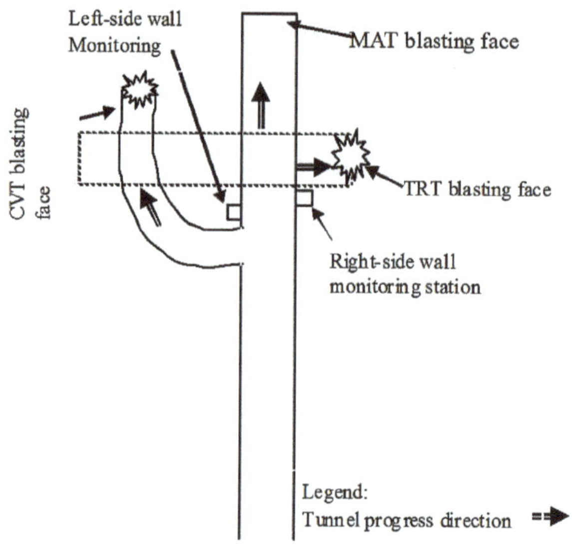

Figure 7. The installation of (i) borehole extensometer and (ii) geophones

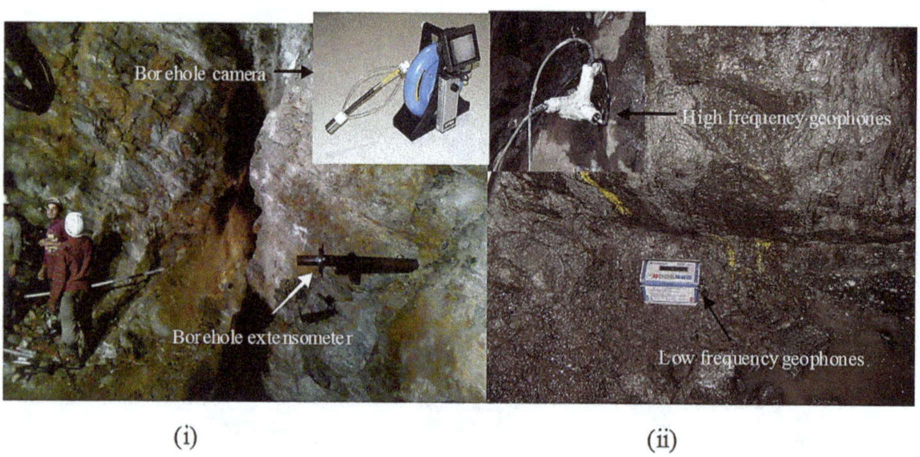

EXPERIMENTATION ON THE EFFECT OF REPEATED BLAST LOADING

Near-Field Blast Damage Assessment

The near-field damage to the rock mass at the experimental site occurred due to the production blast rounds conducted within the tunnel. The near-field damage was assessed by the Holmberg-Persson model (1978) as well as borehole camera survey. The principle of Holmberg-Persson Equation, is to add the contribution of every small portion of the explosives column along the full charge length to derive the peak particle velocity (V_{max}) at a fixed sensor location. More details about the Holmberg-Persson (H-P) model can be known from the sited references (Andrieux et al, 1994; LeBlanc, 1995; McKenzie et al, 1995; Meyer & Dunn, 1996; Liu & Proulx, 1996).

The Holmberg-Persson Equation can be simplified to:

$$V_{max} = K\left[a\right]^{\alpha} \tag{1}$$

where, *a* is here defined as the Holmberg–Persson term and K and α are the rock mass and explosive specific attenuation constants. K and α can be obtained by linear regression from experimental data vibration and distance. The mean values of K and α show the general trend of vibration attenuation in the rock mass. In this study, The Holmberg–Persson approach was applied to determine the site specific constants K and α to model peak particle velocity attenuation across rock mass. Same type of explosive and design parameters were used for all the experiments and analysis of the results are presented in the following paras.

Near-field peak particle velocity (V_{max}) measurements were grouped and analysed separately for two experimental sites. Typical log–log plot of the measured V_{max} values obtained in this study against the Holmberg–Persson term. The V_{max} amplitudes were being experienced at similar distances and for the same design parameters. This plot is used to determine the K and **α** constants by fitting the linear relationship of the form, $\text{Log}(V_{max}) = \boldsymbol{\alpha}\,\text{Log}(a) + \text{Log}(K)$. Results of the analysis are summarized in Table 3. The calculated K and α constants were slightly different for different test sites of the tunnel where the monitoring stations were set up. After determining the site specific attenuation constants, preliminary predictions of the extent of blast damage into wall-rock were made by applying the Holmberg–Persson model and by considering a site specific critical V_{max} or damage threshold (V_{cr})given by the following relationship (Persson et al,1994),

$$V_{cr} = \frac{\sigma_T V_p}{E} \qquad (2)$$

where,

V_{cr} = Critical peak particle velocity before tensile failure (mm/s);
σ_T = Uniaxial tensile strength of rock (Pa);
V_p = Compressional wave velocity in rock mass (mm/s);
E = Young's Modulus of rock (Pa).

From the properties described in Section 4 (Table 2) and by adopting the above relationship (Equation 2), the value of damage threshold peak particle velocities (V_{cr}) calculated were 353, 427 and 407mm/s for the Left-side, Right-side and Crown rock mass, respectively. These threshold values were used to compare the extent of damage caused by the near-field blast rounds, which obviously generated maximum peak particle velocity. The near-field damage analysis was carried out by using H-P model is shown in Figure 8. The above analysis indicates that on average the extent of blast induced damage was 1.40m, 1.30m and 1.35m at Left-side, Right-side and Crown rock mass, respectively. The damage levels assessed by H-P model were cross checked by borehole camera survey before and after each blast round. An initial pre-blasting survey was performed in each hole to take pictures of pre-existing structural features for comparison with post blast surveys. A total of three borehole camera surveys were taken, one each at Left-side, Right-side and Crown rock of tunnel. All the pictures were sorted out first by editing and matching the same images before and after each blast. Consequently, the pictures, which showed differences in fracture existence, were identified by their position coordinates. The images of inspection holes captured by borehole camera clearly indicated that the near-field damage due to production blasts extended up to 1.59m, 1.74m, and 1.53m at Left-side, Right-side and Crown rock mass, respectively. Near-field damage observations of both the methods are given in Table 3. The borehole camera inspection survey results shows contrasting results in comparison to the H-P model analysis. According to the tensile failure criteria the there should be lesser damage to the Right-side wall as it is relatively sound. But the actual damage survey indicates more damage to Right-side wall in comparison to Left-side wall. This might be due to the dominant role of joint orientation over the soundness of the rock. Investigations were further carried out to establish the role of joint orientation in the blast induced damage.

Far-Field Blast Damage Due to Repeated Vibrations

The multiple rounds of blasting activity were carried out at the power house complex as shown in Figure 6. In order to monitor the effect of repeated blasting due to multiple rounds the peak particle velocity of vibrations and plastic deformations were measured continuously by geophones and borehole extensometers respectively. Far-field rock mass damage observations were also carried out by using borehole camera by regular inspection of observation holes. In order to correlate the far field damage with the vibrations, the V_{max} levels were recorded for every blast round uptill the vibration intensity attenuated to about 45mm/s. All the damage observation systems were monitored for 60 rounds of blasts at all the three monitoring locations of MAT i.e. Right-side wall, Left-side wall and Crown rock. As the blast site is moving away from the monitoring point the vibration intensity, obviously, reduced gradually below critical peak particle velocity. The effect of these reduced vibration levels with the repeated number of exposures on the extent of further damage was studied thoroughly by all the three instruments. The blast damage assessment at Left-side, Right-side and Crown rocks of the MAT tunnel is discussed in the following sections separately.

Damage Assessment of Left-Side Rock Mass at MAT

Peak particle velocities versus no. of occurrences of dynamic loading at gneiss rock mass with acute angle joint orientation is shown in Figure 9. The vibration intensity recorded was ranging from 40 to 1753 mm/s at the monitoring location. A typical

Table 3. Extent of predicted rock mass damage into the tunnel wall

Location	Rock Type	K	α	V_{cr} From H-P Model	Damage From H-P Model	Damage From Borehole Camera Survey
Left-side wall	Schistose/Augan gneiss with obtuse joint orientation	671	0.75	353 mm/s	1.40 m	1.59 m
Right-side wall	Schistose/Augan gneiss with acute joint orientation	697	0.99	427 mm/s	1.30 m	1.74 m
Crown rock mass	Schistose/Augan gneiss with right angle joint orientation	687	0.90	407 mm/s	1.35 m	1.53 m

Figure 8. Results of the analysis by using H-P model

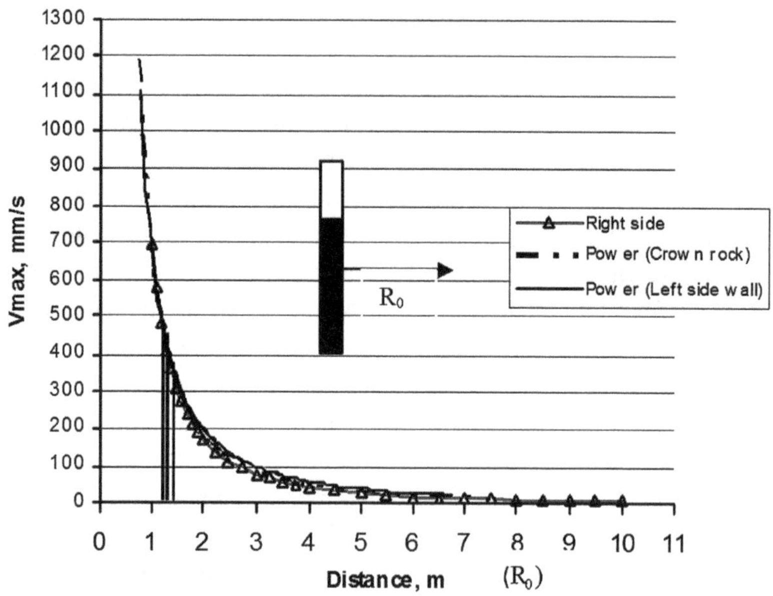

near-field vibration event history recorded during the study is shown in Figure 10. The displacements of extensometer rods of 3 and 4m depth at the Left-side tunnel wall were proportional to the V_{max} for few near-field blast rounds of MAT face blasts, where the V_{max} is above V_{cr} (Figure 11). There was no considerable change in the extensometer readings of both 3 and 4m rods for all the V_{max} levels below V_{cr} (353mm/s) for the MAT and CVT face blasts. After 40 blast rounds, the displacements were again observed for TRT face blasts, even at the vibration levels 233.6 mm/s, which is below the V_{cr}. There were no displacements observed below the V_{max} level of 116mm/s. This effect gives an inference that the damage zone was below 3m from the tunnel perimeter. The exact depth of damage zone inside the tunnel walls could not be found with extensometers. Therefore, borehole camera observations were used for determination of exact depth of damage as well as extent of crack network and fracture frequency. As the range of damage extension was already assessed by means of extensometer, the inspection of exact extension of damage by borehole camera had become much easier. Images were captured by an interval of 5cm, within the range of probable damage extension for precise inspection of rock mass damage. The images of borehole sections which contain the interface of intact and disturbed rock mass of tunnel wall are shown in Figure 12. The damage depth measured by the borehole camera inspection survey at the Left-side wall with acute joint orientation was 2.70m.

Damage Assessment of Right-Side Rock Mass at MAT

Peak particle velocities versus no. of occurrences of dynamic loading at Right-side of the tunnel is shown in Figure 13. The vibration intensity recorded at the monitoring location was ranging from 42 to 1644 mm/s. The displacements indicated by the extensometer rod of 3m depth at the Right-side wall were proportional to the V_{max} for few blast rounds at the MAT whose V_{max} were above V_{cr} (427mm/s), as shown in Figure 14. After 16 blast rounds it was observed that the displacements suddenly dropped to negligible levels even though the V_{max} recorded was above V_{cr}. This effect gives an inference that the damage zone reached the anchor point of the 3.0m extensometer rod after few close field blast loadings. This might be the reason why the 3.0m extensometer rod did not respond to even the vibrations levels which were greater than the critical vibration levels. There were substantial displacements observed to the 4.0m extensometer rod (Figure 14) for the V_{max} levels above V_{cr} for the MAT and CVT blast rounds and no displacements (>1mm) was observed for the V_{max} levels below V_{cr} for the blast rounds of same locations. After 40 rounds of blast occurrences, the displacements were again observed at the vibration level, 254 mm/s, which is much below the V_{cr}. There were no displacements observed below the V_{max} level of 98 mm/s. This result gives an inference that the anchor point of 4.0m rod was in elastic zone and the damage zone was within 4m from the tunnel perimeter. The borehole inspection survey was conducted to cross check the damage assessment by extensometers and for exact damage measurement. The images of borehole sections which contain the interface of intact and disturbed rock mass are shown in Figure 15. The maximum damage depth measured by the borehole camera survey with obtuse joint angle orientation was 3.6m. This was the reason for the roof failure occurred near the test site. The roof failure due to extension of fractures beyond 3m due to repeated blast induced damage is shown in Figure 16. The maximum length of roof bolts designed for that rock mass was also 3m.

Damage Assessment of Crown Rock Mass at MAT

Peak particle velocities versus no. of occurrences of dynamic loading at Right-side of the tunnel is shown in Figure 17. The vibration intensity recorded at the monitoring location was ranging from 43.44 to 1602 mm/s. The displacements indicated by the extensometer rod of 3m depth at the Right-side wall were proportional to the V_{max} for few blast rounds at the MAT whose V_{max} were above V_{cr} (407mm/s), as shown in Figure 18. After 42 blast rounds it was observed that there were no displacements observed even though the V_{max} recorded was above V_{cr}. This effect gives an inference that the damage zone extended beyond the anchor point of the 3.0m extensometer rod after the exposure of 42 close field and far-field blast loadings. This might be

Figure 9. Peak particle velocities vs. no. of dynamic loading cycles at gneiss rock mass with acute joint orientation

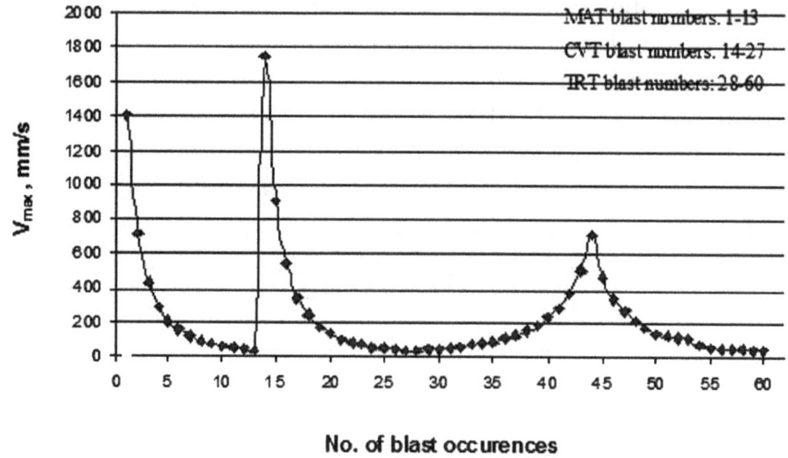

the reason why the 3.0m extensometer rod did not respond to higher vibrations levels than the V_{cr}. There were substantial displacements observed to the 4.0m extensometer rod (Figure 18) for the V_{max} levels above V_{cr} for the MAT and CVT blast rounds and no displacements (>1mm) was observed for the V_{max} levels below V_{cr} for the blast rounds of same locations. After 39 rounds of blast occurrences, the displacements were again observed at the vibration level, 268 mm/s, which is much below the V_{cr}. There were no displacements observed below the V_{max} level of 112 mm/s. This phenomena gives an inference that the anchor point of 4.0m rod of the extensometers was in elastic zone and the damage zone was within 4m from the tunnel perimeter.

The borehole inspection survey was conducted for capturing the damaged rock strata due to repeated blasting at Crown of tunnel. The images of borehole sections which contain the interface of intact and disturbed rock mass are shown in Figure 19. The maximum damage depth measured by the borehole camera survey with joint angle parallel to the tunnel profile was 3.21m, but the maximum length of roof bolts designed for that rock mass was only 3m. This inadequate length of roof bolts led to roof failure due to extension of fractures beyond 3m due to repeated blast induced damage is shown in Figure 20. The far-field damage due to repeated loading at the three monitoring locations is given in Table 4.

Figure 10. Near-field vibration event history recorded during the experimentation

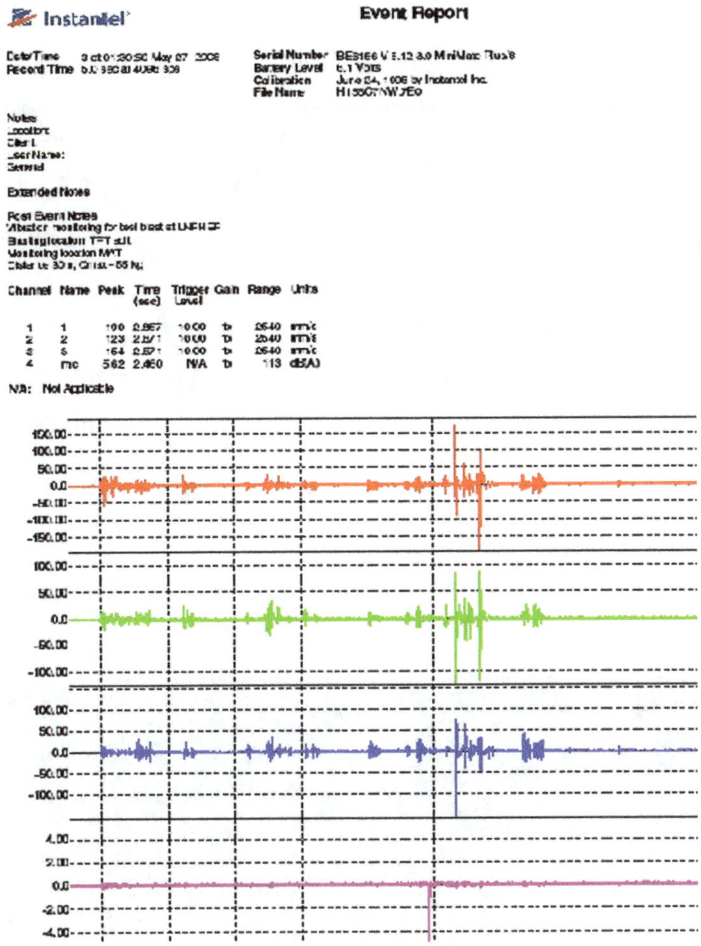

RESULTS AND DISCUSSIONS

The near field blast loading due to main tunnel excavations, could generate displacements in the rock mass, only when the peak particle velocity exceeded the critical vibration levels (V_{cr}). After repeated exposures of vibrations due to blast rounds at MAT, CVT and TRT, plastic displacements observed even at lower levels of vibrations than the V_{cr}. The extra damage due to repeated blast loading was 1.11m, 1.86m and 1.68m at the Left-side, Right-side and Crown portions of the tunnel respectively.

After the occurrences of 53 numbers of blast rounds with the V_{max} levels ranging from 40-1753 mm/s, considerable displacements observed in the extensometer even

Figure 11. Plastic deformations of extensometer rods at Gneiss rock mass with acute joint orientation

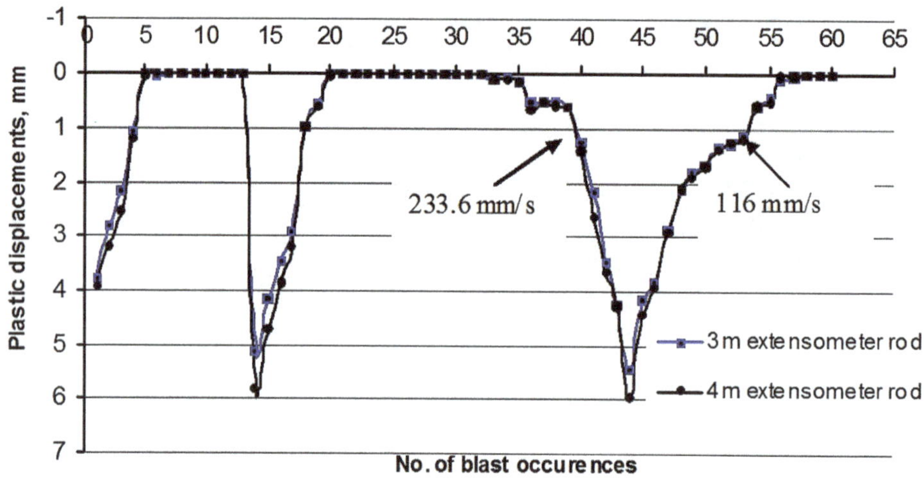

Figure 12. Images of pre-blast and post-blast inspection surveys of rock mass with acute joint orientations captured by borehole camera

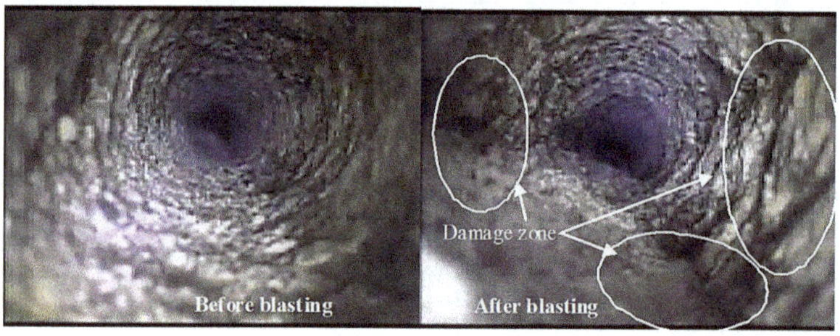

at the V_{max} level of 116 mm/s, which is approximately 33% of V_{cr} in the Gneiss rock mass with acute joint orientation. The displacements observed at the V_{max} level of 98 mm/s i.e. at approximately 23% of V_{cr} in the Gneiss rock mass with obtuse joint orientation, after 54 numbers of occurrences of blast loading. Similarly, displacements observed at the V_{max} level of 112 mm/s i.e. at approximately 27.5% of V_{cr} in the Crown rock mass with parallel joint orientation, after 52 numbers of blast occurrences. These results are well in line with the observations of Dowding & Rosen (1978). The threshold vibration limits with number of cycles of repeated loading for two different rock mass are given in Table 4. The study also reveals that the overall damage was about 33% and 19% more at Right-side and Crown rock mass respectively, in comparison to the Left-side rock mass. The observations also

Figure 13. Peak particle velocities versus no. of occurrences of dynamic loading at Gneiss rock mass with obtuse joint orientation

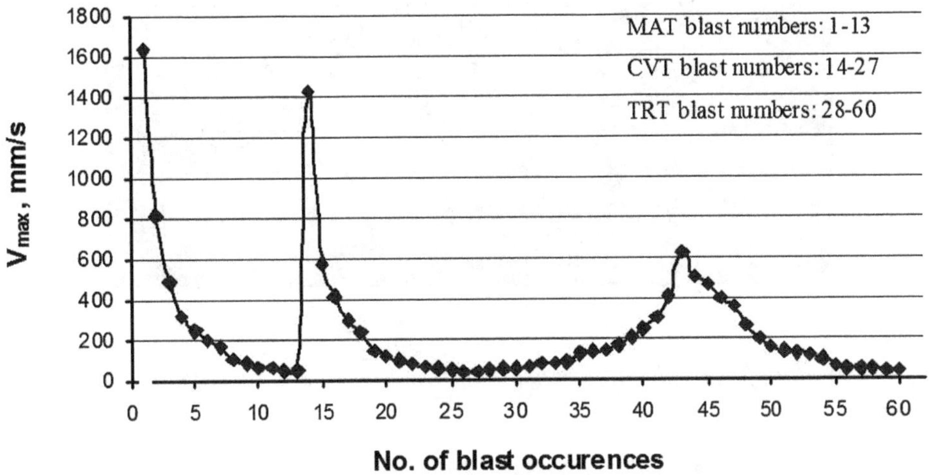

Figure 14. Plastic deformations of extensometer rods at Gneiss rock mass with obtuse joint orientation

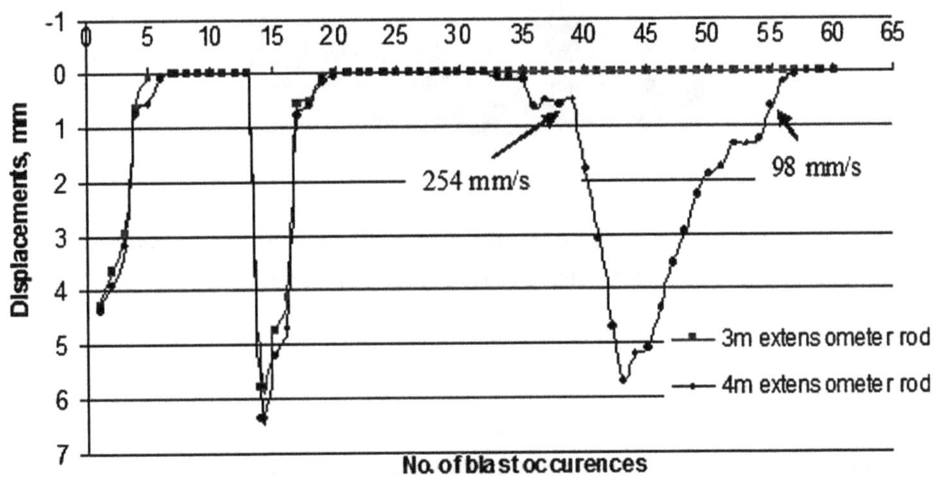

indicate that the repeated dynamic loading resulted in the damage at the vibration levels even at 23% of V_{cr} when the unfavorable joints exist. These observations were almost similar to the findings of Adamson and Scherpenisse (1998), which says that of threshold vibration level falls down to 25% of V_{cr} in repeated loading conditions. The findings of the study clearly indicate that the phenomena of repeated blasting

Figure 15. Images of intact and disturbed rock mass with obtuse joint orientations captured by borehole camera

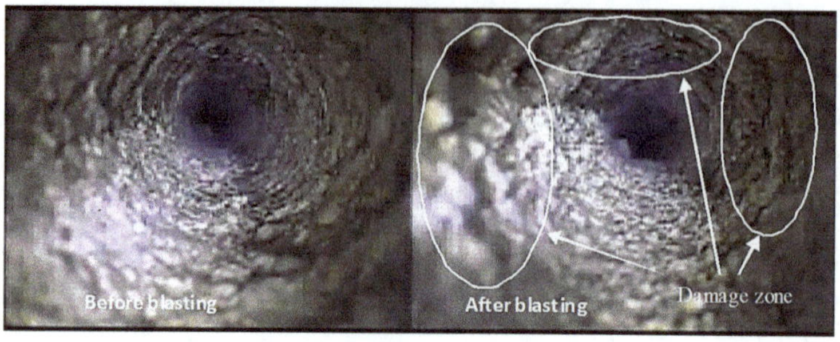

Figure 16. Roof failure due to extension of fractures due to repeated blast induced damage

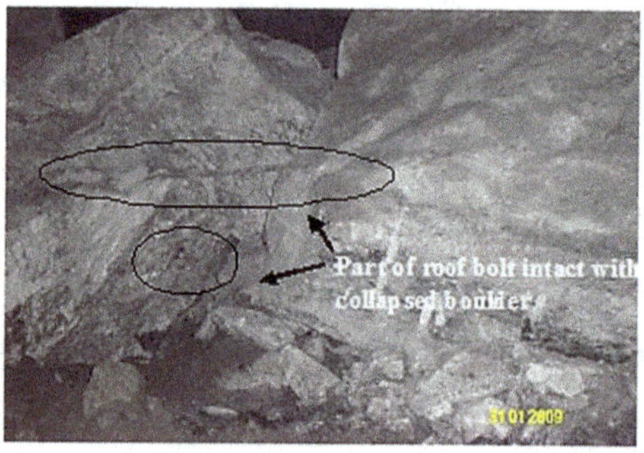

with respect to number of cycles of loading should be taken in to consideration for proper assessment of comprehensive blast induced damage.

CONCLUSION

A comprehensive blast monitoring program aimed at the investigation on the effect of repeated blast induced damage due to near-field and far-field vibrations was conducted in a Gneiss rock mass. The damage levels predicted for near-field blast loading by Holmberg-Persson model were 1.40m, 1.30m and 1.35m for the Left-side, Right-side and Crown rock mass, respectively. The damage levels measured for the same rock mass by borehole camera was 1.59m, 1.74m and 1.53m for the Left-side, Right-side and Crown rock mass, respectively. Estimates of the maximum extent of rock mass damage made through the application of the Holmberg-Persson

Figure 17. Peak particle velocities versus no. of occurrences of dynamic loading at Gneiss rock mass with obtuse joint orientation

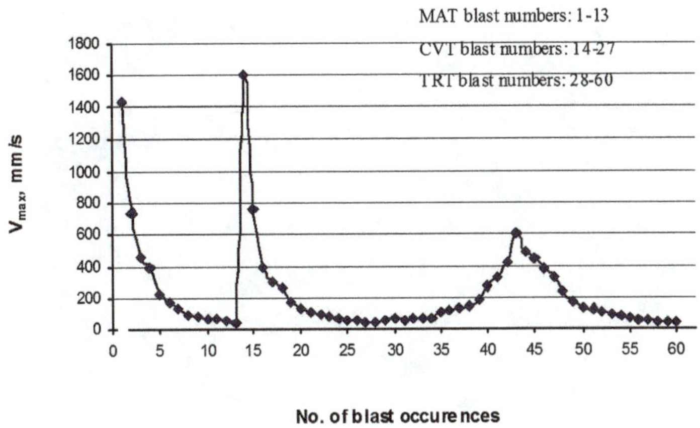

Figure 18. Plastic deformations of extensometer rods at Gneiss rock mass with obtuse angle joint orientation

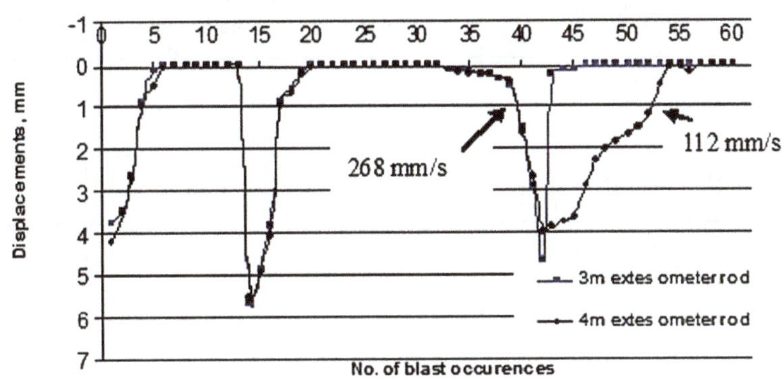

Figure 19. Borehole camera images of intact and disturbed rock mass with obtuse angle joint orientations

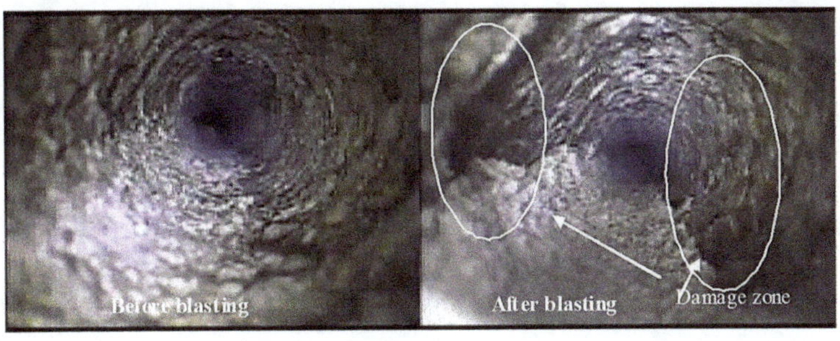

Figure 20. Roof failure due to extension of fractures beyond 3m due to repeated blast induced damage

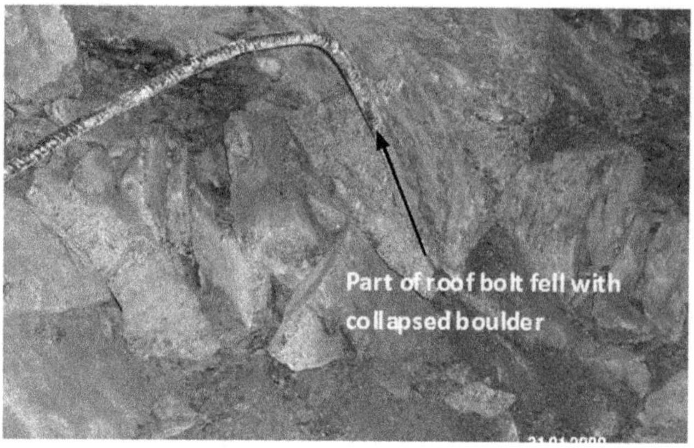

Table 4. Far-field damage due to repeated loading at both the sides of tunnel wall

Location	Rock Type	No of Cycles of Blast Loading	Threshold Vibration Limits, mm/s	Maximum Extent of Damage, m
Left-side wall	Schistose/Augan gneiss with obtuse joint orientation	53	116	2.70
Right-side wall	Schistose/Augan gneiss with acute joint orientation	52	98	3.60
Crown rock	Schistose/Augan gneiss with obtuse joint orientation	54	112	3.21

model compared well with measured results for Left-side wall and Crown rock mass, although the former one was at lower side of damage. The damage results of Holmberg-Persson model were deviating with the measured results for Right-side wall. This might be because of the additional effect of joint orientations, playing a dominant role at Right-side wall rock mass. Although there are deviations, the modelling approaches like Holmberg-Persson model are very useful for the project engineers for preliminary assessment and estimation of the damage and for practical methods to model peak particle velocity attenuation.

The study also found that repeated dynamic loading imparted on the exposed tunnel from subsequent blasts in the vicinity is going to contribute to rock mass weakening and preconditioning. After 53 repeated blast rounds, the threshold vibration level for the Gneiss at Left-side wall was found to be 116 mm/s. The dynamic loading due to repeated blasts resulted in 70%, 107% and 110%of extra damage, in addition to the near-field damage, at Left-side, Right-side and Crown rock mass, respectively. The repeated dynamic loading also resulted in reduced threshold peak particle velocity to 33%, 23% and 28% of critical peak particle velocity at Left-side, Right-side and Crown rock mass, respectively. The study also revealed that the overall damage was about 75% more at tunnel wall with obtuse angle joint orientation in comparison to the acute angle joint orientation. The findings of the study clearly indicate that the repeated blast loading extends damage to a substantially dangerous levels and the effect is severe in case of unfavorable joint orientations. Therefore the phenomena of repeated blasting with respect to number of cycles of loading is an important aspect to be considered for proper assessment of blast induced damage and for planning of safety measures of underground openings.

ACKNOWLEDGMENT

The research work presented in this paper is a part of PhD degree of the first author from Indian Institute of Science and it is also part of CSIR sponsored project No. 70 (0058)/06/ EMR-II-CSIR for which the authors are grateful to CSIR. The authors express their sincere thanks to the Director, CMRI for permitting to publish this paper. The authors express their sense of gratitude to Mr. P.B. Choudhury, Dr. John Loui P., Dr. A.K. Raina, Dr. M.R.Saharan and Mr.B.K.Jha for their cordial help during the field studies. The authors express their sincere thanks to the Director, CIMFR (CMRI) for permitting to publish this paper. The authors also express their sincere thanks to all the scientists of CIMFR regional centre, Nagpur for their valuable assistance during the field work. The authors are also thankful to the executives of M/s Patel Engineering Ltd. and NTPC for their cooperation during the field studies.

REFERENCES

Adamson, W. R., & Scherpenisse, C. R. (1998). The measurement and control of blast induced damage of final pit walls in open pit mining. *Proc. of 24th Annual Conf. on Explosives and Blasting Research*, 539.

Andrieux, P., McKenzie, C., Heilig, J., & Drolet, A. (1994). The Impact of Blasting on Excavation Design – A Geomechanics Approach. *Proceedings of the 10th Symposium on Explosives and Blasting Research,* 107–119.

Anon. (1997). Criteria for safe vibration limits on structures. DGMS.

Atchison, T. C., & Pugliese, J. M. (1964). Comparative Studies of Explosives in Limestone. BuMines Rept. of Inv., 6395.

Barton, N., & Hansteen, H. (1979). Very large span openings at shallow depth: deformation magnitudes from jointed models and finite element analysis. *Proc. 4th Rapid Excavation & Tunneling Conf.*, *2*, 1331-1353.

Beyer, R.R., & Jacobs, A.M. (1986). Borehole television for geotechnical investigations. *Water Power and Dam Construction, 38*(9), 16-28.

Brady, B. H. (1990). Dynamic performance and design of underground excavations in jointed rock. Static and dynamic considerations of Rock Engineering, 1-10.

Brown, E. T., & Hudson, J. A. (1974). Fatigue failure characteristics of some models of jointed rock. *Earthquake Eng. and struct. Dyn.*, *2*, 379–386.

Cameron-Clarke & Mival. (1990). Stabilisation of old workings by blasting: Case study of a failed experiment. In Static and dynamic considerations of Rock Engineering. Academic Press.

Chakraborty, A.K., Raina, A.K., Ramulu, M. Jethwa, J.L., & Gupta, R.N. (1998, November). Lake Tap at Koyna. *World Tunnel. Subsurface Exc.*, 456-460.

Charlie, W. A., Veyera, G. E., Doehring, D. O., & Abt, S. R. (1985). *Blast Induced Liquefaction Potential and Transient Porewater Pressure Response of Saturated Sands.* Final Report to Air Force Office of Scientific Research, Grant No. AFOSR-80-0260.

Connigham, C. V. B., & Goetzsche, A. F. (1996). The specification of blast damage limitations in tunneling contracts. *Tunnelling and Underground Space Technology*, *5*(3), 23–27.

Crandell, F. J. (1949). Ground Vibrations Due to Blasting and Its Effect Upon Structures. *Journal of the Boston Society of Civil Engineers*, 222-245.

Doucet, C., Cameron, A., & Lizotte, Y. (1996). The effects of rock mass characteristics on fragmentation in controlled blasting experiments in small development headings. *ISEE's 22nd Annual Conference*, 1-11.

Dowding, C. H. (1985). *Blast Vibration monitoring*. Prentice Hall.

Dowding, C. H., & Rozen, A. (1978). Damage to rock tunnels from earthquake loading. *J. Geotech.Eng. Div. ASCE, 104*(GT2), 175–191.

Hendron, A. J. (1977). Engineering of Rock Blasting on Civil Projects. In *Structural and Geotechnical Mechanics*. Prentice-Hall, Inc.

Holmberg, R. (1993). Recent developments in control rock damage. In *Rock Fragmentation by Blasting*. Academic Press.

Holmberg, R., & Persson, P. A. (1978). The Swedish Approach to Contour Blasting. In *Conference on Explosives and Blasting Technique*. Society of Explosives Engineers.

Holmberg, R., & Persson, P. A. (1980). Design of Tunnel Perimeter Blast Hole Patterns to Prevent Rock Damage. *Transactions of the Institution of Mining and Metallurgy*, *89*, A37–A40.

ISRM. (1992). *Commission on Testing methods, Suggested methods for installation of borehole extensometer*. International Society for Rock Mechanics.

Jaeger, J. C., & Cook, N. G. W. (1979). *Fundementals of rock mechanics* (3rd ed.). London: Chapman & Hall.

Kim, H.-Y., Lee, K.-W., Mizuta, Y., & Lee, H.-K. (1990). A study on determination of the relaxed zone around an excavated tunnel. Static and Dynamic Considerations of Rock Engineering, 177-182.

Kutter, H. K., & Fairhurst, C. (1971). On the Fracture Process in Blasting. *International Journal of Rock Mechanics and Mining Sciences*, *8*(3), 181–202. doi:10.1016/0148-9062(71)90018-0

Langefors, U., & Kihlstrom, B. (1963). *The modern techniques of rock blasting*. New York: J. Wiley and Sons, Inc.

Law, T. M., May, J., Spathis, A. T., Du Plessis, A. T., & Palmer, A. M. (2001). Blast damage and blast dilution control: The applicaton of bulk emulsion systems at the WMC St Ives junction mine, J. *Rock Fragmentation by Blasting*. *Fragblast*, *5*(1-2), 1–20. doi:10.1076/frag.5.1.1.3317

LeBlanc, T., Heilig, J., & Ryan, J. (1995). Predicting the Envelope of Damage from the Detonation of a Confined Charge. *Proceedings of the Sixth High-Tech Seminar on the State of the Art in Blasting Technology Instrumentation and Explosives Applications*, 225–291.

Leet, L. D. (1946). *Vibrations from Blasting*. Wilmington, DE: Hercules Powder Co.

Lewandowski, T., Mai, L. V. K., & Danell, R. (1996). Influence of discontinuities on presplitting effectiveness. *Proceedings of 5th Int. Symposium on Rock Fragmentation by Blasting*, 217-225.

Liu, Tran, Counter, & Andrieux. (1998). A case study of blast damage evaluation in open stope mining at Kidd Creek mines. *24th Annual Conference on Explosives and Blasting Research*, 323-336.

McGarr, A. (1983). Estimating ground motion for small nearby earthquakes. In Seismic design of embankments and caverns (pp. 113-127). New York: ASCE.

McKenzie, C., Scherpenisse, C., Arriagada, J., & Jones, J. (1995). Application of Computer Assisted Modelling to Final Wall Blast Design. *Proceedings of the EXPLO'95 – A Conference Exploring the Role of Rock Breakage in Mining and Quarrying*, 285–292.

Meyer, T., & Dunn, P. G. (1996). Fragmentation and Rock mass Damage Assessment – Sunburst Excavator and Drill and Blast. *Proceedings of the NARMS 1996 – Rock Mechanics Tools and Techniques*, 609–617.

Niklasson, B. (1985). Vertical crater retreat mining at the Luossavaara research mine, Swedish Detonic Research Foundation. *11th Annual Conference on Explosives and Blasting Research*, 46-61.

Olson, M., & Bergqvist, I. (1996). Crack lengths from explosives in small diameter boreholes. In Rock Fragmentation by Blasting. Academic Press.

Oriard, L. L. (1989). The Scale of Effects in Evaluating Vibration Damage Potential. *Proc. 15th Annual Conference on Explosives and Blasting Research*, 161-176.

Oriord, L. L. (1982). Blasting effects and their control. In Underground Mining Methods Handbook. SME of AIME.

Otuonye, F. O. (1997). Dynamic response of a fully grouted resin roof bolt to blast loading. *Proc. of Annual Conf. on Explosives and Blasting Research.*

Ouchterlony, F. (1997). Prediction of crack lengths in rock after cautious blasting with zero interholedelay,Int. J. *Rock Fragmentation by Blasting, 1*, 417–444.

Ouchterlony, F., Sjoberg, C., & Jonsson, B. A. (1993). *Blast damage prediction from vibration measurements at the SKB underground laboratories at ÄSPÖ in Sweden, Intl.* San Diego, CA: Soc. Explosive Engg.

Paventi, M., Lizotte, Y., Scoble, M., & Mohanty, B. (1996). Measuring rock mass damage in drifting. Rock Fragmentation by Blasting, 131-138.

Persson, P. A., Holmberg, R., & Lee, J. (1994). *Rock Blasting and Explosives Engineering*. Boca Raton, FL: CRC press Inc.

Ramulu, M., Chakraborty, A. K., & Sitharam, T. G. (2009). Damage assessment of basaltic rock mass due to repeated blasting in a railway tunnelling project – a case study. *Tunnelling and Underground Space Technology*, *24*(2), 208–221. doi:10.1016/j.tust.2008.08.002

Rocque, P. (1992). *Techniques utilized for the detection and characterization of blast induced damage* (Unpublished M. Sc. Eng. Thesis). Queen's University at Kingston, Ontario, Canada.

Scott, A., Cocker, A. N., Djordjevic, N., Higgins, M., La Rosa, D., Sarma, K. S., & Wedmaier, R. (1996). *Open Pit Blast Design, Analysis and optimization*. Julius Kruttschnitt Mineral Research Centre, The University of Queensland.

Singh, S. P. (1993). Damage causing potential of different explosives. *Proc, 9th annual symposium on Explosives and Blasting Research*, 325-337.

Singh, S. P., & Xavier, P. (2005). Causes, impact and control of overbreak in underground excavations. *Tunnelling and Underground Space Technology*, *20*(1), 63–71. doi:10.1016/j.tust.2004.05.004

St.John, C. M., & Zahrah, T. F. (1987). A seismic design of underground structures. Tunneling and underground space. *Technology*, *2*(2), 165–197.

Tart, R. J., Oriard, L. L., & Plump, J. H. (1980). Blast Damage Criteria for a Massive Concrete Structure. In Minimizing Detrimental Construction Vibrations. ASCE.

Thoenen, J. R., & Windes, S. L. (1942). Seismic Effects of Quarry Blasting. Bureau of Mines Bull., 442.

Villaescusa, E., Onederra, I., & Scott, C. (2004). Blast Induced Damage and Dynamic Behaviour of Hangingwalls in Bench Stoping. *J. Rock Fragmentation by Blasting. Fragblast*, *8*(1), 23–40. doi:10.1080/13855140512331389614

Wagner, H. (1984). Support requirements for rockburst conditions. In *Rockbursts and seismicity in mines* (pp. 209–218). Academic Press.

Yu & Vongpaisal. (1996). New blast damage criteria for underground blasting. *CIM Bulletin*, *89*(998), 139–145.

Zhang, J. C., & Chang, C. (1999). On damage mechanism of micro crack zone in Rock blasting and its measurements. In *Pro. 6th Int. Symp. of Rock fragmentation by blasting*. The South African Ist. of Mining and Metal.

Compilation of References

Abhishek, S. V., Rajyalakshmi, K., & Madhav, M. R. (2016). Engineering of ground with granular piles: A critical review. *International Journal of Geotechnical Engineering*, *10*(4), 337–357. doi:10.1080/19386362.2016.1145942

Adalier, K., & Elgamal, A. (2004). Mitigation of liquefaction and associated ground deformations by stone columns. *Engineering Geology*, *72*(3-4), 275–291. doi:10.1016/j.enggeo.2003.11.001

Adamson, W. R., & Scherpenisse, C. R. (1998). The measurement and control of blast induced damage of final pit walls in open pit mining. *Proc. of 24th Annual Conf. on Explosives and Blasting Research*, 539.

Ahmed, A. M. L. S., & Wavelet, A. S. (2007). Based Automatic Phase Picking Algorithm for 3-Component Broadband Seismological Data. *Journal of Seismology and Earthquake Engineering*, *9*(1), 15–24.

Ahmed, S., & Al Hussaini, T. M. (1991). Simplified design for vibration screening by open and in-filled trenches. *Journal of Geotechnical Engineering*, *117*(1), 67–88. doi:10.1061/(ASCE)0733-9410(1991)117:1(67)

Alberts, D., Garstka, J., & Stein, F. (2002). *Network Centric Warfare, Developing, and Leveraging Information Superiority*. Washington, DC: Department of Defence Command and Control Research Program.

Al-Homoud, A. S., & Degen, W. S. (2006). Marine Stone Columns to Prevent Earthquake Induced Soil Liquefaction. *Geotechnical and Geological Engineering*, *24*(3), 775–790. doi:10.100710706-005-2783-5

Allen, J. B. (1977). Short-Time Spectral Analysis, Synthesis and Modifications by Discrete Fourier Transform. *IEEE Transactions on Acoustics, Speech, and Signal Processing*, *25*(3), 235–238. doi:10.1109/TASSP.1977.1162950

American Petroleum Institute. (1991). *RP2A: Recommended practice for planning, designing and constructing fixed offshore platforms* (19th ed.). Washington, DC: American Petroleum Institute.

Anderson, D. G., & Stokoe, K. H. (1978). Shear Modulus: A Time-Dependent Soil Property. *ASTM Special Technical Publication*, 66–90.

Andrieux, P., McKenzie, C., Heilig, J., & Drolet, A. (1994). The Impact of Blasting on Excavation Design – A Geomechanics Approach. *Proceedings of the 10th Symposium on Explosives and Blasting Research,* 107–119.

Andrus, R., & Chung, R. (1995). *Ground improvement techniques for liquefaction remediation near existing lifelines. Report, Building and Fire Research Laboratory.* National Institute of Standards and Technology, U.S. Department of Commerce. doi:10.6028/NIST.IR.5714

Anon. (1997). Criteria for safe vibration limits on structures. DGMS.

Asimaki, D., & Jeong, S. (2015). Two-Dimensional Site Effects for Dry Granular Soils. *International Conference on Earthquake Geotechnical Engineering.*

ASTM Standard D 4428-M. (2007). *Standard test methods for seismic cross hole testing*. ASTM International. DOI: 10.1520/D4428_D4428M-07

Atchison, T. C., & Pugliese, J. M. (1964). Comparative Studies of Explosives in Limestone. BuMines Rept. of Inv., 6395.

Augello, A. J., Matasovic, N., Bray, J. D., Kavazanjian, E., Jr., & Seed, R. B. (1995). Geotechnical Special Publication: Vol. 54. Evaluation of solid waste landfill performance during the Northridge earthquake. Academic Press.

Augello, A. J., Bray, J. D., Abrahamson, N. A., & Seed, R. B. (1998). Dynamic properties of solid waste based on back-analysis of OII landfill. *Journal of Geotechnical and Geoenvironmental Engineering*, *124*(3), 211–222. doi:10.1061/(ASCE)1090-0241(1998)124:3(211)

Auger, F., Flandrin, P., Goncalves, P., & Lemoine, O. (1996). *Time-Frequency Toolbox*. Rice University, CNRS France.

Australian Emergency Management. (2002). *Natural disasters in Australia – Reforming relief, recovery and mitigation arrangements, A report to the Council of Australian Governments by a high level officials group, 2002*. Retrieved from http://www.em.gov.au/Documents/Natural%20Disasters%20in%20Australia%20-%20Review.pdf

Baez, J. I. (1995). *A design model for the reduction of soil liquefaction by vibro-stone columns* (PhD thesis). University of Southern California, Los Angeles, CA.

Baez, J. I., & Martin, G. R. (1992). Quantitative evaluation of stone column technique for earthquake liquefaction mitigation. *Proc. 10th World Conference on Earthquake Engineering*, 1477-1483.

Bagherzadeh-Khalkhali, A., & Mirghasemi, A. A. (2009). Numerical and experimental direct shear tests for coarse-grained soils. *Particuology*, *7*(1), 83–91. doi:10.1016/j.partic.2008.11.006

Balaam, N. P., & Booker, J. R. (1981). Analysis of rigid raft supported by granular piles. *International Journal for Numerical and Analytical Methods in Geomechanics*, *5*(4), 379–403. doi:10.1002/nag.1610050405

Barksdale, R. D., & Bachus, R. C. (1983). *Design and Construction of Stone Columns*. Report No. FHWA/RD-83/026, U. S. Department of Transportation, Federal Highway Administration, Washington, DC.

Barton, N., & Hansteen, H. (1979). Very large span openings at shallow depth: deformation magnitudes from jointed models and finite element analysis. *Proc. 4th Rapid Excavation & Tunneling Conf.*, *2*, 1331-1353.

Beskos, D. E., Dasgupta, B., & Vardoulakis, I. G. (1986). Vibration isolation using open or filled trenches Part 1: 2-D homogeneous soil. *Computational Mechanics*, *1*(1), 43–63. doi:10.1007/BF00298637

Beyer, R.R., & Jacobs, A.M. (1986). Borehole television for geotechnical investigations. *Water Power and Dam Construction, 38*(9), 16-28.

Bhattacharya, S. (2003). *Pile instability during earthquake* (Unpublished Ph.D. thesis). University of Cambridge, UK.

Bhattacharya, S., & Madabhushi, S. P. G. (2008). A critical review of methods for pile design in seismically liquefiable soils. *Bulletin of Earthquake Engineering*, *6*(3), 407–446. doi:10.100710518-008-9068-3

Biswas, S. K. (1987). Regional tectonic framework, structure and evolution of western marginal basins of India. *Tectonophysics*, *135*(4), 305–327. doi:10.1016/0040-1951(87)90115-6

Boashash, B. (1988). Note on the Use of the Wigner Distribution for Time Frequency Signal Analysis. *IEEE Transactions on Acoustics, Speech, and Signal Processing*, *36*(9), 1518–1521. doi:10.1109/29.90380

Boominathan, A. (2004). Seismic site characterization for nuclear structures and power plants. *Current Science*, *87*(10), 1388–1397.

Boominathan, A., Dodagoudar, G. R., Suganthi, A., & Uma Maheshwari, R. (2008). Seismic hazard assessment of Chennai city considering local site effects. *Journal of Earth System Science*, *117*(S2), 853–863. doi:10.100712040-008-0072-4

Boore, D. M. (1983). Stochastic simulation of high-frequency ground motions based on seismological models of the radiated spectra. *Bulletin of the Seismological Society of America*, *237*(6), 68–78.

Borcherdt, R. D. (1994). Estimates of site-dependent response spectra for design (methodology and justification). *Earthquake Spectra*, *10*(4), 617–653. doi:10.1193/1.1585791

Bosco, M., Ghersi, A., & Leanza, S. (2008). Force-displacement relationships for R/C members in seismic design. *Proceedings of the 14th World Conference on Earthquake Engineering*.

Bouckovalas, G. D., Papadimitriou, A. G., & Niarchos, D. (2009). Gravel drains for the remediation of liquefiable sites: The Seed & Booker (1977) approach revisited. *Proceedings of the keynote lecture*.

Bouckovalas, G. D., Papadimitriou, A. G., Niarchos, D., & Tsiapas, Y. Z. (2011). Sand fabric evolution effects on drain design for liquefaction mitigation. *Soil Dynamics and Earthquake Engineering*, *31*(10), 1426–1439. doi:10.1016/j.soildyn.2011.05.019

Boulanger, R. W., Curras, C. J., Kutter, B. L., Wilson, D. W., & Abghari, A. (1999). Seismic soil-pile-structure interaction experiments and analyses. *Journal of Geotechnical and Geoenvironmental Engineering*, *125*(9), 750–759. doi:10.1061/(ASCE)1090-0241(1999)125:9(750)

Bracewell, R. N. (1978). *The Fourier Transform and its Applications*. New York: McGraw-Hill.

Bradford, S. C. (2007). *Time-frequency Analysis of Systems with Changing Dynamic Properties* (PhD dissertation). California Institute of Technology.

Brady, B. H. (1990). Dynamic performance and design of underground excavations in jointed rock. Static and dynamic considerations of Rock Engineering, 1-10.

Bray, J. D., & Rathje, E. M. (1998). Earthquake-induced displacements of solid-waste landfills. *Journal of Geotechnical and Geoenvironmental Engineering*, *124*(3), 242–253. doi:10.1061/(ASCE)1090-0241(1998)124:3(242)

Brennan, A. J., Klar, A., & Madabhushi, S. P. G. (2013). Mitigation of seismic accelerations by soft caissons. *International Journal of Geotechnical Earthquake Engineering*, *4*(2), 1–17. doi:10.4018/ijgee.2013070101

Brennan, A. J., & Madabhushi, S. P. G. (2002). Effectiveness of vertical drains in mitigation of liquefaction. *Soil Dynamics and Earthquake Engineering*, *22*(9), 1059–1065. doi:10.1016/S0267-7261(02)00131-8

Brennan, A. J., Madabhushi, S. P. G., & Houghton, N. E. (2006) Comparing laminar and equivalent shear beam (ESB) containers for dynamic centrifuge modelling. In *Proceedings of 2nd Int. Conf. on Physical Modelling in Geotechnics* (pp.171-176). Hong Kong. Taylor and Francis/Balkema. 10.1201/NOE0415415866.ch17

Brennan, A. J., Thusyanthan, N. I., & Madabhushi, S. P. G. (2005). Evaluation of shear modulus and damping in dynamic centrifuge tests. *Journal of Geotechnical and Geoenvironmental Engineering*, *131*(12), 1488–1497. doi:10.1061/(ASCE)1090-0241(2005)131:12(1488)

Brown, E. T., & Hudson, J. A. (1974). Fatigue failure characteristics of some models of jointed rock. *Earthquake Eng. and struct. Dyn.*, *2*, 379–386.

BSSC. (2001). NEHRP recommended provisions for seismic regulations for new buildings and other structures 2000 edition, part 1: Provisions. Report no. FEMA 368, Building seismic safety council for the federal emergency management agency.

Cameron-Clarke & Mival. (1990). Stabilisation of old workings by blasting: Case study of a failed experiment. In Static and dynamic considerations of Rock Engineering. Academic Press.

Campbell, K. W. (1981). Near source attenuation of peak horizontal acceleration. *Bulletin of the Seismological Society of America*, *71*, 2039–2070.

Carey, P. J., Koragappa, N., & Gurda, J. J. (1993). *A case study of Brookhaven landfill. In WasteTech Proceedings.* Washington, DC: National Solid Waste Management Association.

Carlos, S. O., Antoni, R., & Xavier, G. (2006). *Assessing and managing earthquake risk*. Springer.

Castro, G. (1975). Liquefaction and cyclic mobility of saturated sands. *J. of the Geotechnical Engineering Division, ASCE.*, *101*(GT6), 551–569.

Cauzzi, C., Kalogeras, I., Melis, N., Stupazzini, M., Mazzieri, I., & Clinton, J. (2015). Seismic Response of the Acropolis of Athens (Greece) through Recorded Earthquake Data and Numerical Simulations. *International Conference on Earthquake Geotechnical Engineering*.

Celebi, E., Firat, S., Beyhan, G., Cankaya, I., Vural, I., & Kirtel, O. (2009). Field experiments on wave propagation and vibration isolation by using barriers. *Soil Dynamics and Earthquake Engineering*, *29*(5), 824–833. doi:10.1016/j.soildyn.2008.08.007

Chakraborty, A.K., Raina, A.K., Ramulu, M. Jethwa, J.L., & Gupta, R.N. (1998, November). Lake Tap at Koyna. *World Tunnel. Subsurface Exc.*, 456-460.

Chakraborty, A., & Okaya, D. (1995). Frequency-Time Decomposition of Seismic Data using Wavelet-Based Method. *Geophysics*, *60*(6), 1906–1916. doi:10.1190/1.1443922

Chan, C. L., & Low, B. K. (2012). Probabilistic analysis of laterally loaded piles using response surface and neural network approaches. *Computers and Geotechnics*, *43*, 101–110. doi:10.1016/j.compgeo.2012.03.001

Chang, F.K., & Krinitzsky, E.L. (1977). *Duration, spectral content and predominant period of strong motion earthquake records from Western United States.* Miscellaneous paper 5-73-1, U.S. Army Corps Engineers Waterways Experiment Station.

Charlie, W. A., Veyera, G. E., Doehring, D. O., & Abt, S. R. (1985). *Blast Induced Liquefaction Potential and Transient Porewater Pressure Response of Saturated Sands.* Final Report to Air Force Office of Scientific Research, Grant No. AFOSR-80-0260.

Cheung, Y. K., Tharn, L. G., & Lie, Z. X. (1995). Transient response of single pile under harmonic excitation. *Earthquake Engineering & Structural Dynamics*, *24*(7), 1017–1038. doi:10.1002/eqe.4290240707

Choi, H., & Williams, W. J. (1989). Improved Time-Frequency Representation of Multicomponent Signals using Exponential kernels. *IEEE Transactions on Acoustics, Speech, and Signal Processing*, *37*(6), 862–871. doi:10.1109/ASSP.1989.28057

Chopra, A. K. (2001). *Dynamics of structures*. Delhi: Pearson Education.

Choudhury, D., & Savoikar, P. (2009). Simplified method to characterize municipal solid waste properties under seismic conditions. *Waste Management (New York, N.Y.)*, *29*(2), 924–933. doi:10.1016/j.wasman.2008.05.008 PMID:18682319

Chui, C. K. (1992). Wavelet Transforms and Time-Frequency Analysis. In C. K. Chui (Ed.), *An Introduction to Wavelets* (pp. 49–74). Academic Press Inc.

Clayton, C. R. I., & Heymann, G. (2001). Stiffness of geomaterials at very small strains. *Geotechnique*, *51*(3), 245–255. doi:10.1680/geot.2001.51.3.245

Cohen, L. (1989). Time-Frequency Distributions: A Review. *Proceedings of the IEEE*, *77*(7), 94–981. doi:10.1109/5.30749

Cohen, L. (1995). *Time-Frequency Signal Analysis*. New York: Prentice Hall.

Coleman, D. J. (2010). Volunteered Geographic Information in Spatial Data Infrastructure: an Early Look at Opportunities and Constraints. In A. Rajabifard, J. Crompvoets, M. Kalantari, & B. Kok (Eds.), *Spatially Enabling Society: Research, emerging trends and critical assessments* (pp. 131–147). Leuven University Press.

Connigham, C. V. B., & Goetzsche, A. F. (1996). The specification of blast damage limitations in tunneling contracts. *Tunnelling and Underground Space Technology*, *5*(3), 23–27.

Cramer, C.H., & Wheeler, R.L. (2001). The 2001 Gujarat, India earthquake and seismic hazard in central and Eastern North America. *Abstract in Seismological Research Letters*, *72*, 396.

Crandell, F. J. (1949). Ground Vibrations Due to Blasting and Its Effect Upon Structures. *Journal of the Boston Society of Civil Engineers*, 222-245.

Cubrinovski, M., Uzuoka, R., Sugita, H., Tokimatsu, K., Sato, M., Ishihara, K., & Kamata, T. (2008). Prediction of pile response to lateral spreading by 3-D soil-water coupled dynamic analysis: Shaking in the direction of ground flow. *Soil Dynamics and Earthquake Engineering*, *28*(6), 421–435. doi:10.1016/j.soildyn.2007.10.015

Cundall, P. A. (1971). A computer model for simulating progressive large scale movements in Blocky Rock systems. *Proceedings of the Symposium of the International Society of Rock Mechanics*, 1, Paper No II-8.

Cundall, P. A., & Strack, O. D. L. (1979a). A Discrete Numerical Model for Granular Assemblies. *Geotechnique*, *29*(1), 47–65. doi:10.1680/geot.1979.29.1.47

Dash, S. R., Bhattacharya, S., & Blakeborough, A. (2008). *Bending-buckling interaction: Investigation of this failure mechanism of pile foundations in seismically liquefiable deposits. Technical Report-2302/08*. Department of Engineering Science, University of Oxford.

Datye, K. R., & Nagaraju, S. S. (1981). Design approach and field control for stone columns, *Proc.10th International Conf. on Soil Mechanics and Foundation Engineering, 3*, 637-640.

Daubechies, I. (1990). The Wavelet Transform, Time-Frequency Localization and Signal Analysis. *IEEE Transactions on Information Theory, 36*(5), 395–457. doi:10.1109/18.57199

Department of Defence (DoD). (2005). *NCW Roadmap*. Canberra, ACT: Defence Publishing Service.

Department of Defence (DoD). (2009). *NCW Integration and Implementation Strategy 2010*. Canberra, ACT: Defence Publishing Service.

Desai, C. S., Park, I. J., & Shao, C. (1998). Fundamental yet simplified model for liquefaction instability. *International Journal for Numerical and Analytical Methods in Geomechanics, 22*(7), 721–748. doi:10.1002/(SICI)1096-9853(199809)22:9<721::AID-NAG942>3.0.CO;2-F

Devi, V., & Sharma, M.L. (2016a). Recent Spectral Decomposition Techniques and its Application in analysis of Seismological Data: A review. *IJIRSET, 5*.

Devi, V., & Sharma, M.L. (2016b). Spectral Estimation of Noisy Seismogram using Time-Frequency Analyses. *International Journal of Geotechnical Earthquake Engineering, 7*, 19-32.

Dinesh, S.V., Sitharam, T. G., & Vinod, J.S. (2004). Dynamic properties and liquefaction behaviour of granular materials using discrete element method. *Current Science, Special Section: Geotechnics and Earthquake Hazards, 87*(10), 1379-1387.

Ding, J.J. (n.d.). *Time Frequency Analysis Tutorial Gabor Feature and its Applications*. Academic Press.

Dobry, R., & Ng, T. T. (1992). Discrete modeling of stress-strain behaviour of granular media at small and large strains. *Engineering Computations, 9*(2), 129–143. doi:10.1108/eb023853

Dobson, T. (1987). Case Histories of the Vibro Systems to Minimize the Risk of Liquefaction, Soil Improvement–A Ten Year Update. *ASCE Symposium*, 167-183.

Doucet, C., Cameron, A., & Lizotte, Y. (1996). The effects of rock mass characteristics on fragmentation in controlled blasting experiments in small development headings. *ISEE's 22nd Annual Conference*, 1-11.

Dowding, C. H. (1985). *Blast Vibration monitoring*. Prentice Hall.

Dowding, C. H., & Rozen, A. (1978). Damage to rock tunnels from earthquake loading. *J. Geotech.Eng. Div. ASCE, 104*(GT2), 175–191.

Drnevich, V. P., & Richart, F. E. (1970). Dynamic pre straining of dry sand. *Journal of Soil Mechanics and Foundation*, *96*, 453–469.

Emani, P. K., & Maheshwari, B. K. (2009). Dynamic impedances of pile groups with embedded caps in homogeneous elastic soils using CIFECM. *Soil Dynamics and Earthquake Engineering*, *29*(6), 963–973. doi:10.1016/j.soildyn.2008.11.003

Engelhardt, K., & Golding, H. C. (1975). Field Testing to Evaluate Stone column Performance in a Seismic Area. *Geotechnique*, *25*(1), 61–69. doi:10.1680/geot.1975.25.1.61

Fassett, J. B., Leonards, G. A., & Repetto, P. C. (1994). Geotechnical properties of municipal solid wastes and their use in landfill design. *Waste Tech*, *94*, 13–14.

Finn, W. D. L., & Fujita, N. (2002). Piles in liquefiable soils: Seismic analysis and design issues. *Soil Dynamics and Earthquake Engineering*, *22*(9-12), 731–742. doi:10.1016/S0267-7261(02)00094-5

Fleischhauer, M., Greiving, S., & Wanczura, S. (2005). Spatial planning in the focus of hazard and risk assessment/management in Europe. EURO-RIOB Conference, Wroclaw, Poland.

Flora, A., Lombardi, D., Nappa, V., & Bilotta, E. (2018). Numerical Analyses of the Effectiveness of Soft Barriers into the Soil for the Mitigation of Seismic Risk. *Journal of Earthquake Engineering*, *22*(1), 63–93. doi:10.1080/13632469.2016.1217802

Frankel, A., Mueller, C., Barnhard, T., Perkins, D., Leyendecker, E.V., Dickman, N., Hanson, S., & Hopper, M. (1996). *National seismic hazard maps: documentation June 1996*. Open-File Report (United States Geological Survey) 96-532.

Garcia, X., Latham, J. P., Xiang, J., & Harrison, J. P. (2009). A clustered overlapping sphere algorithm to represent real particles in distinct element modelling. *Geotechnique*, *59*(9), 1–6. doi:10.1680/geot.8.T.037

Gasparini, D. A., & Vanmarcke, E. H. (1976). *Simulated earthquakes compatible with prescribed response spectra. Publication No. R76-4*. Cambridge, MA: Massachusetts Institute of Technology.

Gazetas, G., & Mylonakis, G. (1998) Seismic soil-structure interaction: new evidence and emerging issues. In Proceedings of Geotechnical Earthquake Engineering and Soil Dynamics III (pp. 1119-1174). ASCE Geotechnical Special Publication 75.

Genes, M. C., & Kocak, S. (2005). Dynamic soil-structure interaction analysis of layered unbounded media via a coupled finite element/boundary element/scaled boundary finite element model. *International Journal for Numerical Methods in Engineering*, *62*(6), 798–823. doi:10.1002/nme.1212

Geosyntec Consultants, Inc. (1996). *Operating Industries, Inc. Landfill, Monterey Park, CA. Technical Reports SWP-1 through SWP-9*. Huntington Beach, CA: GeoSyntec Consultants, Inc.

Ghosh, B., & Madabhushi, S. P. G. (2005). Comparison of free-field ground response of layered soil under liquefiable conditions. *Indian Geotechnical Journal*, *35*(2), 199–219.

Gong, G., & Zha, X.-X. (2012). DEM simulation of liquefaction for cohesionless media at grain scale. *Journal of Central South University*, *19*(9), 2643–2649. doi:10.100711771-012-1322-9

Greiving, S., Fleischhauer, M., & Wanczura, S. (2006). Management of natural hazards in Europe: The role of spatial planning in selected EU member states. *Journal of Environmental Planning and Management*, *49*(5), 739–757. doi:10.1080/09640560600850044

GSI. (2000). *Seismotechnic Atlas of India*. New Delhi: Geology Society of India.

Hanumantharao, C., & Ramana, G. V. (2008). Dynamic soil properties for microzonation of Delhi India. *Journal of Earth System Science*, *117*(S2), 719–730. doi:10.100712040-008-0066-2

Hanumantharao, C., & Ramana, G. V. (2009). Site Specific Ground Response Analyses at Delhi, India. *The Electronic Journal of Geotechnical Engineering*, 14.

Harder, L. S. Jr. (1991). Performance of Earth Dams During the Loma Prieta Earthquake. *Proceedings Second International Conference on Recent Advances in Geotechnical Earthquake Engineering and Soil Dynamics*, 1673.

Hashash, Y. M., & Park, D. (2001). Non-linear one-dimensional seismic ground motion propagation in the Mississippi embayment. *Engineering Geology*, *62*(1), 185–206. doi:10.1016/S0013-7952(01)00061-8

Hayward Baker. (2004). *Vibro systems*. Hayward Baker Inc. Retrieved from www.HaywardBaker.com

Hazarika, H., & Boominathan, A. (2009). Liquefaction and ground failures during the 2001 Bhuj Earthquake India. *Proceedings of International Conference on Performance-Based Design in Earthquake Geotechnical Engineering from case history to practice*. 10.1201/NOE0415804844.ch13

Heidari, A., & Salajegheh, E. (2008). *Wavelet Analysis for Processing of Earthquake Records. Asian Journal of Civil Engineering*, *9*(5), 513–524.

Hendron, A. J. (1977). Engineering of Rock Blasting on Civil Projects. In *Structural and Geotechnical Mechanics*. Prentice-Hall, Inc.

Henke, K. F. (1985). Standsicherheit von Deponien aus Mull-und Abfallstoffen. *Mull-und Abfall besitigung*. Kennzahl 4546.

Hlawatsch, F. (1998). *Time-Frequency Analysis and Synthesis of Linear Signal Spaces*. Boston: Kluwer Academic Publishers; doi:10.1007/978-1-4757-2815-6

Hlawatsch, F., & Boudreaux-Bartels, G. F. (1992). Linear and Quadratic Time-Frequency Signal Representations. *IEEE SP Magazine*, *9*(2), 21–67. doi:10.1109/79.127284

Holmberg, R. (1993). Recent developments in control rock damage. In *Rock Fragmentation by Blasting*. Academic Press.

Holmberg, R., & Persson, P. A. (1978). The Swedish Approach to Contour Blasting. In *Conference on Explosives and Blasting Technique*. Society of Explosives Engineers.

Holmberg, R., & Persson, P. A. (1980). Design of Tunnel Perimeter Blast Hole Patterns to Prevent Rock Damage. *Transactions of the Institution of Mining and Metallurgy*, *89*, A37–A40.

Housner, G. W. (1959). Behavior of structures during earthquakes. *Journal of the Engineering Mechanics Division, 85*(4), 109-130.

Huerta, C. I., Shin, Y., Powers, E., & Roesset, J. (2000). Time-Frequency Analysis of Earthquake Records. Pro. of the 12th WCEE.

Hughes, J. M. O., & Withers, N. J. (1974). Reinforcing of Soft Cohesive Soils with Stone Columns. *Ground Engineering*, *7*(3), 42–49.

Hunt, R. E. (2005). *Geotechnical Investigation Handbook* (2nd ed.). Boca Raton, FL: Taylor & Francis. doi:10.1201/9781420039153

Hushmand Associates. (1994). *Landfill response to seismic events, Report prepared for USEPA Region XI*. Laguna Niguel, CA: Hushmand Associates.

Iai, S., & Koizumi, K. (1986). Estimation of earthquake induced excess pore water pressure for gravel drains. *Proc. 7th Japan Earthquake Engg. Symposium*, 679-684.

Iai, S., Koizumi, K., Noda, S., & Tsuchida, H. (1988). Large scale model tests and analysis of gravel drains. *Proc., 9th World Conference on Earthquake. Engg.*, *3*.

Idriss, I. M. (1990). Response of soft soil sites during earthquakes. *Proceedings Symposium to Honor H. B. Seed*, 273–289.

Idriss, I. M., & Boulanger, R. W. (2005). Evaluation of Liquefaction Potential, Consequences and Mitigation. *Proc. Indian Geotechnical Conference*, 3-25.

Idriss, I. M., & Sun, J. I. (1992). *SHAKE91: A computer program for conducting equivalent linear seismic response analyses of horizontally layered soil deposits*. Davis, CA: Center for Geotechnical Modeling, Department of Civil and Environmental Engineering, University of California.

Indian Standards. (2009). *IS 1893(Part 1):2002. Criteria for Earthquake Resistant Design of Structures*. General Provisions and Buildings.

IS 15284 (Part 1). (2003). *Design and construction for ground improvement — Guidelines*. Part 1 Stone Columns, analysis. Bureau of Indian Standards, New Delhi.

IS 1893 (Part 1). 2002 *Indian Standard Criteria for Earthquake Resistant Design of structures: Part 1 General Provisions and Buildings* (5th Revision).

Ishibashi, I., & Zhang, X. (1993). Unified dynamic shear moduli and damping ratios of sand and clay. *Soil and Foundation*, *33*(1), 182–191. doi:10.3208andf1972.33.182

Ishihara, K. (1993). Liquefaction and flow failure during earthquakes. *Geotechnique*, *42*(3), 351–415. doi:10.1680/geot.1993.43.3.351

Ishihara, K., Tatsuoka, F., & Yasuda, S. (1975). Undrained deformation and liquefaction of sand under cyclic stresses. *Soil and Foundation*, *15*(1), 29–44. doi:10.3208andf1972.15.29

ISRM. (1992). *Commission on Testing methods, Suggested methods for installation of borehole extensometer*. International Society for Rock Mechanics.

Itasca. (2002). *FLAC Ver. 4.0 user manual*. Itasca Group.

Iwasaki, T., Tatsuoka, F., Tokida, K., & Yasuda, S. (1978) A practical method for assessing soil liquefaction potential based on case studies at various sites in Japan. *Proceedings of 2nd Int. Conf. on Microzonation for Safer Construction – Research and Application*, 885-896.

Iyama, J., & Kuwamura, H. (1999). Application of Wavelets to Analysis and Simulation of Earthquake Motions. *Earthquake Engineering & Structural Dynamics*, *28*(3), 255–272. doi:10.1002/(SICI)1096-9845(199903)28:3<255::AID-EQE815>3.0.CO;2-C

Iyengar, R. N., & Ghosh, S. (2004). Microzonation of earthquake hazard in Greater Delhi area. *Current Science*, *87*(9), 1193–1202.

Jaeger, J. C., & Cook, N. G. W. (1979). *Fundementals of rock mechanics* (3rd ed.). London: Chapman & Hall.

Jeong, J., & Williams, W. J. (1992). Kernel Design for Reduced Interference Distributions. *IEEE Transactions on Signal Processing*, *40*(2), 402-412.

JGS. (1998). Remedial Measures against Soil Liquefaction. The Japanese Geotechnical Society.

Kaila, K. L., Krishna, V. G., & Mall, D. M. (1981). Crustal structure along Mehmedabad-Billimora profile in the Cambay basin, India, from deep seismic sounding. *Tectonophysics*, *76*(1-2), 99–130. doi:10.1016/0040-1951(81)90255-9

Kaila, K. L., Tewari, H. C., Krishna, V. G., Dixit, M. M., Sarkar, D., & Reddy, M. S. (1990). Deep seismic sounding studies in the north Cambay and Sanchor basins, India. *Geophysical Journal International*, *103*(3), 621–637. doi:10.1111/j.1365-246X.1990.tb05676.x

Kareem, A., & Gurley, K. (1999). Applications of Wavelet Transforms in Earthquake, Wind and Ocean Engineering. *Engineering Structures*, *26*(3).

Kavazanjian, J. E., & Matasovic, N. (1995). Seismic analysis of solid waste landfills. Geoenvironment 200 Proceedings, 1066-1080.

Kavazanjian, E., Matasovic, N., Stokoe, K. H., & Bray, J. D. (1996). In situ shear wave velocity of solid waste from surface wave measurements. In M. Kamon (Ed.), *Proceedings of Environmental Geotechnics,* 97–102.

Kerwin, S., & Stone, J. (1997). Liquefaction failure and remediation: King Harbor Redondo Beach, California. *Journal of Geotechnical and Geoenvironmental Engineering*, *123*(8), 760–769. doi:10.1061/(ASCE)1090-0241(1997)123:8(760)

Kijko, A., & Graham, G. (1998). Parametric-historic procedure for probabilistic seismic hazard analysis, Part I: Estimation of maximum regional magnitude Mmax. *Pure and Applied Geophysics*, *152*(3), 413–442. doi:10.1007000240050161

Kim, H.-Y., Lee, K.-W., Mizuta, Y., & Lee, H.-K. (1990). A study on determination of the relaxed zone around an excavated tunnel. Static and Dynamic Considerations of Rock Engineering, 177-182.

Kirtas, E., & Pitilakis, K. (2009). Subsoil Interventions Effect on Structural Seismic Response. Part II: Parametric Investigation. *Journal of Earthquake Engineering*, *13*(3), 328–344. doi:10.1080/13632460802347471

Komatitsch, D., Liu, Q., Tromp, J., Suss, P., Stidham, C., & Shaw, J. H. (2004). Simulations of ground motion in the Los Angeles basin based upon the spectral-element method. *Bulletin of the Seismological Society of America*, *94*(1), 187–206. doi:10.1785/0120030077

Komatitsch, D., & Vilotte, J. P. (1998). The spectral-element method: An efficient tool to simulate the seismic response of 2D and 3D geological structures. *Bulletin of the Seismological Society of America*, *88*, 368–392.

Kramer, S. L. (1996). *Geotechnical earthquake engineering*. Prentice Hall.

Kramer, S. L. (1996). *Geotechnical Earthquake Engineering*. Prentice-Hall, Inc.

Kramer, S. L., & Stewart, J. P. (2004). Geotechnical aspects of seismic hazards. In *Earthquake Engineering from Engineering Seismology to Performance Based Engineering*. CRC Press.

Krinitzsky, E. L. (2002). How to obtain earthquake ground motions for engineering design. *Engineering Geology*, *65*(1), 1–16. doi:10.1016/S0013-7952(01)00098-9

Krinitzsky, E. L. (2003). How to combine deterministic and probabilistic methods for assessing earthquake hazards. *Engineering Geology*, *70*(1-2), 157–163. doi:10.1016/S0013-7952(02)00269-7

Krishna, A. M., Kumar, A. M., & Baruah, A. K. (2018). Stone Columns/Granular Piles for Improving Liquefiable Sites: Case studies. *Geotechnical Engineering Journal of the SEAGS & AGSSEA*, *49*(1), 136–142.

Krishna, A. M., & Madhav, M. R. (2008). Densification and dilation effects of granular piles in liquefaction mitigation. *Indian Geotechnical Journal*, *38*(3), 295–316.

Krishna, A. M., Madhav, M. R., & Kumar, K. (2014). Ground Engineering with Granular Inclusions for Loose Saturated Sands subjected to Seismic Loadings. *Indian Geotechnical Journal*, *44*(2), 205–217. doi:10.100740098-013-0085-z

Krishna, A. M., Madhav, M. R., & Latha, G. M. (2006). Liquefaction mitigation of ground treated with granular piles: Densification effect. *ISET Journal of Earthquake Technology*, *43*(4), 105–120.

Krishna, A. M., Madhav, M. R., & Latha, G. M. (2007). Densification effect of granular piles on settlement response of treated ground. *Ground Improvement*, *11*(3), 127–136. doi:10.1680/grim.2007.11.3.127

Kumar, A., Gupta, S. C., Sharma, M. L., Pandey, A. D., Jindal, A. K., & Jain, S. K. Goswami, Neetu, (2010). Attributes of local seismicity of The Garhwal Himalaya. *Proceedings of the14th Symp. Earthquake Engineering*, 90-99.

Kumar, R., Sumathi, P., & Kumar, A. (2015). A Time–Frequency Approach for Generation of Synthetic Time-Histories of Earthquake Signals. *Acta Geodaetica et Geophysica*. doi:10.100740328-015-0115-7

Kutter, H. K., & Fairhurst, C. (1971). On the Fracture Process in Blasting. *International Journal of Rock Mechanics and Mining Sciences*, *8*(3), 181–202. doi:10.1016/0148-9062(71)90018-0

Landva, A. O., & Clark, J. I. (1986, August). Geotechnical testing of waste fill. *Proceedings, 39th Canadian Geotechnical Conference,* 3714385.

Langefors, U., & Kihlstrom, B. (1963). *The modern techniques of rock blasting*. New York: J. Wiley and Sons, Inc.

Law, T. M., May, J., Spathis, A. T., Du Plessis, A. T., & Palmer, A. M. (2001). Blast damage and blast dilution control: The applicaton of bulk emulsion systems at the WMC St Ives junction mine, J. *Rock Fragmentation by Blasting*. *Fragblast*, *5*(1-2), 1–20. doi:10.1076/frag.5.1.1.3317

LeBlanc, T., Heilig, J., & Ryan, J. (1995). Predicting the Envelope of Damage from the Detonation of a Confined Charge. *Proceedings of the Sixth High-Tech Seminar on the State of the Art in Blasting Technology Instrumentation and Explosives Applications*, 225–291.

Lee, S. J., Chen, H. W., Liu, Q., Komatitsch, D., Huang, B. S., & Tromp, J. (2008). Three Dimensional Simulations of Seismic-Wave Propagation in the Taipei Basin with Realistic Topography Based upon the Spectral-Element Method. *Bulletin of the Seismological Society of America*, *98*(1), 253–264. doi:10.1785/0120070033

Leet, L. D. (1946). *Vibrations from Blasting*. Wilmington, DE: Hercules Powder Co.

Lehmann, L. (2005). An effective finite element approach for soil-structure analysis in the time-domain. *Structural Engineering and Mechanics*, *21*(4), 1–14. doi:10.12989em.2005.21.4.437

Leonards, G. A. (Ed.). (1962). *Foundation Engineering*. New York: McGraw-Hill.

Le, T., Lee, V. W., & Luo, H. (2016). Out of plane (SH) soil-structure interaction: A shear wall with rigid and flexible ring foundation. *Earth Science*, *29*(1), 44–55. doi:10.100711589-016-0139-2

Lew. (2001). *The Seismic Design Handbook*. Kluwer Academic Publishers.

Lewandowski, T., Mai, L. V. K., & Danell, R. (1996). Influence of discontinuities on presplitting effectiveness. *Proceedings of 5th Int. Symposium on Rock Fragmentation by Blasting*, 217-225.

Lindsay, B. R. (2011). *Social Media and Disasters: Current Uses, Future Options, and Policy considerations*. Washington, DC: Congressional Research Service.

Lin, X., & Ng, T. T. (1997). A three-dimensional discrete element model using arrays of ellipsoids. *Geotechnique*, *47*(2), 319–329. doi:10.1680/geot.1997.47.2.319

Liu, Tran, Counter, & Andrieux. (1998). A case study of blast damage evaluation in open stope mining at Kidd Creek mines. *24th Annual Conference on Explosives and Blasting Research*, 323-336.

Liu, G., Fomel, S., & Chen, X. (2011). Time-frequency Analysis of Seismic Data using Local Attributes. *Geophysics*, *76*(6), 23–34. doi:10.1190/geo2010-0185.1

Liyanapathirana D. S., & Poulos H. G. (2005). Seismic lateral response of piles in liquefying soil. *Journal of Geotechnical and Geo-Environmental Engineering*, *131*(12), 1466e79.

Lockwood, O. G., & Kanamori, H. (2006). Wavelet Analysis of the Seismograms of the 2004 Sumatra-Andaman Earthquake and its Application to Tsunami Early Warning. *Geochemistry Geophysics Geosystems*, *7*(9). doi:.10.1029/2006GC001272

Lombardi, D., Dash, S. R., Bhattacharya, S., Ibraim, E., Wood, D. M., & Taylor, C. A. (2017). Construction of simplified design p-y curves for liquefied soils. *Geotechnique*, *67*(3), 216–227. doi:10.1680/jgeot.15.P.116

Lombardi, D., Flora, A., Lirer, S., & Silvestri, F. (2014). Soil grouting for the seismic protection of existing buildings. *Ground Improvement*, *167*(3GI3), 206–218. doi:10.1680/grim.13.00010

Lopez, R. A., & Hayden, R. F. (1992). The Use of Vibro Systems in Seismic Design. Grouting, Soil Improvement and Geosynthetics, 30(2), 1433-1445.

Luehring, R., Snorteland, N., Mejia, L., & Stevens, M. (2001). Liquefaction Mitigation of a Silty Dam Foundation Using Vibro-stone Columns and Drainage Wicks: A Case History at Salmon Lake Dam. *Proc. 21st USSD annual meeting and lecture.*

Madabhushi, G., Knappett, J., & Haigh, S. (2010). *Design of pile foundations in liquefiable soils*. London: Imperial College Press.

Madabhushi, S. P. G. (2007). Ground improvement methods for liquefaction remediation. *Ground Improvement*, *11*(4), 195–206. doi:10.1680/grim.2007.11.4.195

Madabhushi, S. P. G., Schofield, A. N., & Lesley, S. (1998). A new stored angular momentum (SAM) based actuator. *Proceedings of Centrifuge*, *98*, 111–116.

Madhav, M. R., & Arlekar, J. N. (2000).Dilation of granular piles in mitigating liquefaction of sand deposits. *Proc. 12th World Conference Earthquake Engineering.*

Madhav, M. R., Iyengar, N. G. R., Vitkar, R. P., & Nandia, A. (1979). Increased Bearing Capacity and Reduced Settlements Due to Inclusions in Soil. *Proc. International Conference on Soil Reinforcement: Reinforced Earth and Other Techniques*, *2*, 329-333.

Madhav, M. R., & Thiruselvam, K. (1988). Effect of Installation Methods on Granular Pile Response. *Proc. Int. Symp. on Theory and Practice of Earth Reinforcement*, 221-226.

Maheshwari, B. K., & Emani, P. K. (2014). Three-dimensional nonlinear seismic analysis of pile groups using FE-CIFECM coupling in a hybrid domain and HiSS plasticity model. *International Journal of Geomechanics, 15*(3), 04014055-1-12.

Maheshwari, B. K., Truman, K. Z., Gould, P. L., & El Naggar, M. H. (2005). Three-dimensional nonlinear seismic analysis of single piles using finite element model: Effects of plasticity of soil. *International Journal of Geomechanics*, *5*(1), 35–44.

Mallat, S. (1999). *A Wavelet Tour of Signal Processing*. New York: Academi Press.

Marks, J. R. II. (2009). *Handbook of Fourier Analysis & its Applications*. Oxford University Press.

Martin, G. R. (1975). Fundamentals of Liquefaction Under Cyclic Loading. *Journal of Geotechnical Engineering*, *101*(5), 423–438.

Martin, G. R., Finn, W. D. L., & Seed, H. B. (1975). Fundamentals of liquefaction under cyclic loading. *J. of the Geotechnical Engineering Division, ASCE*, *101*(GT5), 425–438.

Matasovic, N., & Kavazanjian, E. Jr. (1998). Cyclic characterization of OII landfill solid waste. *Journal of Geotechnical and Geoenvironmental Engineering*, *124*(3), 197–210. doi:10.1061/(ASCE)1090-0241(1998)124:3(197)

Matasovic, N., & Vucetic, M. (1995). Seismic response of soil deposits composed of fully-saturated clay and sand layers. *First International Conference on Earthquake Geotechnical Engineering*.

May, T. W., & Bolt, B. A. (1982). Effectiveness of trenches in reducing seismic motion. *Earthquake Engineering & Structural Dynamics, 10*(2), 195–210. doi:10.1002/eqe.4290100203

McGarr, A. (1983). Estimating ground motion for small nearby earthquakes. In Seismic design of embankments and caverns (pp. 113-127). New York: ASCE.

McKenzie, C., Scherpenisse, C., Arriagada, J., & Jones, J. (1995). Application of Computer Assisted Modelling to Final Wall Blast Design. *Proceedings of the EXPLO'95 – A Conference Exploring the Role of Rock Breakage in Mining and Quarrying*, 285–292.

Mejia, L. H. (2005). Seismic Retrofit of Embankment Dams to Mitigate Liquefaction: Selected Case Histories. In *Proc. Geo-Frontiers 2005 Congress*. ASCE Geotechnical Special Publication.

Mener, A. S. (2007). *Disaster response in the United States of America: An analysis of the bureaucratic and political history of a failing system* (PhD thesis). University of Pennsylvania.

Meyer, T., & Dunn, P. G. (1996). Fragmentation and Rock mass Damage Assessment – Sunburst Excavator and Drill and Blast. *Proceedings of the NARMS 1996 – Rock Mechanics Tools and Techniques*, 609–617.

Miao, X. G., & Moon, W. M. (1999). Application of Wavelet Transform in Reflection Seismic Data Analysis. *Geosciences Journal, 3*(3), 171–179. doi:10.1007/BF02910273

Mitchell, J. K., & Wentz, F. K. (1991). *Performance of improved ground during Loma Prieta earthquake*. Report No. EERC91/12. Earthquake Engineering Research Center, University of California, Berkeley, CA.

Mitrani, H., & Madabhushi, S. P. G. (2008). Centrifuge modelling of inclined micro-piles for liquefaction remediation of existing buildings. *Geomechanics and Geoengineering, 3*(4), 245–256. doi:10.1080/17486020802483730

Moghtar, A. S., Abdel-Motaal, M. A., & Wahidy, M. (2014). Lateral displacement and pile instability due to soil liquefaction using numerical model. *Ain Shams Engineering Journal*, *5*(4), 1019–1032. doi:10.1016/j.asej.2014.05.002

Munfakh, G. A., Abramson, L. W., Barksdale, R. D., & Juran, I. (1987). In-Situ Ground Reinforcement. In J. P. Welsh (Ed.), Geotechnical Special Publication: Vol. 12. *Soil Improvement – A Ten Year Update* (pp. 1–17). ASCE.

Mylonakis, G., & Gazetas, G. (2000). Seismic soil-structure interaction: Detrimental or beneficial. *Journal of Earthquake Engineering*, *4*(3), 277–301. doi:10.1080/13632460009350372

Naveen, B. P., Sitharam, T. G., & Sivapullaiah, P. V. (2014). Evaluating the Dynamic Characteristics of Municipal Solid Waste for Geotechnical Purpose. *Current Advances in Civil Engineering, 2*(1).

Naveen, B.P., Sitharam, T.G., & Sivapullaiah, P.V. (2015). Seismic Analysis of Municipal Solid Waste Landfill in India. *International Journal of Geotechnical Earthquake Engineering*, *6*(2), 35-55.

Nawab, S. N., & Quatieri, T. F. (1988). *Short Time Fourier Transform, Chapter in Advanced Topics in Signal Processing* (J. S. Lim & A. V. Oppenheim, Eds.). Englewood Cliffs, NJ: Prentice Hall.

Nazarian, S., & Stokoe, K.H. (1983). *Use of spectral analysis of surface waves for determination of moduli and thicknesses of pavement systems*. Transportation Research Record No. 954.

Neelima Satyam, D., & Rao, K. S. (2009). Dynamic site characterization in Delhi region using MASW testing. *International Journal of Earth Sciences and Engineering*, *2*(1), 32–42.

Newmark, N. M. (1967). Design criteria for nuclear reactors subjected to earthquake hazards. *Proc., IAEA Panel on a Seismic Design and Testing of Nuclear Facilities.*

Newmark, N. M., & Hall, W. J. (1973). *Procedures and criteria for earthquake resistant design. building practice for disaster mitigation*. Washington, DC: National Bureau of Standards.

Newmark, N. M., & Riddell, R. (1980, September). Inelastic spectra for seismic design. *Proc. 7th Worm Conf. on Earthquake Engineering*, 4, 129-136.

Nguyen, H. B. K., Rahman, M. M., & Fourie, A. B. (2017). Undrained behaviour of granular material and the role of fabric in isotropic and K0 consolidations: DEM approach. *Geotechnique*, *67*(2), 153–167. doi:10.1680/jgeot.15.P.234

Niklasson, B. (1985). Vertical crater retreat mining at the Luossavaara research mine, Swedish Detonic Research Foundation. *11th Annual Conference on Explosives and Blasting Research*, 46-61.

Nogami, T., Otani, J., Konagai, K., & Chen, H. L. (1992). Nonlinear soil-pile interaction model for dynamic lateral motion. *Journal of Geotechnical Engineering*, *118*(1), 89–106. doi:10.1061/(ASCE)0733-9410(1992)118:1(89)

Noorzad, A., Poorooshasb, H. B., & Madhav, M. R. (2007). Performance of partially penetrating stone columns during an earthquake. *Proc. Tenth Symposium on Numerical models in Geomechanics (NUMOG X)*, 503-508.

NSDR. (2011). *National Strategy for Disaster Resilience*, retrieved from http://www.em.gov.au/Documents/1National%20Strategy%20for%20Disaster%20Resilience%20-%20pdf.PDF

Olsen, K. B., Pechmann, J. C., & Schuster, G. T. (1996). An Analysis of Simulated and Observed Blast Records in the Salt Lake Basin. *Bulletin of the Seismological Society of America*, *86*(4), 1061–1076.

Olson, M., & Bergqvist, I. (1996). Crack lengths from explosives in small diameter boreholes. In Rock Fragmentation by Blasting. Academic Press.

Onoue, A. (1988). Diagrams considering well resistance for designing spacing ratio of gravel drains. *Soil and Foundation*, *28*(3), 160–168. doi:10.3208andf1972.28.3_160

Onoue, A., Mori, N., & Takano, S. (1987). In-situ experiment and analysis on well resistance of gravel drains. *Soil and Foundation*, *27*(2), 42–60. doi:10.3208andf1972.27.2_42

Oriard, L. L. (1989). The Scale of Effects in Evaluating Vibration Damage Potential. *Proc. 15th Annual Conference on Explosives and Blasting Research*, 161-176.

Oriord, L. L. (1982). Blasting effects and their control. In Underground Mining Methods Handbook. SME of AIME.

Otuonye, F. O. (1997). Dynamic response of a fully grouted resin roof bolt to blast loading. *Proc. of Annual Conf. on Explosives and Blasting Research.*

Ouchterlony, F. (1997). Prediction of crack lengths in rock after cautious blasting with zero interholedelay,Int. J. *Rock Fragmentation by Blasting*, *1*, 417–444.

Ouchterlony, F., Sjoberg, C., & Jonsson, B. A. (1993). *Blast damage prediction from vibration measurements at the SKB underground laboratories at ÄSPÖ in Sweden, Intl.* San Diego, CA: Soc. Explosive Engg.

Oweis, I. S., & Khera, R. P. (1998). *Geotechnology of waste management*. PWS Pub.

Park, R., & Paulay, T. (1975). *Reinforced concrete structures*. John Wiley & Sons, Inc. doi:10.1002/9780470172834

Paventi, M., Lizotte, Y., Scoble, M., & Mohanty, B. (1996). Measuring rock mass damage in drifting. Rock Fragmentation by Blasting, 131-138.

Persson, P. A., Holmberg, R., & Lee, J. (1994). *Rock Blasting and Explosives Engineering*. Boca Raton, FL: CRC press Inc.

Pestana, J. M., Hunt, C. E., & Goughnour, R. R. (1997). *FEQDrain: a finite element computer program for the analysis of the earthquake generation and dissipation of pore water pressure in layered sand deposits with vertical drains.* Report No. UCB/EERC-97-15.

Pestana, J. M., Hunt, C. E., Goughnour, R. R., & Kammerer, A. M. (1998). Effect of storage capacity on vertical drain performance in liquefiable sand deposits. *Proc. Second International Conference on Ground Improvement Techniques*, 373-380.

Petersen, M. D., Rastogi, B. K., Schweig, E. S., Harmsen, S. C., & Gomberg, J. S. (2004). Sensitivity analysis of seismic hazard for the northwestern portion of the state of Gujarat. Academic Press.

Phanikanth, V. S., Choudhury, D., & Rami Reddy, G. (2011). Equivalent-Linear Seismic Ground Response Analysis of Some Typical Sites in Mumbai. *Geotechnical and Geological Engineering*, *29*(6), 1109–1126. doi:10.100710706-011-9443-8

Phanikanth, V. S., Choudhury, D., & Reddy, G. R. (2013). Behavior of single pile in liquefied soil deposits during earthquakes. *International Journal of Geomechanics*, *13*(4), 454–462. doi:10.1061/(ASCE)GM.1943-5622.0000224

Phoon, K. K., & Kulhawy, F. H. (1999). Evaluation of geotechnical property variability. *Canadian Geotechnical Journal*, *36*(4), 625–639. doi:10.1139/t99-039

PHRI. (1997). Handbook on liquefaction remediation of reclaimed land. Port and Harbor Research Institute.

Poorooshasb, H. B., Madhav, M. R., & Noorzad, A. (2006). Performance of Stone Columns subjected to a seismic Base Excitation. *Proc. Int. Symp. on Lowland Technology*, 1189-1194.

Poorooshasb, H. B., & Noorzad, A. (2000). Miura, N. and Madhav, M. R. Prevention of Earthquake Induced Liquefaction of Sandy Deposits using Stone Columns. *Proc. Int. Symp. on Lowland Technology*, 213-220.

Price-Robertson, R., & Knight, K. (2012). *Natural disasters and community resilience: A framework for support, Communities and Families.* CFCA paper no. 3. Retrieved from http://www.aifs.gov.au/cfca/topics/communities.php

Priebe, H. J. (1995). Design of vibro replacement. *Ground Engineering*, *28*(10), 31–37.

Priebe, H. J. (1998). Vibro Replacement to Prevent Earthquake Induced Liquefaction. *Proceedings of the 5th Geotechnique-Colloquium.*

Qiang, Z., Wang, Y., & Shen, G. (2011). Comparison and Application of Time-Frequency Analysis Methods for Non-stationary Signal Processing. In *Advanced Research on Computer Education, Simulation and Modeling* (Vol. 175, pp. 286–291). CCIS. doi:10.1007/978-3-642-21783-8_47

Qian, S., & Chen, D. (1993). Discrete Gabor Transform. IEEE Trans. *Signal Processing*, *41*, 2429–2438.

Qian, S., & Chen, D. (1994). Decomposition of the Wigner-Ville Distribution and Time-Frequency Distribution Series. *IEEE Transactions on Signal Processing*, *42*(10), 2836–2842. doi:10.1109/78.324750

Qian, S., & Chen, D. (1996). *Joint Time-Frequency Analysis, Methods and Applications*. Prentice Hall PTR.

Qian, S., & Chen, D. (1999). Understanding the Nature of the Signals whose Power Spectrum Change with Time Joint Analysis. *IEEE Signal Processing Magazine*, 53–67.

Qian, S., Hui, S., & Wei, J. (2003). *Discrete Gabor Expansion for Order Tracking. IEEE Trans. of Instrumentation and Measurements.*

Rahimi, M., Chan, D., & Nouri, A. (2017). Constitutive model for cyclic behaviour of cohesionless sands. *Geomechanics and Geoengineering*, *12*(1), 36–47. doi:10.1080/17486025.2016.1156168

Rahmani, A., & Pak, A. (2012). Dynamic behavior of pile foundations under cyclic loading in liquefiable soils. *Computers and Geotechnics*, *40*, 114–126. doi:10.1016/j.compgeo.2011.09.002

Rajabifard, A., Crompvoets, J., Kalantari, M., & Kok, B. (2010b). Spatially enabled societies. In A. Rajabifard, J. Crompvoets, M. Kalantari, & B. Kok (Eds.), *Spatially Enabling Society: Research, emerging trends and critical assessments* (pp. 15–25). Leuven University Press.

Rajabifard, A., Williamson, I., & Vaez, S. (2010a). Spatially enabled societies: Asia and the Pacific. In A. Rajabifard, J. Crompvoets, M. Kalantari, & B. Kok (Eds.), *Spatially Enabling Society: Research, emerging trends and critical assessments* (pp. 87–94). Leuven University Press.

Raju, V. R., & Sondermann, W. (2005). Ground Improvement using Deep Vibro Techniques. In Ground Improvement — Case Histories. Elsevier Publishers (UK).

Ramulu, M., Chakraborty, A. K., & Sitharam, T. G. (2009). Damage assessment of basaltic rock mass due to repeated blasting in a railway tunnelling project – a case study. *Tunnelling and Underground Space Technology*, *24*(2), 208–221. doi:10.1016/j.tust.2008.08.002

Rastogi, B. K. (2001). Ground deformation study of Mw 7.7 Bhuj earthquake of 2001. *Episodes*, *24*, 160–165.

Reddy, K. R., Kumar, G., & Giri, R. K. (2017). Modelling Coupled Processes in Municipal Solid Waste Landfills: An Overview with Key Engineering Challenges. *Int. J. of Geosynthetic. and Ground Engineering*, *3*(6), 1–21.

Reiter, L. (1990). *Earthquake hazard analysis – Issues and Insights*. New York: Columbia University Press.

Riddell, R., & Newmark, N. M. (1979). *Statistical analysis of the response of nonlinear systems subjected to earthquakes. University of Illinois Engineering Experiment Station. College of Engineering*. University of Illinois at Urbana-Champaign.

Riks, E. (1972). The application of Newton's method to the problem of elastic stability. *Journal of Applied Mechanics*, *39*(4), 1060–1065. doi:10.1115/1.3422829

Riks, E. (1979). An incremental approach to the solution of snapping and buckling problems. *International Journal of Solids and Structures*, *15*(7), 529–551. doi:10.1016/0020-7683(79)90081-7

Rioul, O., & Vetterli, M. (1991). Wavelets and Signal Processing. *IEEE Signal Processing Magazine*, *14*(38).

Rocque, P. (1992). *Techniques utilized for the detection and characterization of blast induced damage* (Unpublished M. Sc. Eng. Thesis). Queen's University at Kingston, Ontario, Canada.

Rothenburg, L., & Bathurst, R. J. (1992). Micromechanical features of granular assemblies with planar elliptical particles. *Geotechnique*, *42*(4), 79–95. doi:10.1680/geot.1992.42.1.79

Saito, A., Taghawa, K., Tamura, T., Oishi, H., Nagayama, H., & Shimaoka, H. (1987). A countermeasure for sand liquefaction: gravel drain method. Nippon Kokan Technical Report, Overseas No. 51.

Sanchez-Alciturri, J. M., Palma, J., Sagaseta, C., & Canizal, J. (1993). Mechanical properties of wastes in a sanitary landfill. *Proceedings of the Green '93, Waste Disposal by Landfill,* 357-363.

Sasaki, Y., & Taniguchi, E. (1982). Shaking table tests on gravel drains to prevent liquefaction of sand deposits. *Soil and Foundation, 22*(3), 1–14. doi:10.3208andf1972.22.3_1

Saurez, L. E., & Montejo, L. A. (2005). Generation of Artificial Earthquakes via the Wavelet Transform. *International Journal of Solids and Structures, 21-22*(21-22), 5905–5919. doi:10.1016/j.ijsolstr.2005.03.025

Sawicki, A., & Mierczynski, J. (2006). Developments in Modeling Liquefaction of Granular Soils, Caused by Cyclic Loads. *Applied Mechanics Reviews, Transactions of the ASME, 59*(3), 91-106.

Schnabel, P. B., Lysmer, J., & Seed, H. B. (1972) *SHAKE: A computer program for earthquake response analysis of horizontally layered sites* (Report No. UCB/EERC-72/12). Earthquake Engineering Research Center, University of California, Berkeley, CA.

Schnabel, P. M., Lysmer, J., & Seed, H. B. (1972). *SHAKE: A computer program for earthquake response analysis of horizontally layered sites. Report No. EERC 72/12.* Berkeley, CA: Earthquake Engineering Research Centre, University of California.

Schofield, A. N. (1980). Cambridge geotechnical centrifuge operations. *Geotechnique, 30*(3), 227–268. doi:10.1680/geot.1980.30.3.227

Scott, A., Cocker, A. N., Djordjevic, N., Higgins, M., La Rosa, D., Sarma, K. S., & Wedmaier, R. (1996). *Open Pit Blast Design, Analysis and optimization.* Julius Kruttschnitt Mineral Research Centre, The University of Queensland.

Seed, H. B., & Idriss, I. M. (1971). Simplified procedure for evaluating soil liquefaction potential. *Journal of the Soil Mechanics and Foundations Division*, *97*(9), 1249-1273.

Seed, H. B., Idriss, I. M., Makdisi, F., & Bannerjee, N. (1975). *Representation of irregular stress time histories by equivalent uniform stress series in liquefaction analyses.* Report No. UCB/EERC/75-29, Earthquake Engineering Research Center, University of California, Berkeley, CA.

Seed, H. B. (1979). Soil liquefaction and cyclic mobility evaluation for level ground during earthquakes. *J. of the Geotechnical Engineering Division, ASCE, 105*(2), 201–255.

Seed, H. B., & Booker, J. R. (1977). Stabilization of potentially liquefiable sand deposits using gravel drains. *J. of the Geotechnical Engineering Division, ASCE, 103*(7), 757–768.

Seed, H. B., & Idriss, I. M. (1970). *Soil modules and damping factors for dynamic response analyses. Report EERC 70–10*. Berkeley, CA: Earthquake Engineering Research Center, University of California.

Seed, H. B., & Idriss, I. M. (1971). Simplified Procedure for Evaluating Soil Liquefaction Potential. *Journal of the Soil Mechanics and Foundations Division*, *97*, 1249–1273.

Seed, H. B., & Idriss, I. M. (1982). *Ground Motions and Soil Liquefaction during Earthquakes*. Oakland, CA: Earthquake Engineering Research Institute.

Seed, H. B., & Lee, K. L. (1966). Liquefaction of Saturated Sands during Cyclic Loading. *Journal of the Soil Mechanics and Foundations Division*, *92*, 105–134.

Seed, H. B., Martin, P. P., & Lysmer, J. (1976). Pore-water pressure changes during soil liquefaction. *Journal of Geotechnical Engineering*, *102*(4), 323–346.

Seed, H. B., Wong, R. T., Idriss, I. M., & Tokimatsu, K. (1986). Moduli and damping factors for dynamic analyses of cohesionless soils. *Journal of Geotechnical Engineering*, *112*(11), 1016–1032. doi:10.1061/(ASCE)0733-9410(1986)112:11(1016)

Seed, R. B., Cetin, K. O., Moss, R. E. S., Kammerer, A. M., Wu, J., Pestana, J. M., & Riemer, M. F. (2001). Recent Advances in Soil Liquefaction Engineering and Seismic Site Response Evaluation. *Proc. 4th Int. Conf. Recent Advances in Geotechnical Earthquake Engineering and Soil Dynamics*, 1-45.

Sejdic, E., Djurovic, I., & Jiang, J. (2009). Time-Frequency Feature Representation using Energy Concentration: An Overview of Recent Advances. *Digital Signal Processing*, *19*(1), 153–183. doi:10.1016/j.dsp.2007.12.004

SHAKE. (2000). *A computer program for conducting equivalent-linear seismic response analyses for horizontally layered soil deposits. A modified PC version of the original SHAKE program published in 1972 by Schnabel, Lysmer and Seed (modifications made by Idriss IM, Sum JI). EERI*. Berkeley, CA: University of California.

Sharma, M. L. (1992b). *Procedures for Analysis of Digital Telemetered Seismic Array Data* (Ph.D. thesis). EQD, Univ. of Roorkee, India.

Sharma, M. L., Kumar, A., Gupta, S.C., Jindal, A.K., Sen, A., Jain, S.K., … Vandana. (2014). Earthquake source parameters and focal mechanism of local earthquakes around Tehri region. *Proceedings of SEE '10*.

Sharma, H. D., Dukes, M. T., & Olsen, D. M. (1990). *Field measurements of dynamic moduli and Poisson's ratio of refuse and underlying soils at a landfill site. Geotechnics of Waste Fills – Theory and Practice, STP 1070*. Philadelphia, PA: ASTM.

Sharma, H. D., Dukes, M. T., & Olsen, D. M. (1991). Field measurements of dynamic moduli and Poisson's ratios of refuse and underlying soils at a landfill site. *International Journal of Rock Mechanics and Mining Sciences & Geomechanics Abstracts*, *28*(6), A356. doi:10.1016/0148-9062(91)91335-O

Sharma, M. L. (1992a). *Sample Telemetered Digital Seismic Array in Garhwal Himalaya-Software and Data Management*. Geo. Soc. of India.

Sharma, M. L., Gupta, S. C., & Wason, H. R. (1990). An Automatic Phase Picker for Earthquakes. *Proc. International Conference on Automatic Robotics and Computer Vision*, 247-250.

Sharma, M. L., Wason, H. R., & Dimri, R. (2003). Seismic zonation of the Delhi region for bedrock ground motion. *Pure and Applied Geophysics, 160*(12), 2381–2398. doi:10.100700024-003-2400-6

Shokrollahi, E., Zargar, G., & Riahi, M.A. (2013). Using Continuous Wavelet Transform and Short Timc Fouricr Transform as Spectral Decomposition Methods to Detect of Stratigraphic Channel in One of the Iranian South-West Oil Fields. *Int. J Sci. Emerging Tech, 5*(5).

Shumann, H. (1989). Standdestigkeitsberechnung fur Hausmulldeponien–Praktische Erfahrungen. *Fortschritte der Deponietechnik, 89*, 153–179.

Sifuzzaman, M., Islam, M. R., & Ali, M. Z. (2009). Application of Wavelet Transform and its Advantages Compared to Fourier Transform. *The Journal of Physiological Sciences; JPS, 13*, 121–134.

Simons, F. J., Dando, B., & Allen, R. (2006). Automatic Detection and Rapid Determination of Earthquake Magnitude by Wavelet Multi Scale Analysis of the Primary Arrival. *Earth and Planetary Science Letters, 250*(1-2), 214–223. doi:10.1016/j.epsl.2006.07.039

Singh, S. P. (1993). Damage causing potential of different explosives. *Proc, 9th annual symposium on Explosives and Blasting Research*, 325-337.

Singh, S. P., & Xavier, P. (2005). Causes, impact and control of overbreak in underground excavations. *Tunnelling and Underground Space Technology, 20*(1), 63–71. doi:10.1016/j.tust.2004.05.004

Sitharam, T.G., & Anbazhagan, P. (2007). Seismic hazard analysis for Bangalore region. *Journal of Natural Hazards, 40*, 261-278.

Sitharam, T. G. (2003). Discrete element modeling of cyclic behaviour of granular materials. *Geotechnical and Geological Engineering, 21*(4), 297–329. doi:10.1023/B:GEGE.0000006036.00597.0b

Sitharam, T. G., & Govindaraju, L. (2004). Geotechnical aspects and ground response studies in Bhuj earthquake, India. *Geotechnical and Geological Engineering*, *22*(3), 439–455. doi:10.1023/B:GEGE.0000025045.90576.d3

So, M. M. L., Mote, T. I., Pappin, J. W., Lau, V. S. Y., & Yim, I. P. H. (2012). 3D GIS technique for seismic hazard assessment for a subduction zone region. *Proc. of. The Annual Technical Conference: Australian Earthquake Engineering Society*.

Sokolov, Y. V., Loh, C. H., & Wen, K. L. (2001). Empirical models for site- and region-dependent ground motion parameters in the Taepei area: A unified approach. *Earthquake Spectra*, *17*(2), 313–332. doi:10.1193/1.1586177

Sondermann, W., & Wehr, W. (2004). Deep vibro techniques. In Ground Improvement (2nd ed.). Spon Press, Taylor & Francis Group.

St.John, C. M., & Zahrah, T. F. (1987). A seismic design of underground structures. Tunneling and underground space. *Technology*, *2*(2), 165–197.

ST060088. (2013). *AISRF additional report: Spatially enabled nations: A meta-integrative framework for improving disaster management practices in India and Australia.* A report submitted to the Governments of Australia and India.

Steeghs, P., & Drijkoningen, G. G. (1996). Time-Frequency Analysis of Seismic Reflection Signals. *Proceedings of the IEEE International Conference on Acoustics, Speech, and Signal Processing*, *5*, 2972–2975.

Sun, J. I., Golesorkhi, R., & Seed, H. B. (1988). *Dynamic moduli and damping ratios for cohesive soils. EERC 88-15*. Berkeley, CA: University of California.

Sutanta, H., Rajabifard, A., & Bishop, I. D. (2010). Integrating spatial planning and disaster risk reduction at the local level in the context of spatially enabled government. In A. Rajabifard, J. Crompvoets, M. Kalantari, & B. Kok (Eds.), *Spatially Enabling Society: Research, emerging trends and critical assessments* (pp. 205–218). Leuven University Press.

Suto, K. (2007). Multichannel analysis of surface waves (MASW) for investigation of ground competence: an introduction in Engineering Advances in Earthworks. In *Proceedings of the Sydney Chapter 2007 Symposium* (pp. 71-81). Australian Geomechanics Society.

Tang, L., Maula, B. H., Ling, X., & Su, L. (2014). Numerical simulations of shake-table experiment for dynamic soil-pile-structure interaction in liquefiable soils. *Earthquake Engineering and Engineering Vibration, 13*(1), 171–180. doi:10.100711803-014-0221-5

Tarquis, F., & Roesset, J. M. (1992). Inelastic response spectra for narrow-banded earthquakes. *Proceedings of 10th World Conference on Earthquake Engineering,* 2, 645-650.

Tart, R. J., Oriard, L. L., & Plump, J. H. (1980). Blast Damage Criteria for a Massive Concrete Structure. In Minimizing Detrimental Construction Vibrations. ASCE.

Tary, J. B., Herrera, R. H., Han, J., & Baan, M. (2014). Spectral Estimation—what is new? what is next? *Reviews of Geophysics, 52*(4), 723–749. doi:10.1002/2014RG000461

Tewari, H. C., Dixit, M. M., Sarkar, D., & Kaila, K. L. (1991). A crustal density model across Cambay basin, India and its relationship with the Aravallis. *Tectonophysics, 194*(1-2), 123–130. doi:10.1016/0040-1951(91)90276-X

The Time Frequency Toolbox Tutorial. (n.d.). Retrieved from http://tftb.nongnu.org/tutorial.pdf

Thoenen, J. R., & Windes, S. L. (1942). Seismic Effects of Quarry Blasting. Bureau of Mines Bull., 442.

Ting, J. M., Khawaja, M., Meachum, L. R., & Rowell, J. D. (1993). An ellipse-based discrete element model for granular materials. *International Journal for Numerical and Analytical Methods in Geomechanics, 17*(9), 603–623. doi:10.1002/nag.1610170902

Tokimatsu, K., Mizuno, H., & Kakurai, M. (1996) Building damage associated with geotechnical problems. In Soils and Foundations special edition January 1996, on geotechnical aspects of the January 17th 1995 Hyogoken-Nambu earthquake (pp. 219-234). Academic Press. doi:10.3208andf.36.Special_219

Tokimatsu, K., & Yoshimi, Y. (1980). Effects of vertical drains on the bearing capacity of saturated sand during earthquakes. *Proc., International Conference on Engineering for Protection from Natural Disasters*, 643–655.

Tseng, C. (2012). *Tutorial: Time-Frequency Analysis for Seismology*. R99942044.

UBC. (1997). Uniform Building Code. *Int. Conf. Building Officials*, 2.

Uma Maheshwari, R., Boominathan, A., & Dodagoudar, G. R. (2008). Development of empirical correltation between shear wave velocity and standard penetration resistance in soils of Chennai. *Proceedings of 14th World Conference on Earthquake Engineering*.

Uma Maheswari, R., Boominathan, A., & Dodagoudar, G. R. (2010a). Use of Surface Waves in Statistical Correlations of Shear Wave Velocity and Penetration Resistance of Chennai Soils. *Geotechnical and Geological Engineering, 28,* 119-137.

Uma Maheswari, R., Boominathan, A., & Dodagoudar, G. R. (2010b). Seismic site classification and site period mapping of Chennai City using geophysical and geotechnical data. *Journal of Applied Geophysics, 72*, 152–168.

Upegui-Botero, F. M., Huerta-Lopez, C. I., Caro-Cortes, J. A., Martinez-Cruzado, J. A., & Suarez-Colche, L. E. (2012). Joint Time- Frequency Analysis of Seismic Records. *Proceedings of the 15 WCEE.*

Vaid, Y. P., Byrne, P. M., & Hughes, J. M. O. (1981). Dilation angle and liquefaction potential. *Journal of the Geotechnical Engineering Division. Proceedings of the ASCE, 107*(GT7), 1003–1008.

Vaid, Y. P., & Thomas, J. (1995). Liquefaction and post liquefaction behaviour of sand. *Journal of Geotechnical Engineering, 121*(2), 163–173. doi:10.1061/(ASCE)0733-9410(1995)121:2(163)

Villaescusa, E., Onederra, I., & Scott, C. (2004). Blast Induced Damage and Dynamic Behaviour of Hangingwalls in Bench Stoping. *J. Rock Fragmentation by Blasting. Fragblast*, *8*(1), 23–40. doi:10.1080/13855140512331389614

Ville, J. (1948). Theory and Applications of the Notion of the Analytic Signal. *Câbles et Transmissions*, *2*, 61–74.

Wagner, H. (1984). Support requirements for rockburst conditions. In *Rockbursts and seismicity in mines* (pp. 209–218). Academic Press.

Wang, J. G., Sun, W., & Anand, S. (2009). Numerical investigation on active isolation of ground shock by soft porous layers. *Journal of Sound and Vibration*, *321*(3-5), 492–509. doi:10.1016/j.jsv.2008.09.047

Wang, J., Yu, H. S., Langston, P., & Fraige, E. (2011). Particle shape effects in discrete element modelling of cohesive angular particles. *Granular Matter*, *13*(1), 1–12. doi:10.100710035-010-0217-4

Wang, L. M., Wu, Z. J., & Xia, K. (2015). Study on the Effect of Loess Sites on Seismic Ground Motion and Its Application in Seismic Design. *International Conference on Earthquake Geotechnical Engineering*.

Wang, S., Kutter, B. L., Chacko, M. J., Wilson, D. W., Boulanger, R. W., & Abghari, A. (1998). Nonlinear seismic soil-pile structure interaction. *Earthquake Spectra*, *14*(2), 377–396. doi:10.1193/1.1586006

Wang, W. (1990). *Bounding surface hypoplasticity model for granular soils and its applications*. Davis, CA: University of California.

Watts, K. S., & Charles, J. A. (1999). Settlement characteristics of landfill wastes. *Proceedings of ICE. Geotechnical Engineering*, *137*(4), 225–233. doi:10.1680/gt.1999.370408

Wei, J., & Wang, G. (2017). Discrete-element method analysis of initial fabric effects on pre- and post-liquefaction behavior of sands. *Géotechnique Letters*, *7*(2), 161–166. doi:10.1680/jgele.16.00147

Wensrich, C. M., & Katterfeld, A. (2012). Rolling friction as a technique for modelling particle shape in DEM. *Powder Technology*, *217*, 409–417. doi:10.1016/j.powtec.2011.10.057

Wiegel, L. R. (1970). *Earthquake Engineering*. Englewood Cliffs, NJ: Prentice-Hall Inc.

Wiemer, K. (1982). *Qualitative and quantitative Kriterien zur Bestimmung der Dichte von Abfällen in geordneten Deponien*. Academic Press.

Wigner, E. (1932). On the Quantum Correction for Thermodynamic Equilibrium. *Physical Review*, *40*(5), 749–759. doi:10.1103/PhysRev.40.749

Wijewickreme, D., & Atukorala, U. (2005). Ground improvement for mitigating liquefaction-induced geotechnical hazards. In *Ground Improvement — Case Histories*. UK: Elsevier Publishers.

Wolf, J. P., & Deeks, A. (2004). *Foundation vibration analysis: A strength-of-materials approach*. Oxford, UK: Elsevier.

Woodward Clyde Consultants. (1987). *Seismic stability evaluation operating industries Islandfill. A Report Prepared for Camp Dresser Mckee*. Santa Ana, CA: Woodward Clyde Consultants.

Wu, G., & Finn, W. D. L. (1997). Dynamic nonlinear analysis of pile foundations using finite element method in the time domain. *Canadian Geotechnical Journal*, *34*(1), 44–52. doi:10.1139/t96-088

Yang, C., Huang, M. S., & Cui, Y. J. (2011). Constitutive model of unsaturated structured soils under cyclic loading. In Alonso & Gens (Eds.), Unsaturated Soils (pp. 987-992). Taylor & Francis Group.

Yang, Y. B., & Hung, H. H. (1997). Parametric study of wave barriers for reduction of train-induced vibrations. *International Journal for Numerical Methods in Engineering*, *40*(20), 3729–3747. doi:10.1002/(SICI)1097-0207(19971030)40:20<3729::AID-NME236>3.0.CO;2-8

Yasuda, S., Ishihara, K., Harada, K., & Shinkawa, N. (1996). Effect of soil improvement on ground subsidence due to liquefaction. *Soils and Foundations*, 99-107.

Yazdchi, M., Khalili, N., & Valliappan, S. (1999). Dynamic soil-structure interaction analysis via coupled finite-element-boundary-element method. *Soil Dynamics and Earthquake Engineering*, *18*(7), 499–517. doi:10.1016/S0267-7261(99)00019-6

Yegian, M. K., & Catan, U. (2004). S*oil isolation for seismic protection using a smooth synthetic liner. Journal of Geotechnical and Geoenvironmental Engineering*, *130*(11), 1131–1139. doi:10.1061/(ASCE)1090-0241(2004)130:11(1131)

Yegian, M. K., & Kadakal, U. (2004). Foundation isolation for seismic protection using a smooth synthetic liner. *Journal of Geotechnical and Geoenvironmental Engineering*, *130*(11), 1121–1130. doi:10.1061/(ASCE)1090-0241(2004)130:11(1121)

Ye, L., & Tan, Y. (2011). Mitigation of building responses to DDC impacts by soft and stiff barriers. *Journal of Vibration and Control*, *17*(2), 259–277. doi:10.1177/1077546309355297

Yilmaz, S., & Bayrak, Y. (2013). Time Frequency Analysis of Georgia Earthquake. *Proceedings of the 7th Congress of Balkan Geophysical Society*. doi:10.3997/2214-4609.20131753

Youd, T. L., Idriss, I. M., Ronald, D. A., Arango, I., Castro, G., Christian, J. T., ... Stokoe, K. H. II. (2001). Liquefaction Resistance of Soils: Summary Report from the 1996 NCEER and 1998 NCEER/NSF Workshops on Evaluation of Liquefaction Resistance of Soils. *Journal of Geotechnical and Geoenvironmental Engineering*, *124*(10), 817–833. doi:10.1061/(ASCE)1090-0241(2001)127:10(817)

Yu & Vongpaisal. (1996). New blast damage criteria for underground blasting. *CIM Bulletin*, *89*(998), 139–145.

Zabihi, N. E., & Siahkoohi, H. R. (2006). Single Frequency Seismic Attribute Based on Short Time Fourier Transform, Continuous Wavelet Transform and S Transform. *Proceedings of the 6th International Conference & Exposition on Petroleum Geophysics*, 662-669.

Zekkos, D. P. (2005). Evaluation of static and dynamic properties of municipal solid waste, *University of California, Berkeley* (Ph.D. thesis). Academic Press.

Zekkos, D., Bray, J. D., & Riemer, M. F. (2008). Shear modulus and material damping of municipal solid waste based on large-scale cyclic triaxial testing. *Canadian Geotechnical Journal, 45*(1), 45–58. doi:10.1139/T07-069

Zhang, J. C., & Chang, C. (1999). On damage mechanism of micro crack zone in Rock blasting and its measurements. In *Pro. 6th Int. Symp. of Rock fragmentation by blasting*. The South African Ist. of Mining and Metal.

Zhang, R. (2008). Spectral Decomposition of Seismic Data via Wigner-Ville Distribution and Chirplet Transform. *Proceedings of the70th EAGE Conference & Exhibition.* 10.3997/2214-4609.20147874

Zhao, Y., Atlas, L. E., & Marks, R. J. (1990). The Use of Cone-Shape Kernels for Generalized Time-Frequency Representations of Nonstationary Signals. *IEEE Transactions on Acoustics, Speech, and Signal Processing, 38*(7), 1084–1091. doi:10.1109/29.57537

Zienkiewicz, O. C., Chan, A. H. C., Pastor, M., Schrefler, B. A., & Shiomi, T. (1999). *Computational geomechanics*. Chichester, UK: John Wiley & Sons.

Related References

To continue our tradition of advancing academic research, we have compiled a list of recommended IGI Global readings. These references will provide additional information and guidance to further enrich your knowledge and assist you with your own research and future publications.

Abresch, J. (2003). Geographic Information Systems Research and Data Centers. In A. Hanson & B. Levin (Eds.), *Building a Virtual Library* (pp. 52–64). Hershey, PA: IGI Global. doi:10.4018/978-1-59140-106-3.ch004

Ajiboye, O. E., & Yusuff, O. S. (2017). Foreign Land Acquisition: Food Security and Food Chains – The Nigerian Experience. In I. Management Association (Ed.), Natural Resources Management: Concepts, Methodologies, Tools, and Applications (pp. 1524-1545). Hershey, PA: IGI Global. doi:10.4018/978-1-5225-0803-8.ch072

Alapiki, H. E., & Amadi, L. A. (2018). Sustainable Food Consumption in the Neoliberal Order: Challenges and Policy Implications. In A. Obayelu (Ed.), *Food Systems Sustainability and Environmental Policies in Modern Economies* (pp. 90–123). Hershey, PA: IGI Global. doi:10.4018/978-1-5225-3631-4.ch005

Altaş, A. (2018). Geographical Information System Applications Utilized in Museums in Turkey Within the Scope of the Cultural Heritage Tourism: A Case Study of Mobile Application of Müze Asist. In S. Chaudhuri & N. Ray (Eds.), *GIS Applications in the Tourism and Hospitality Industry* (pp. 42–60). Hershey, PA: IGI Global. doi:10.4018/978-1-5225-5088-4.ch002

Andreea, I. R. (2018). Beyond Macroeconomics of Food and Nutrition Security. *International Journal of Sustainable Economies Management*, *7*(1), 13–22. doi:10.4018/IJSEM.2018010102

Anwar, J. (2017). Reproductive and Mental Health during Natural Disaster: Implications and Issues for Women in Developing Nations – A Case Example. In I. Management Association (Ed.), Gaming and Technology Addiction: Breakthroughs in Research and Practice (pp. 446-472). Hershey, PA: IGI Global. doi:10.4018/978-1-5225-0778-9.ch021

Awadh, H., Aksissou, M., Benhardouze, W., Darasi, F., & Snaiki, J. (2018). Socioeconomic Status of Artisanal Fishers in the West Part of Moroccan Mediterranean. *International Journal of Social Ecology and Sustainable Development*, *9*(1), 40–52. doi:10.4018/IJSESD.2018010104

Aye, G. C., & Haruna, R. F. (2018). Effect of Climate Change on Crop Productivity and Prices in Benue State, Nigeria: Implications for Food Security. In V. Erokhin (Ed.), *Establishing Food Security and Alternatives to International Trade in Emerging Economies* (pp. 244–268). Hershey, PA: IGI Global. doi:10.4018/978-1-5225-2733-6.ch012

Azizan, S. A., & Suki, N. M. (2017). Consumers' Intentions to Purchase Organic Food Products. In T. Esakki (Ed.), *Green Marketing and Environmental Responsibility in Modern Corporations* (pp. 68–82). Hershey, PA: IGI Global. doi:10.4018/978-1-5225-2331-4.ch005

Barakabitze, A. A., Fue, K. G., Kitindi, E. J., & Sanga, C. A. (2017). Developing a Framework for Next Generation Integrated Agro Food-Advisory Systems in Developing Countries. In I. Management Association (Ed.), Agri-Food Supply Chain Management: Breakthroughs in Research and Practice (pp. 47-67). Hershey, PA: IGI Global. doi:10.4018/978-1-5225-1629-3.ch004

Beachcroft-Shaw, H., & Ellis, D. (2018). Using Successful Cases to Promote Environmental Sustainability: A Social Marketing Approach. In I. Management Association (Ed.), Sustainable Development: Concepts, Methodologies, Tools, and Applications (pp. 936-953). Hershey, PA: IGI Global. doi:10.4018/978-1-5225-3817-2.ch042

Behnassi, M., Kahime, K., Boussaa, S., Boumezzough, A., & Messouli, M. (2017). Infectious Diseases and Climate Vulnerability in Morocco: Governance and Adaptation Options. In I. Management Association (Ed.), Public Health and Welfare: Concepts, Methodologies, Tools, and Applications (pp. 91-109). Hershey, PA: IGI Global. doi:10.4018/978-1-5225-1674-3.ch005

Bekele, F., & Bekele, I. (2017). Social and Environmental Impacts on Agricultural Development. In W. Ganpat, R. Dyer, & W. Isaac (Eds.), *Agricultural Development and Food Security in Developing Nations* (pp. 21–56). Hershey, PA: IGI Global. doi:10.4018/978-1-5225-0942-4.ch002

Benaouda, A., & García-Peñalvo, F. J. (2018). Towards an Intelligent System for the Territorial Planning: Agricultural Case. In F. García-Peñalvo (Ed.), *Global Implications of Emerging Technology Trends* (pp. 158–178). Hershey, PA: IGI Global. doi:10.4018/978-1-5225-4944-4.ch010

Bhaskar, A., Rao, G. B., & Vencatesan, J. (2017). Characterization and Management Concerns of Water Resources around Pallikaranai Marsh, South Chennai. In P. Rao & Y. Patil (Eds.), *Reconsidering the Impact of Climate Change on Global Water Supply, Use, and Management* (pp. 102–121). Hershey, PA: IGI Global. doi:10.4018/978-1-5225-1046-8.ch007

Bianchini, E., & Sicurella, S. (2010). GIS: A New Tool for Criminology and Victimology's Studies. In R. Sette (Ed.), *Cases on Technologies for Teaching Criminology and Victimology: Methodologies and Practices* (pp. 87–110). Hershey, PA: IGI Global. doi:10.4018/978-1-60566-872-7.ch006

Bogataj, D., & Drobne, D. (2017). Control of Perishable Goods in Cold Logistic Chains by Bionanosensors. In S. Joo (Ed.), *Applying Nanotechnology for Environmental Sustainability* (pp. 376–402). Hershey, PA: IGI Global. doi:10.4018/978-1-5225-0585-3.ch016

Bogataj, D., & Drobne, D. (2017). Control of Perishable Goods in Cold Logistic Chains by Bionanosensors. In I. Management Association (Ed.), Materials Science and Engineering: Concepts, Methodologies, Tools, and Applications (pp. 471-497). Hershey, PA: IGI Global. doi:10.4018/978-1-5225-1798-6.ch019

Bogueva, D., & Marinova, D. (2018). What Is More Important: Perception of Masculinity or Personal Health and the Environment? In D. Bogueva, D. Marinova, & T. Raphaely (Eds.), *Handbook of Research on Social Marketing and Its Influence on Animal Origin Food Product Consumption* (pp. 148–162). Hershey, PA: IGI Global. doi:10.4018/978-1-5225-4757-0.ch010

Bouzid, M. (2017). Waterborne Diseases and Climate Change: Impact and Implications. In M. Bouzid (Ed.), *Examining the Role of Environmental Change on Emerging Infectious Diseases and Pandemics* (pp. 89–108). Hershey, PA: IGI Global. doi:10.4018/978-1-5225-0553-2.ch004

Bowles, D. C. (2017). Climate Change-Associated Conflict and Infectious Disease. In M. Bouzid (Ed.), *Examining the Role of Environmental Change on Emerging Infectious Diseases and Pandemics* (pp. 68–88). Hershey, PA: IGI Global. doi:10.4018/978-1-5225-0553-2.ch003

Buck, J. J., & Lowry, R. K. (2017). Oceanographic Data Management: Quills and Free Text to the Digital Age and "Big Data". In P. Diviacco, A. Leadbetter, & H. Glaves (Eds.), *Oceanographic and Marine Cross-Domain Data Management for Sustainable Development* (pp. 1–22). Hershey, PA: IGI Global. doi:10.4018/978-1-5225-0700-0.ch001

Buse, C. G. (2017). Are Climate Change Adaptation Policies a Game Changer?: A Case Study of Perspectives from Public Health Officials in Ontario, Canada. In M. Bouzid (Ed.), *Examining the Role of Environmental Change on Emerging Infectious Diseases and Pandemics* (pp. 230–257). Hershey, PA: IGI Global. doi:10.4018/978-1-5225-0553-2.ch010

Calderon, F. A., Giolo, E. G., Frau, C. D., Rengel, M. G., Rodriguez, H., Tornello, M., ... Gallucci, R. (2018). Seismic Microzonation and Site Effects Detection Through Microtremors Measures: A Review. In N. Ceryan (Ed.), *Handbook of Research on Trends and Digital Advances in Engineering Geology* (pp. 326–349). Hershey, PA: IGI Global. doi:10.4018/978-1-5225-2709-1.ch009

Carfì, D., Donato, A., & Panuccio, D. (2018). A Game Theory Coopetitive Perspective for Sustainability of Global Feeding: Agreements Among Vegan and Non-Vegan Food Firms. In I. Management Association (Ed.), Game Theory: Breakthroughs in Research and Practice (pp. 71-104). Hershey, PA: IGI Global. doi:10.4018/978-1-5225-2594-3.ch004

Carr, D. A., & Carr, T. R. (2007). Geographic Information System Applications in the Public Sector. In G. Garson (Ed.), *Modern Public Information Technology Systems: Issues and Challenges* (pp. 293–311). Hershey, PA: IGI Global. doi:10.4018/978-1-59904-051-6.ch015

Castagnolo, V. (2018). Analyzing, Classifying, Safeguarding: Drawing for the Borgo Murattiano Neighbourhood of Bari. In G. Carlone, N. Martinelli, & F. Rotondo (Eds.), *Designing Grid Cities for Optimized Urban Development and Planning* (pp. 93–108). Hershey, PA: IGI Global. doi:10.4018/978-1-5225-3613-0.ch006

Chekima, B. (2018). The Dilemma of Purchase Intention: A Conceptual Framework for Understanding Actual Consumption of Organic Food. *International Journal of Sustainable Economies Management*, *7*(2), 1–13. doi:10.4018/IJSEM.2018040101

Chen, Y. (2017). Sustainable Supply Chains and International Soft Landings: A Case of Wetland Entrepreneurship. In B. Christiansen & F. Kasarcı (Eds.), *Corporate Espionage, Geopolitics, and Diplomacy Issues in International Business* (pp. 232–247). Hershey, PA: IGI Global. doi:10.4018/978-1-5225-1031-4.ch013

Çıtak, L., Akel, V., & Ersoy, E. (2018). Investors' Reactions to the Announcement of New Constituents of BIST Sustainability Index: An Analysis by Event Study and Mean-Median Tests. In M. Risso & S. Testarmata (Eds.), *Value Sharing for Sustainable and Inclusive Development* (pp. 270–289). Hershey, PA: IGI Global. doi:10.4018/978-1-5225-3147-0.ch012

Crossland, M. D. (2009). Geographic Information Systems as Decision Tools. In M. Khosrow-Pour, D.B.A. (Ed.), Encyclopedia of Information Science and Technology, Second Edition (pp. 1630-1633). Hershey, PA: IGI Global. doi:10.4018/978-1-60566-026-4.ch257

Crossland, M. D., Herschel, R. T., Perkins, W. C., & Scudder, J. N. (2002). Geographic Information Systems: How Cognitive Style Impacts Decision-Making Effectiveness. In M. Mahmood (Ed.), *Advanced Topics in End User Computing* (Vol. 1, pp. 285–301). Hershey, PA: IGI Global. doi:10.4018/978-1-930708-42-6.ch017

D'Aleo, V., D'Aleo, F., & Bonanno, R. (2018). New Food Industries Toward a New Level of Sustainable Supply: Success Stories, Business Models, and Strategies. In V. Erokhin (Ed.), *Establishing Food Security and Alternatives to International Trade in Emerging Economies* (pp. 74–97). Hershey, PA: IGI Global. doi:10.4018/978-1-5225-2733-6.ch004

Dagevos, H., & Reinders, M. J. (2018). Flexitarianism and Social Marketing: Reflections on Eating Meat in Moderation. In D. Bogueva, D. Marinova, & T. Raphaely (Eds.), *Handbook of Research on Social Marketing and Its Influence on Animal Origin Food Product Consumption* (pp. 105–120). Hershey, PA: IGI Global. doi:10.4018/978-1-5225-4757-0.ch007

Deenapanray, P. N., & Ramma, I. (2017). Adaptations to Climate Change and Climate Variability in the Agriculture Sector in Mauritius: Lessons from a Technical Needs Assessment. In I. Management Association (Ed.), Natural Resources Management: Concepts, Methodologies, Tools, and Applications (pp. 655-680). Hershey, PA: IGI Global. doi:10.4018/978-1-5225-0803-8.ch030

Deenapanray, P. N., & Ramma, I. (2017). Adaptations to Climate Change and Climate Variability in the Agriculture Sector in Mauritius: Lessons from a Technical Needs Assessment. In I. Management Association (Ed.), Natural Resources Management: Concepts, Methodologies, Tools, and Applications (pp. 655-680). Hershey, PA: IGI Global. doi:10.4018/978-1-5225-0803-8.ch030

Deshpande, S., Basu, S. K., Li, X., & Chen, X. (2017). Smart, Innovative and Intelligent Technologies Used in Drug Designing. In I. Management Association (Ed.), Pharmaceutical Sciences: Breakthroughs in Research and Practice (pp. 1175-1191). Hershey, PA: IGI Global. doi:10.4018/978-1-5225-1762-7.ch045

Dlamini, P. N. (2017). Use of Information and Communication Technologies Tools to Capture, Store, and Disseminate Indigenous Knowledge: A Literature Review. In P. Ngulube (Ed.), *Handbook of Research on Theoretical Perspectives on Indigenous Knowledge Systems in Developing Countries* (pp. 225–247). Hershey, PA: IGI Global. doi:10.4018/978-1-5225-0833-5.ch010

Dolejsova, M., & Kera, D. (2017). The Fermentation GutHub Project and the Internet of Microbes. In S. Konomi & G. Roussos (Eds.), *Enriching Urban Spaces with Ambient Computing, the Internet of Things, and Smart City Design* (pp. 25–46). Hershey, PA: IGI Global. doi:10.4018/978-1-5225-0827-4.ch002

Dolunay, O. (2018). A Paradigm Shift: Empowering Farmers to Eliminate the Waste in the Form of Fresh Water and Energy Through the Implementation of 4R+T. In I. Management Association (Ed.), Sustainable Development: Concepts, Methodologies, Tools, and Applications (pp. 882-892). Hershey, PA: IGI Global. doi:10.4018/978-1-5225-3817-2.ch039

Dube, P., Heijman, W. J., Ihle, R., & Ochieng, J. (2018). The Potential of Traditional Leafy Vegetables for Improving Food Security in Africa. In V. Erokhin (Ed.), *Establishing Food Security and Alternatives to International Trade in Emerging Economies* (pp. 220–243). Hershey, PA: IGI Global. doi:10.4018/978-1-5225-2733-6.ch011

Duruji, M. M., & Urenma, D. F. (2017). The Environmentalism and Politics of Climate Change: A Study of the Process of Global Convergence through UNFCCC Conferences. In I. Management Association (Ed.), *Natural Resources Management: Concepts, Methodologies, Tools, and Applications* (pp. 77-108). Hershey, PA: IGI Global. 10.4018/978-1-5225-0803-8.ch004

Dutta, U. (2017). Agro-Geoinformatics, Potato Cultivation, and Climate Change. In S. Londhe (Ed.), *Sustainable Potato Production and the Impact of Climate Change* (pp. 247–271). Hershey, PA: IGI Global. doi:10.4018/978-1-5225-1715-3.ch012

Edirisinghe, R., Stranieri, A., & Wickramasinghe, N. (2017). A Taxonomy for mHealth. In N. Wickramasinghe (Ed.), *Handbook of Research on Healthcare Administration and Management* (pp. 596–615). Hershey, PA: IGI Global. doi:10.4018/978-1-5225-0920-2.ch036

Ekpeni, N. M., & Ayeni, A. O. (2018). Global Natural Hazard and Disaster Vulnerability Management. In A. Eneanya (Ed.), *Handbook of Research on Environmental Policies for Emergency Management and Public Safety* (pp. 83–104). Hershey, PA: IGI Global. doi:10.4018/978-1-5225-3194-4.ch005

Ene, C., Voica, M. C., & Panait, M. (2017). Green Investments and Food Security: Opportunities and Future Directions in the Context of Sustainable Development. In M. Mieila (Ed.), *Measuring Sustainable Development and Green Investments in Contemporary Economies* (pp. 163–200). Hershey, PA: IGI Global. doi:10.4018/978-1-5225-2081-8.ch007

Escamilla, I., Ruíz, M. T., Ibarra, M. M., Soto, V. L., Quintero, R., & Guzmán, G. (2018). Geocoding Tweets Based on Semantic Web and Ontologies. In M. Lytras, N. Aljohani, E. Damiani, & K. Chui (Eds.), *Innovations, Developments, and Applications of Semantic Web and Information Systems* (pp. 372–392). Hershey, PA: IGI Global. doi:10.4018/978-1-5225-5042-6.ch014

Escribano, A. J. (2018). Marketing Strategies for Trendy Animal Products: Sustainability as a Core. In F. Quoquab, R. Thurasamy, & J. Mohammad (Eds.), *Driving Green Consumerism Through Strategic Sustainability Marketing* (pp. 169–203). Hershey, PA: IGI Global. doi:10.4018/978-1-5225-2912-5.ch010

Eudoxie, G., & Roopnarine, R. (2017). Climate Change Adaptation and Disaster Risk Management in the Caribbean. In W. Ganpat & W. Isaac (Eds.), *Environmental Sustainability and Climate Change Adaptation Strategies* (pp. 97–125). Hershey, PA: IGI Global. doi:10.4018/978-1-5225-1607-1.ch004

Farmer, L. S. (2017). Data Analytics for Strategic Management: Getting the Right Data. In V. Wang (Ed.), *Encyclopedia of Strategic Leadership and Management* (pp. 810–822). Hershey, PA: IGI Global. doi:10.4018/978-1-5225-1049-9.ch056

Fattal, L. R. (2017). Catastrophe: An Uncanny Catalyst for Creativity. In R. Shin (Ed.), *Convergence of Contemporary Art, Visual Culture, and Global Civic Engagement* (pp. 244–262). Hershey, PA: IGI Global. doi:10.4018/978-1-5225-1665-1.ch014

Forti, I. (2017). A Cross Reading of Landscape through Digital Landscape Models: The Case of Southern Garda. In A. Ippolito (Ed.), *Handbook of Research on Emerging Technologies for Architectural and Archaeological Heritage* (pp. 532–561). Hershey, PA: IGI Global. doi:10.4018/978-1-5225-0675-1.ch018

Gharbi, A., De Runz, C., & Akdag, H. (2017). Urban Development Modelling: A Survey. In S. Faiz & K. Mahmoudi (Eds.), *Handbook of Research on Geographic Information Systems Applications and Advancements* (pp. 96–124). Hershey, PA: IGI Global. doi:10.4018/978-1-5225-0937-0.ch004

Ghosh, I., & Ghoshal, I. (2018). Implications of Trade Liberalization for Food Security Under the ASEAN-India Strategic Partnership: A Gravity Model Approach. In V. Erokhin (Ed.), *Establishing Food Security and Alternatives to International Trade in Emerging Economies* (pp. 98–118). Hershey, PA: IGI Global. doi:10.4018/978-1-5225-2733-6.ch005

Glaves, H. M. (2017). Developing a Common Global Framework for Marine Data Management. In P. Diviacco, A. Leadbetter, & H. Glaves (Eds.), *Oceanographic and Marine Cross-Domain Data Management for Sustainable Development* (pp. 47–68). Hershey, PA: IGI Global. doi:10.4018/978-1-5225-0700-0.ch003

Godulla, A., & Wolf, C. (2018). Future of Food: Transmedia Strategies of National Geographic. In R. Gambarato & G. Alzamora (Eds.), *Exploring Transmedia Journalism in the Digital Age* (pp. 162–182). Hershey, PA: IGI Global. doi:10.4018/978-1-5225-3781-6.ch010

Gomes, P. P. (2018). Food and Environment: A Review on the Sustainability of Six Different Dietary Patterns. In A. Obayelu (Ed.), *Food Systems Sustainability and Environmental Policies in Modern Economies* (pp. 15–31). Hershey, PA: IGI Global. doi:10.4018/978-1-5225-3631-4.ch002

Gonzalez-Feliu, J. (2018). Sustainability Evaluation of Green Urban Logistics Systems: Literature Overview and Proposed Framework. In A. Paul, D. Bhattacharyya, & S. Anand (Eds.), *Green Initiatives for Business Sustainability and Value Creation* (pp. 103–134). Hershey, PA: IGI Global. doi:10.4018/978-1-5225-2662-9.ch005

Goodland, R. (2017). A Fresh Look at Livestock Greenhouse Gas Emissions and Mitigation. In I. Management Association (Ed.), Natural Resources Management: Concepts, Methodologies, Tools, and Applications (pp. 124-139). Hershey, PA: IGI Global. doi:10.4018/978-1-5225-0803-8.ch006

Goundar, S., & Appana, S. (2018). Mainstreaming Development Policies for Climate Change in Fiji: A Policy Gap Analysis and the Role of ICTs. In I. Management Association (Ed.), Sustainable Development: Concepts, Methodologies, Tools, and Applications (pp. 402-432). Hershey, PA: IGI Global. doi:10.4018/978-1-5225-3817-2.ch020

Granell-Canut, C., & Aguilar-Moreno, E. (2018). Geospatial Influence in Science Mapping. In M. Khosrow-Pour, D.B.A. (Ed.), Encyclopedia of Information Science and Technology, Fourth Edition (pp. 3473-3483). Hershey, PA: IGI Global. doi:10.4018/978-1-5225-2255-3.ch302

Grigelis, A., Blažauskas, N., Gelumbauskaitė, L. Ž., Gulbinskas, S., Suzdalev, S., & Ferrarin, C. (2017). Marine Environment Data Management Related to the Human Activity in the South-Eastern Baltic Sea (The Lithuanian Sector). In P. Diviacco, A. Leadbetter, & H. Glaves (Eds.), *Oceanographic and Marine Cross-Domain Data Management for Sustainable Development* (pp. 282–302). Hershey, PA: IGI Global. doi:10.4018/978-1-5225-0700-0.ch012

Guma, I. P., Rwashana, A. S., & Oyo, B. (2018). Food Security Policy Analysis Using System Dynamics: The Case of Uganda. *International Journal of Information Technologies and Systems Approach*, *11*(1), 72–90. doi:10.4018/IJITSA.2018010104

Guma, I. P., Rwashana, A. S., & Oyo, B. (2018). Food Security Indicators for Subsistence Farmers Sustainability: A System Dynamics Approach. *International Journal of System Dynamics Applications*, *7*(1), 45–64. doi:10.4018/IJSDA.2018010103

Gupta, P., & Goyal, S. (2017). Wildlife Habitat Evaluation. In A. Santra & S. Mitra (Eds.), *Remote Sensing Techniques and GIS Applications in Earth and Environmental Studies* (pp. 258–264). Hershey, PA: IGI Global. doi:10.4018/978-1-5225-1814-3.ch013

Hanson, T., & Hildebrand, E. (2018). GPS Travel Diaries in Rural Transportation Research: A Focus on Older Drivers. In I. Management Association (Ed.), Intelligent Transportation and Planning: Breakthroughs in Research and Practice (pp. 609-625). Hershey, PA: IGI Global. doi:10.4018/978-1-5225-5210-9.ch027

Hartman, M. B. (2017). Research-Based Climate Change Public Education Programs. In I. Management Association (Ed.), Natural Resources Management: Concepts, Methodologies, Tools, and Applications (pp. 992-1003). Hershey, PA: IGI Global. doi:10.4018/978-1-5225-0803-8.ch046

Hashim, N. (2017). Zanzibari Seaweed: Global Climate Change and the Promise of Adaptation. In I. Management Association (Ed.), Natural Resources Management: Concepts, Methodologies, Tools, and Applications (pp. 365-391). Hershey, PA: IGI Global. doi:10.4018/978-1-5225-0803-8.ch019

Herrera, J. E., Argüello, L. V., Gonzalez-Feliu, J., & Jaimes, W. A. (2017). Decision Support System Design Requirements, Information Management, and Urban Logistics Efficiency: Case Study of Bogotá, Colombia. In G. Jamil, A. Soares, & C. Pessoa (Eds.), *Handbook of Research on Information Management for Effective Logistics and Supply Chains* (pp. 223–238). Hershey, PA: IGI Global. doi:10.4018/978-1-5225-0973-8.ch012

Huizinga, T., Ayanso, A., Smoor, M., & Wronski, T. (2017). Exploring Insurance and Natural Disaster Tweets Using Text Analytics. *International Journal of Business Analytics*, *4*(1), 1–17. doi:10.4018/IJBAN.2017010101

Hung, K., Kalantari, M., & Rajabifard, A. (2017). An Integrated Method for Assessing the Text Content Quality of Volunteered Geographic Information in Disaster Management. *International Journal of Information Systems for Crisis Response and Management*, *9*(2), 1–17. doi:10.4018/IJISCRAM.2017040101

Husnain, A., & Avdic, A. (2018). Identifying the Contemporary Status of E-Service Sustainability Research. In I. Management Association (Ed.), Sustainable Development: Concepts, Methodologies, Tools, and Applications (pp. 467-485). Hershey, PA: IGI Global. doi:10.4018/978-1-5225-3817-2.ch022

Iarossi, M. P., & Ferro, L. (2017). "The Past is Never Dead. It's Not Even Past": Virtual Archaeological Promenade. In A. Ippolito & M. Cigola (Eds.), *Handbook of Research on Emerging Technologies for Digital Preservation and Information Modeling* (pp. 228–255). Hershey, PA: IGI Global. doi:10.4018/978-1-5225-0680-5.ch010

Ignjatijević, S., & Cvijanović, D. (2018). Analysis of Serbian Honey Production and Exports. In *Exploring the Global Competitiveness of Agri-Food Sectors and Serbia's Dominant Presence: Emerging Research and Opportunities* (pp. 109–139). Hershey, PA: IGI Global. doi:10.4018/978-1-5225-2762-6.ch005

Jana, S. K., & Karmakar, A. K. (2017). Globalization, Governance, and Food Security: The Case of BRICS. In I. Management Association (Ed.), Natural Resources Management: Concepts, Methodologies, Tools, and Applications (pp. 692-712). Hershey, PA: IGI Global. doi:10.4018/978-1-5225-0803-8.ch032

Jana, S. K., & Karmakar, A. K. (2017). Food Security in Asia: Is There Convergence? In I. Management Association (Ed.), Natural Resources Management: Concepts, Methodologies, Tools, and Applications (pp. 109-123). Hershey, PA: IGI Global. doi:10.4018/978-1-5225-0803-8.ch005

John, J., & Kumar, S. (2018). A Locational Decision Making Framework for Shipbreaking Under Multiple Criteria. In I. Management Association (Ed.), Operations and Service Management: Concepts, Methodologies, Tools, and Applications (pp. 504-527). Hershey, PA: IGI Global. doi:10.4018/978-1-5225-3909-4.ch024

John, J., & Srivastava, R. K. (2018). Decision Insights for Shipbreaking using Environmental Impact Assessment: Review and Perspectives. *International Journal of Strategic Decision Sciences*, *9*(1), 45–62. doi:10.4018/IJSDS.2018010104

Joshi, Y., & Rahman, Z. (2018). Determinants of Sustainable Consumption Behaviour: Review and Conceptual Framework. In A. Paul, D. Bhattacharyya, & S. Anand (Eds.), *Green Initiatives for Business Sustainability and Value Creation* (pp. 239–262). Hershey, PA: IGI Global. doi:10.4018/978-1-5225-2662-9.ch011

Juma, D. W., Reuben, M., Wang, H., & Li, F. (2018). Adaptive Coevolution: Realigning the Water Governance Regime to the Changing Climate. In I. Management Association (Ed.), Hydrology and Water Resource Management: Breakthroughs in Research and Practice (pp. 346-357). Hershey, PA: IGI Global. doi:10.4018/978-1-5225-3427-3.ch014

K., S., & Tripathy, B. K. (2018). Neighborhood Rough-Sets-Based Spatial Data Analytics. In M. Khosrow-Pour, D.B.A. (Ed.), *Encyclopedia of Information Science and Technology, Fourth Edition* (pp. 1835-1844). Hershey, PA: IGI Global. doi:10.4018/978-1-5225-2255-3.ch160

Kabir, F. (2018). Towards a More Gender-Inclusive Climate Change Policy. In N. Mahtab, T. Haque, I. Khan, M. Islam, & I. Wahid (Eds.), *Handbook of Research on Women's Issues and Rights in the Developing World* (pp. 354–369). Hershey, PA: IGI Global. doi:10.4018/978-1-5225-3018-3.ch022

Kabir, F. (2018). Towards a More Gender-Inclusive Climate Change Policy. In I. Management Association (Ed.), Climate Change and Environmental Concerns: Breakthroughs in Research and Practice (pp. 525-540). Hershey, PA: IGI Global. doi:10.4018/978-1-5225-5487-5.ch027

Kanyamuka, J. S., Jumbe, C. B., & Ricker-Gilbert, J. (2018). Making Agricultural Input Subsidies More Effective and Profitable in Africa: The Role of Complementary Interventions. In A. Obayelu (Ed.), *Food Systems Sustainability and Environmental Policies in Modern Economies* (pp. 172–187). Hershey, PA: IGI Global. doi:10.4018/978-1-5225-3631-4.ch008

Karimi, H., & Gholamrezafahimi, F. (2017). Study of Integrated Coastal Zone Management and Its Environmental Effects: A Case of Iran. In R. Singh, A. Singh, & V. Srivastava (Eds.), *Environmental Issues Surrounding Human Overpopulation* (pp. 64–88). Hershey, PA: IGI Global. doi:10.4018/978-1-5225-1683-5.ch004

Kaya, I. R., Hutabarat, J., & Bambang, A. N. (2018). "Sasi": A New Path to Sustain Seaweed Farming From Up-Stream to Down-Stream in Kotania Bay, Molucass. *International Journal of Social Ecology and Sustainable Development*, *9*(2), 28–36. doi:10.4018/IJSESD.2018040103

Khader, V. (2018). Technologies for Food, Health, Livelihood, and Nutrition Security. In I. Management Association (Ed.), Food Science and Nutrition: Breakthroughs in Research and Practice (pp. 94-112). Hershey, PA: IGI Global. doi:10.4018/978-1-5225-5207-9.ch005

Kocadağlı, A. Y. (2017). The Temporal and Spatial Development of Organic Agriculture in Turkey. In W. Ganpat, R. Dyer, & W. Isaac (Eds.), *Agricultural Development and Food Security in Developing Nations* (pp. 130–156). Hershey, PA: IGI Global. doi:10.4018/978-1-5225-0942-4.ch006

Koundouri, P., Giannouli, A., & Souliotis, I. (2017). An Integrated Approach for Sustainable Environmental and Socio-Economic Development Using Offshore Infrastructure. In I. Management Association (Ed.), Renewable and Alternative Energy: Concepts, Methodologies, Tools, and Applications (pp. 1581-1601). Hershey, PA: IGI Global. doi:10.4018/978-1-5225-1671-2.ch056

Kumar, A., & Dash, M. K. (2017). Sustainability and Future Generation Infrastructure on Digital Platform: A Study of Generation Y. In N. Ray (Ed.), *Business Infrastructure for Sustainability in Developing Economies* (pp. 124–142). Hershey, PA: IGI Global. doi:10.4018/978-1-5225-2041-2.ch007

Kumar, A., Mukherjee, A. B., & Krishna, A. P. (2017). Application of Conventional Data Mining Techniques and Web Mining to Aid Disaster Management. In A. Kumar (Ed.), *Web Usage Mining Techniques and Applications Across Industries* (pp. 138–167). Hershey, PA: IGI Global. doi:10.4018/978-1-5225-0613-3.ch006

Kumar, C. P. (2017). Impact of Climate Change on Groundwater Resources. In I. Management Association (Ed.), Natural Resources Management: Concepts, Methodologies, Tools, and Applications (pp. 1094-1120). Hershey, PA: IGI Global. doi:10.4018/978-1-5225-0803-8.ch052

Kumari, S., & Patil, Y. (2017). Achieving Climate Smart Agriculture with a Sustainable Use of Water: A Conceptual Framework for Sustaining the Use of Water for Agriculture in the Era of Climate Change. In P. Rao & Y. Patil (Eds.), *Reconsidering the Impact of Climate Change on Global Water Supply, Use, and Management* (pp. 122–143). Hershey, PA: IGI Global. doi:10.4018/978-1-5225-1046-8.ch008

Kumari, S., & Patil, Y. (2018). Achieving Climate Smart Agriculture With a Sustainable Use of Water: A Conceptual Framework for Sustaining the Use of Water for Agriculture in the Era of Climate Change. In I. Management Association (Ed.), Climate Change and Environmental Concerns: Breakthroughs in Research and Practice (pp. 111-133). Hershey, PA: IGI Global. doi:10.4018/978-1-5225-5487-5.ch006

Kursah, M. B. (2017). Least-Cost Pipeline using Geographic Information System: The Limit to Technicalities. *International Journal of Applied Geospatial Research*, *8*(3), 1–15. doi:10.4018/ijagr.2017070101

Lahiri, S., Ghosh, D., & Bhakta, J. N. (2017). Role of Microbes in Eco-Remediation of Perturbed Aquatic Ecosystem. In J. Bhakta (Ed.), *Handbook of Research on Inventive Bioremediation Techniques* (pp. 70–107). Hershey, PA: IGI Global. doi:10.4018/978-1-5225-2325-3.ch004

Lallo, C. H., Smalling, S., Facey, A., & Hughes, M. (2018). The Impact of Climate Change on Small Ruminant Performance in Caribbean Communities. In I. Management Association (Ed.), Climate Change and Environmental Concerns: Breakthroughs in Research and Practice (pp. 193-218). Hershey, PA: IGI Global. doi:10.4018/978-1-5225-5487-5.ch010

Laurini, R. (2017). Nature of Geographic Knowledge Bases. In S. Faiz & K. Mahmoudi (Eds.), *Handbook of Research on Geographic Information Systems Applications and Advancements* (pp. 29–60). Hershey, PA: IGI Global. doi:10.4018/978-1-5225-0937-0.ch002

Lawrence, J., Simpson, L., & Piggott, A. (2017). Protected Agriculture: A Climate Change Adaptation for Food and Nutrition Security. In I. Management Association (Ed.), Natural Resources Management: Concepts, Methodologies, Tools, and Applications (pp. 140-158). Hershey, PA: IGI Global. doi:10.4018/978-1-5225-0803-8.ch007

Leadbetter, A., Cheatham, M., Shepherd, A., & Thomas, R. (2017). Linked Ocean Data 2.0. In P. Diviacco, A. Leadbetter, & H. Glaves (Eds.), *Oceanographic and Marine Cross-Domain Data Management for Sustainable Development* (pp. 69–99). Hershey, PA: IGI Global. doi:10.4018/978-1-5225-0700-0.ch004

Leipnik, M. R., & Mehta, S. S. (2005). Geographic Information Systems (GIS) in E-Marketing. In I. Clarke III & T. Flaherty (Eds.), *Advances in Electronic Marketing* (pp. 193–210). Hershey, PA: IGI Global. doi:10.4018/978-1-59140-321-0.ch011

Lucas, M. R., Rego, C., Vieira, C., & Vieira, I. (2017). Proximity and Cooperation for Innovative Regional Development: The Case of the Science and Technology Park of Alentejo. In L. Carvalho (Ed.), *Handbook of Research on Entrepreneurial Development and Innovation Within Smart Cities* (pp. 199–228). Hershey, PA: IGI Global. doi:10.4018/978-1-5225-1978-2.ch010

Ma, X., Beaulieu, S. E., Fu, L., Fox, P., Di Stefano, M., & West, P. (2017). Documenting Provenance for Reproducible Marine Ecosystem Assessment in Open Science. In P. Diviacco, A. Leadbetter, & H. Glaves (Eds.), *Oceanographic and Marine Cross-Domain Data Management for Sustainable Development* (pp. 100–126). Hershey, PA: IGI Global. doi:10.4018/978-1-5225-0700-0.ch005

Mabe, L. K., & Oladele, O. I. (2017). Application of Information Communication Technologies for Agricultural Development through Extension Services: A Review. In T. Tossy (Ed.), *Information Technology Integration for Socio-Economic Development* (pp. 52–101). Hershey, PA: IGI Global. doi:10.4018/978-1-5225-0539-6.ch003

Malomo, B. I. (2018). A Review of Psychological Resilience as a Response to Natural Hazards in Nigeria. In A. Eneanya (Ed.), *Handbook of Research on Environmental Policies for Emergency Management and Public Safety* (pp. 147–165). Hershey, PA: IGI Global. doi:10.4018/978-1-5225-3194-4.ch008

Manchiraju, S. (2018). Predicting Behavioral Intentions Toward Sustainable Fashion Consumption: A Comparison of Attitude-Behavior and Value-Behavior Consistency Models. In I. Management Association (Ed.), Fashion and Textiles: Breakthroughs in Research and Practice (pp. 1-21). Hershey, PA: IGI Global. doi:10.4018/978-1-5225-3432-7.ch001

Manzella, G. M., Bartolini, R., Bustaffa, F., D'Angelo, P., De Mattei, M., Frontini, F., ... Spada, A. (2017). Semantic Search Engine for Data Management and Sustainable Development: Marine Planning Service Platform. In P. Diviacco, A. Leadbetter, & H. Glaves (Eds.), *Oceanographic and Marine Cross-Domain Data Management for Sustainable Development* (pp. 127–154). Hershey, PA: IGI Global. doi:10.4018/978-1-5225-0700-0.ch006

Mbonigaba, J. (2018). Comparing the Effects of Unsustainable Production and Consumption of Food on Health and Policy Across Developed and Less Developed Countries. In A. Obayelu (Ed.), *Food Systems Sustainability and Environmental Policies in Modern Economies* (pp. 124–158). Hershey, PA: IGI Global. doi:10.4018/978-1-5225-3631-4.ch006

McKeown, A. E. (2017). Nurses, Healthcare, and Environmental Pollution and Solutions: Breaking the Cycle of Harm. In I. Management Association (Ed.), Natural Resources Management: Concepts, Methodologies, Tools, and Applications (pp. 392-415). Hershey, PA: IGI Global. doi:10.4018/978-1-5225-0803-8.ch020

Mennecke, B. E., & West, L. A. Jr. (2001). Geographic Information Systems in Developing Countries: Issues in Data Collection, Implementation and Management. *Journal of Global Information Management*, *9*(4), 44–54. doi:10.4018/jgim.2001100103

Mili, B., Barua, A., & Katyaini, S. (2017). Climate Change and Adaptation through the Lens of Capability Approach: A Case Study from Darjeeling, Eastern Himalaya. In I. Management Association (Ed.), Natural Resources Management: Concepts, Methodologies, Tools, and Applications (pp. 1351-1365). Hershey, PA: IGI Global. doi:10.4018/978-1-5225-0803-8.ch064

Mir, S. A., Shah, M. A., Mir, M. M., & Iqbal, U. (2017). New Horizons of Nanotechnology in Agriculture and Food Processing Industry. In B. Nayak, A. Nanda, & M. Bhat (Eds.), *Integrating Biologically-Inspired Nanotechnology into Medical Practice* (pp. 230–258). Hershey, PA: IGI Global. doi:10.4018/978-1-5225-0610-2.ch009

Mujere, N., & Moyce, W. (2017). Climate Change Impacts on Surface Water Quality. In W. Ganpat & W. Isaac (Eds.), *Environmental Sustainability and Climate Change Adaptation Strategies* (pp. 322–340). Hershey, PA: IGI Global. doi:10.4018/978-1-5225-1607-1.ch012

Mukherjee, A. B., Krishna, A. P., & Patel, N. (2018). Geospatial Technology for Urban Sciences. In *Geospatial Technologies in Urban System Development: Emerging Research and Opportunities* (pp. 99–120). Hershey, PA: IGI Global. doi:10.4018/978-1-5225-3683-3.ch005

Nagarajan, S. K., & Sangaiah, A. K. (2017). Vegetation Index: Ideas, Methods, Influences, and Trends. In N. Kumar, A. Sangaiah, M. Arun, & S. Anand (Eds.), *Advanced Image Processing Techniques and Applications* (pp. 347–386). Hershey, PA: IGI Global. doi:10.4018/978-1-5225-2053-5.ch016

Naraine, L., & Meehan, K. (2017). Strengthening Food Security with Sustainable Practices by Smallholder Farmers in Lesser Developed Economies. In W. Ganpat, R. Dyer, & W. Isaac (Eds.), *Agricultural Development and Food Security in Developing Nations* (pp. 57–81). Hershey, PA: IGI Global. doi:10.4018/978-1-5225-0942-4.ch003

Naraine, L., & Meehan, K. (2017). Strengthening Food Security with Sustainable Practices by Smallholder Farmers in Lesser Developed Economies. In W. Ganpat, R. Dyer, & W. Isaac (Eds.), *Agricultural Development and Food Security in Developing Nations* (pp. 57–81). Hershey, PA: IGI Global. doi:10.4018/978-1-5225-0942-4.ch003

Naraine, L., & Meehan, K. (2017). Strengthening Food Security with Sustainable Practices by Smallholder Farmers in Lesser Developed Economies. In W. Ganpat, R. Dyer, & W. Isaac (Eds.), *Agricultural Development and Food Security in Developing Nations* (pp. 57–81). Hershey, PA: IGI Global. doi:10.4018/978-1-5225-0942-4.ch003

Neufeld, D. J., & Griffith, S. (2006). Isobord's Geographic Information System (GIS) Solution. In M. Khosrow-Pour, D.B.A. (Ed.), Cases on Database Technologies and Applications (pp. 64-80). Hershey, PA: IGI Global. doi:10.4018/978-1-59904-399-9.ch004

Nikolaou, K., Tsakiridou, E., Anastasiadis, F., & Mattas, K. (2017). Exploring Alternative Distribution Channels of Agricultural Products. *International Journal of Food and Beverage Manufacturing and Business Models*, *2*(2), 36–66. doi:10.4018/IJFBMBM.2017070103

Nishat, K. J., & Rahman, M. S. (2018). Disaster, Vulnerability, and Violence Against Women: Global Findings and a Research Agenda for Bangladesh. In N. Mahtab, T. Haque, I. Khan, M. Islam, & I. Wahid (Eds.), *Handbook of Research on Women's Issues and Rights in the Developing World* (pp. 235–250). Hershey, PA: IGI Global. doi:10.4018/978-1-5225-3018-3.ch014

O'Hara, S., Jones, D., & Trobman, H. B. (2018). Building an Urban Food System Through UDC Food Hubs. In A. Burtin, J. Fleming, & P. Hampton-Garland (Eds.), *Changing Urban Landscapes Through Public Higher Education* (pp. 116–143). Hershey, PA: IGI Global. doi:10.4018/978-1-5225-3454-9.ch006

Obayelu, A. E. (2018). Integrating Environment, Food Systems, and Sustainability in Feeding the Growing Population in Developing Countries. In A. Obayelu (Ed.), *Food Systems Sustainability and Environmental Policies in Modern Economies* (pp. 1–14). Hershey, PA: IGI Global. doi:10.4018/978-1-5225-3631-4.ch001

Othman, R., Nath, N., & Laswad, F. (2018). Environmental Reporting and Accounting: Sustainability Hybridisation. In G. Azevedo, J. da Silva Oliveira, R. Marques, & A. Ferreira (Eds.), *Handbook of Research on Modernization and Accountability in Public Sector Management* (pp. 130–158). Hershey, PA: IGI Global. doi:10.4018/978-1-5225-3731-1.ch007

Padigala, B. S. (2018). Traditional Water Management System for Climate Change Adaptation in Mountain Ecosystems. In I. Management Association (Ed.), Climate Change and Environmental Concerns: Breakthroughs in Research and Practice (pp. 630-655). Hershey, PA: IGI Global. doi:10.4018/978-1-5225-5487-5.ch033

Panda, C. K. (2018). Mobile Phone Usage in Agricultural Extension in India: The Current and Future Perspective. In F. Mtenzi, G. Oreku, D. Lupiana, & J. Yonazi (Eds.), *Mobile Technologies and Socio-Economic Development in Emerging Nations* (pp. 1–21). Hershey, PA: IGI Global. doi:10.4018/978-1-5225-4029-8.ch001

Pandian, S. L., Yarrakula, K., & Chaudhury, P. (2018). GIS-Based Decision Support System for Village Level: A Case Study in Andhra Pradesh. In S. Chaudhuri & N. Ray (Eds.), *GIS Applications in the Tourism and Hospitality Industry* (pp. 275–295). Hershey, PA: IGI Global. doi:10.4018/978-1-5225-5088-4.ch012

Quaranta, G., & Salvia, R. (2017). Social-Based Product Innovation and Governance in The Milk Sector: The Case of Carciocacio and Innonatura. In T. Tarnanidis, M. Vlachopoulou, & J. Papathanasiou (Eds.), *Driving Agribusiness With Technology Innovations* (pp. 293–310). Hershey, PA: IGI Global. doi:10.4018/978-1-5225-2107-5.ch015

Rahman, M. K., Schmidlin, T. W., Munro-Stasiuk, M. J., & Curtis, A. (2017). Geospatial Analysis of Land Loss, Land Cover Change, and Landuse Patterns of Kutubdia Island, Bangladesh. *International Journal of Applied Geospatial Research*, *8*(2), 45–60. doi:10.4018/IJAGR.2017040104

Rajack-Talley, T. A. (2017). Agriculture, Trade Liberalization and Poverty in the ACP Countries. In W. Ganpat, R. Dyer, & W. Isaac (Eds.), *Agricultural Development and Food Security in Developing Nations* (pp. 1–20). Hershey, PA: IGI Global. doi:10.4018/978-1-5225-0942-4.ch001

Rajamanickam, S. (2018). Exploring Landscapes in Regional Convergence: Environment and Sustainable Development in South Asia. In I. Management Association (Ed.), Sustainable Development: Concepts, Methodologies, Tools, and Applications (pp. 1051-1087). Hershey, PA: IGI Global. doi:10.4018/978-1-5225-3817-2.ch047

Rizvi, S. M., & Dearden, A. (2018). KHETI: ICT Solution for Agriculture Extension and Its Replication in Open and Distance Learning. In U. Pandey & V. Indrakanti (Eds.), *Open and Distance Learning Initiatives for Sustainable Development* (pp. 163–174). Hershey, PA: IGI Global. doi:10.4018/978-1-5225-2621-6.ch008

Roşu, L., & Macarov, L. I. (2017). Management of Drought and Floods in the Dobrogea Region. In I. Management Association (Ed.), Agri-Food Supply Chain Management: Breakthroughs in Research and Practice (pp. 372-403). Hershey, PA: IGI Global. doi:10.4018/978-1-5225-1629-3.ch016

Rouzbehani, K., & Rouzbehani, S. (2018). Mapping Women's World: GIS and the Case of Breast Cancer in the US. *International Journal of Public Health Management and Ethics*, *3*(1), 14–25. doi:10.4018/IJPHME.2018010102

Roy, D. (2018). Success Factors of Adoption of Mobile Applications in Rural India: Effect of Service Characteristics on Conceptual Model. In M. Khosrow-Pour, D.B.A. (Ed.), Green Computing Strategies for Competitive Advantage and Business Sustainability (pp. 211-238). Hershey, PA: IGI Global. doi:10.4018/978-1-5225-5017-4.ch010

Sajeva, M., Lemon, M., & Sahota, P. S. (2017). Governance for Food Security: A Framework for Social Learning and Scenario Building. *International Journal of Food and Beverage Manufacturing and Business Models*, *2*(2), 67–84. doi:10.4018/IJFBMBM.2017070104

Sallam, G. A., Youssef, T., Embaby, M. E., & Shaltot, F. (2011). Using Geographic Information System to Infollow the Fertilizers Pollution Migration. In I. Management Association (Ed.), Green Technologies: Concepts, Methodologies, Tools and Applications (pp. 564-586). Hershey, PA: IGI Global. doi:10.4018/978-1-60960-472-1.ch312

Sambhanthan, A., & Potdar, V. (2017). A Study of the Parameters Impacting Sustainability in Information Technology Organizations. *International Journal of Knowledge-Based Organizations*, *7*(3), 27–39. doi:10.4018/IJKBO.2017070103

Sanga, C., Kalungwizi, V. J., & Msuya, C. P. (2017). Bridging Gender Gaps in Provision of Agricultural Extension Service Using ICT: Experiences from Sokoine University of Agriculture (SUA) Farmer Voice Radio (FVR) Project in Tanzania. In I. Management Association (Ed.), Discrimination and Diversity: Concepts, Methodologies, Tools, and Applications (pp. 682-697). Hershey, PA: IGI Global. doi:10.4018/978-1-5225-1933-1.ch031

Santra, A., & Mitra, D. (2017). Role of Remote Sensing in Potential Fishing Zone Forecast. In A. Santra & S. Mitra (Eds.), *Remote Sensing Techniques and GIS Applications in Earth and Environmental Studies* (pp. 243–257). Hershey, PA: IGI Global. doi:10.4018/978-1-5225-1814-3.ch012

Schaap, D. (2017). SeaDataNet: Towards a Pan-European Infrastructure for Marine and Ocean Data Management. In P. Diviacco, A. Leadbetter, & H. Glaves (Eds.), *Oceanographic and Marine Cross-Domain Data Management for Sustainable Development* (pp. 155–177). Hershey, PA: IGI Global. doi:10.4018/978-1-5225-0700-0.ch007

Seckin-Celik, T. (2017). Sustainability Reporting and Sustainability in the Turkish Business Context. In U. Akkucuk (Ed.), *Ethics and Sustainability in Global Supply Chain Management* (pp. 115–132). Hershey, PA: IGI Global. doi:10.4018/978-1-5225-2036-8.ch006

Segbefia, A. Y., Barnes, V. R., Akpalu, L. A., & Mensah, M. (2018). Environmental Location Assessment for Seaweed Cultivation in Ghana: A Spatial Multi-Criteria Approach. *International Journal of Applied Geospatial Research*, *9*(1), 51–64. doi:10.4018/IJAGR.2018010104

Sen, Y. (2018). How to Manage Sustainability: A Framework for Corporate Sustainability Tools. In I. Management Association (Ed.), Sustainable Development: Concepts, Methodologies, Tools, and Applications (pp. 568-589). Hershey, PA: IGI Global. doi:10.4018/978-1-5225-3817-2.ch026

Shamshiry, E., Abdulai, A. M., Mokhtar, M. B., & Komoo, I. (2015). Regional Landfill Site Selection with GIS and Analytical Hierarchy Process Techniques: A Case Study of Langkawi Island, Malaysia. In P. Thomas, M. Srihari, & S. Kaur (Eds.), *Handbook of Research on Cultural and Economic Impacts of the Information Society* (pp. 248–282). Hershey, PA: IGI Global. doi:10.4018/978-1-4666-8598-7.ch011

Sharma, Y. K., Mangla, S. K., Patil, P. P., & Uniyal, S. (2018). Analyzing Sustainable Food Supply Chain Management Challenges in India. In M. Ram & J. Davim (Eds.), *Soft Computing Techniques and Applications in Mechanical Engineering* (pp. 162–180). Hershey, PA: IGI Global. doi:10.4018/978-1-5225-3035-0.ch008

Silvestrelli, P. (2018). The Impact of Events: To Which Extent Are They Sustainable for Tourist Destinations? Some Evidences From Expo Milano 2015. In M. Risso & S. Testarmata (Eds.), *Value Sharing for Sustainable and Inclusive Development* (pp. 185–204). Hershey, PA: IGI Global. doi:10.4018/978-1-5225-3147-0.ch008

Silvius, G. (2018). Sustainability Evaluation of IT/IS Projects. In I. Management Association (Ed.), Sustainable Development: Concepts, Methodologies, Tools, and Applications (pp. 26-40). Hershey, PA: IGI Global. doi:10.4018/978-1-5225-3817-2.ch002

Singh, R., Srivastava, P., Singh, P., Upadhyay, S., & Raghubanshi, A. S. (2017). Human Overpopulation and Food Security: Challenges for the Agriculture Sustainability. In R. Singh, A. Singh, & V. Srivastava (Eds.), *Environmental Issues Surrounding Human Overpopulation* (pp. 12–39). Hershey, PA: IGI Global. doi:10.4018/978-1-5225-1683-5.ch002

Singh, R., Srivastava, P., Singh, P., Upadhyay, S., & Raghubanshi, A. S. (2017). Human Overpopulation and Food Security: Challenges for the Agriculture Sustainability. In R. Singh, A. Singh, & V. Srivastava (Eds.), *Environmental Issues Surrounding Human Overpopulation* (pp. 12–39). Hershey, PA: IGI Global. doi:10.4018/978-1-5225-1683-5.ch002

Srivastava, N. (2017). Climate Change Mitigation: Collective Efforts and Responsibly. In I. Management Association (Ed.), Natural Resources Management: Concepts, Methodologies, Tools, and Applications (pp. 64-76). Hershey, PA: IGI Global. doi:10.4018/978-1-5225-0803-8.ch003

Stanganelli, M., & Gerundo, C. (2017). Understanding the Role of Urban Morphology and Green Areas Configuration During Heat Waves. *International Journal of Agricultural and Environmental Information Systems*, *8*(2), 50–64. doi:10.4018/IJAEIS.2017040104

Stewart, M. K., Hagood, D., & Ching, C. C. (2017). Virtual Games and Real-World Communities: Environments that Constrain and Enable Physical Activity in Games for Health. *International Journal of Game-Based Learning*, *7*(1), 1–19. doi:10.4018/IJGBL.2017010101

Stone, R. J. (2017). Modelling the Frequency of Tropical Cyclones in the Lower Caribbean Region. In W. Ganpat & W. Isaac (Eds.), *Environmental Sustainability and Climate Change Adaptation Strategies* (pp. 341–349). Hershey, PA: IGI Global. doi:10.4018/978-1-5225-1607-1.ch013

Syed, A., & Jabeen, U. A. (2018). Climate Change Impact on Agriculture and Food Security. In A. Eneanya (Ed.), *Handbook of Research on Environmental Policies for Emergency Management and Public Safety* (pp. 223–237). Hershey, PA: IGI Global. doi:10.4018/978-1-5225-3194-4.ch012

Tam, G. C. (2017). The Global View of Sustainability. In *Managerial Strategies and Green Solutions for Project Sustainability* (pp. 1–24). Hershey, PA: IGI Global. doi:10.4018/978-1-5225-2371-0.ch001

Tam, G. C. (2017). Understanding Project Sustainability. In *Managerial Strategies and Green Solutions for Project Sustainability* (pp. 110–139). Hershey, PA: IGI Global. doi:10.4018/978-1-5225-2371-0.ch005

Tam, G. C. (2017). Perspectives on Sustainability. In *Managerial Strategies and Green Solutions for Project Sustainability* (pp. 53–76). Hershey, PA: IGI Global. doi:10.4018/978-1-5225-2371-0.ch003

Tang, M., & Karunanithi, A. T. (2018). Visual Logic Maps (vLms). In *Advanced Concept Maps in STEM Education: Emerging Research and Opportunities* (pp. 108–149). Hershey, PA: IGI Global. doi:10.4018/978-1-5225-2184-6.ch005

Taşçıoğlu, M., & Yener, D. (2018). The Value and Scope of GIS in Marketing and Tourism Management. In S. Chaudhuri & N. Ray (Eds.), *GIS Applications in the Tourism and Hospitality Industry* (pp. 189–211). Hershey, PA: IGI Global. doi:10.4018/978-1-5225-5088-4.ch009

Tianming, G. (2018). Food Security and Rural Development on Emerging Markets of Northeast Asia: Cases of Chinese North and Russian Far East. In V. Erokhin (Ed.), *Establishing Food Security and Alternatives to International Trade in Emerging Economies* (pp. 155–176). Hershey, PA: IGI Global. doi:10.4018/978-1-5225-2733-6.ch008

Tiftikçigil, B. Y., Yaşgül, Y. S., & Güriş, B. (2017). Sustainability of Foreign Trade Deficit in Energy: The Case of Turkey. In N. Ray (Ed.), *Business Infrastructure for Sustainability in Developing Economies* (pp. 94–109). Hershey, PA: IGI Global. doi:10.4018/978-1-5225-2041-2.ch005

Tiwari, S., Vaish, B., & Singh, P. (2018). Population and Global Food Security: Issues Related to Climate Change. In I. Management Association (Ed.), Climate Change and Environmental Concerns: Breakthroughs in Research and Practice (pp. 41-64). Hershey, PA: IGI Global. doi:10.4018/978-1-5225-5487-5.ch003

Toujani, A., & Achour, H. (2018). A Data Mining Framework for Forest Fire Mapping. In I. Management Association (Ed.), Information Retrieval and Management: Concepts, Methodologies, Tools, and Applications (pp. 771-794). Hershey, PA: IGI Global. doi:10.4018/978-1-5225-5191-1.ch033

Tripathy, B., & K., S. B. (2017). Rough Fuzzy Set Theory and Neighbourhood Approximation Based Modelling for Spatial Epidemiology. In I. Management Association (Ed.), *Public Health and Welfare: Concepts, Methodologies, Tools, and Applications* (pp. 1257-1268). Hershey, PA: IGI Global. doi:10.4018/978-1-5225-1674-3.ch058

Trukhachev, A. (2017). New Approaches to Regional Branding through Green Production and Utilization of Existing Natural Advantages. In I. Management Association (Ed.), Advertising and Branding: Concepts, Methodologies, Tools, and Applications (pp. 1758-1778). Hershey, PA: IGI Global. doi:10.4018/978-1-5225-1793-1.ch081

Tsobanoglou, G. O., & Vlachopoulou, E. I. (2017). Social-Ecological Systems in Local Fisheries Communities. In G. Korres, E. Kourliouros, & M. Michailidis (Eds.), *Handbook of Research on Policies and Practices for Sustainable Economic Growth and Regional Development* (pp. 306–316). Hershey, PA: IGI Global. doi:10.4018/978-1-5225-2458-8.ch026

Tsou, M. (2010). Geographic Information Retrieval and Text Mining on Chinese Tourism Web Pages. *International Journal of Information Technology and Web Engineering*, *5*(1), 56–75. doi:10.4018/jitwe.2010010104

Tuydes-Yaman, H., & Karatas, P. (2018). Evaluation of Walkability and Pedestrian Level of Service. In I. Management Association (Ed.), Intelligent Transportation and Planning: Breakthroughs in Research and Practice (pp. 264-291). Hershey, PA: IGI Global. doi:10.4018/978-1-5225-5210-9.ch012

Uddin, S., Chakravorty, S., Ray, A., & Sherpa, K. S. (2018). Optimal Location of Sub-Station Using Q-GIS and Multi-Criteria Decision Making Approach. *International Journal of Decision Support System Technology*, *10*(2), 65–79. doi:10.4018/IJDSST.2018040104

Uddin, S., Chakravorty, S., Sherpa, K. S., & Ray, A. (2018). Power Distribution System Planning Using Q-GIS. *International Journal of Energy Optimization and Engineering*, *7*(2), 61–75. doi:10.4018/IJEOE.2018040103

Uzun, F. V. (2018). Natural Resources Management. In A. Eneanya (Ed.), *Handbook of Research on Environmental Policies for Emergency Management and Public Safety* (pp. 1–21). Hershey, PA: IGI Global. doi:10.4018/978-1-5225-3194-4.ch001

V, M., Agrawal, R., Sharma, V., & T.N., K. (2018). Supply Chain Social Sustainability and Manufacturing. In I. Management Association (Ed.), *Technology Adoption and Social Issues: Concepts, Methodologies, Tools, and Applications* (pp. 226-252). Hershey, PA: IGI Global. doi:10.4018/978-1-5225-5201-7.ch011 van der Vliet-Bakker, J. M. (2017). Environmentally Forced Migration and Human Rights. In C. Akrivopoulou (Ed.), *Defending Human Rights and Democracy in the Era of Globalization* (pp. 146–180). Hershey, PA: IGI Global. doi:10.4018/978-1-5225-0723-9.ch007

Vaskov, A. G., Lin, Z. Y., Tyagunov, M. G., Shestopalova, T. A., & Deryugina, G. V. (2018). Design of Renewable Sources GIS for ASEAN Countries. In V. Kharchenko & P. Vasant (Eds.), *Handbook of Research on Renewable Energy and Electric Resources for Sustainable Rural Development* (pp. 1–25). Hershey, PA: IGI Global. doi:10.4018/978-1-5225-3867-7.ch001

Vázquez, D. G., & Gil, M. T. (2017). Sustainability in Smart Cities: The Case of Vitoria-Gasteiz (Spain) – A Commitment to a New Urban Paradigm. In L. Carvalho (Ed.), *Handbook of Research on Entrepreneurial Development and Innovation Within Smart Cities* (pp. 248–268). Hershey, PA: IGI Global. doi:10.4018/978-1-5225-1978-2.ch012

Wahab, I. N., & Soonthodu, S. (2018). Geographical Information System in Eco-Tourism. In S. Chaudhuri & N. Ray (Eds.), *GIS Applications in the Tourism and Hospitality Industry* (pp. 61–75). Hershey, PA: IGI Global. doi:10.4018/978-1-5225-5088-4.ch003

Weiss-Randall, D. (2018). Cultivating Environmental Justice. In *Utilizing Innovative Technologies to Address the Public Health Impact of Climate Change: Emerging Research and Opportunities* (pp. 110–143). Hershey, PA: IGI Global. doi:10.4018/978-1-5225-3414-3.ch004

Weiss-Randall, D. (2018). Cultivating Resilience. In *Utilizing Innovative Technologies to Address the Public Health Impact of Climate Change: Emerging Research and Opportunities* (pp. 204–235). Hershey, PA: IGI Global. doi:10.4018/978-1-5225-3414-3.ch007

Weiss-Randall, D. (2018). Climate Change Solutions: Where Do We Go From Here? In *Utilizing Innovative Technologies to Address the Public Health Impact of Climate Change: Emerging Research and Opportunities* (pp. 236–268). Hershey, PA: IGI Global. doi:10.4018/978-1-5225-3414-3.ch008

Whyte, K. P., List, M., Stone, J. V., Grooms, D., Gasteyer, S., Thompson, P. B., . . . Bouri, H. (2018). Uberveillance, Standards, and Anticipation: A Case Study on Nanobiosensors in U.S. Cattle. In I. Management Association (Ed.), Biomedical Engineering: Concepts, Methodologies, Tools, and Applications (pp. 577-596). Hershey, PA: IGI Global. doi:10.4018/978-1-5225-3158-6.ch025

Wulff, E. (2017). Data and Operational Oceanography: A Review in Support of Responsible Fisheries and Aquaculture. In P. Diviacco, A. Leadbetter, & H. Glaves (Eds.), *Oceanographic and Marine Cross-Domain Data Management for Sustainable Development* (pp. 303–324). Hershey, PA: IGI Global. doi:10.4018/978-1-5225-0700-0.ch013

Yener, D. (2017). Geographic Information Systems and Its Applications in Marketing Literature. In S. Faiz & K. Mahmoudi (Eds.), *Handbook of Research on Geographic Information Systems Applications and Advancements* (pp. 158–172). Hershey, PA: IGI Global. doi:10.4018/978-1-5225-0937-0.ch006

About the Contributors

T. G. Sitharam has Masters from Indian Institute of Science, Bangalore in 1986 and Ph.D. from University of Waterloo, Waterloo, Ontario, Canada in 1991. At University of Texas at Austin, Texas, USA, he was working until 1994. Since 1994 he is at Indian Institute of Science and presently he is a PROFESSOR in the department of Civil Engineering. He is the founder Chairman of Centre for infrastructure, Sustainable Transportation and Urban Planning (CiSTUP) at Indian Institute of Science from Jan, 2009 to March 2014, Bangalore. He is the Chief Editor of an International journal of Geotechnical Earthquake Engineering published by IGI Global, PA, USA. Professor Sitharam has guided 25 Ph.D students and has published more than 150 papers in International / national journals and 250 publications in conferences. H-index of 25 and i-10 index of 50. He has received many awards.

* * *

A. Boominathan is currently professor at Civil Department, IIT Madras. His research interests include Soil Structure Interaction, Ground Improvement Techniques, Liquefaction, Seismic Hazard Analysis, Ground Response Analysis and Base Isolation.

Andrew Brennan graduated in Engineering Science from the University of Oxford in 1999 before defecting to University of Cambridge where he undertook his PhD on centrifuge modelling of vertical drains as remediation against earthquake-induced soil liquefaction. On completion of his thesis he remained at Cambridge as a Junior Research Fellow at Wolfson College, undertaking various dynamic centrifuge projects under the umbrella of the EU FP5 funded NEMISREF project, some of which have led to the research presented here. In 2005 he left Cambridge for the University of Dundee in Scotland, where he has commissioned an earthquake shaker for the Dundee 3 m radius beam centrifuge so he can continue researching problems in earthquake geotechnical engineering. He also has a new-found interest in problems related to offshore geotechnics, particularly in cases where dynamic/cyclic loads are expected.

Kumar Emani is a 2001 Civil Engineering graduate from Nagarjuna University of the state of Andhra Pradesh in India. He worked for the Department of Atomic energy as a Scientific officer/ Engineer. Then completed his Ph. D from Indian Institute of Technology, Roorkee. For the last decade he has been teaching under-graduate and post-graduate students of Civil Engineering. He is also carrying out two sponsored research projects funded by the Government of India under different schemes.

Emad Gad is the Head of Civil Engineering Discipline in the Faculty of Engineering and Industrial Sciences at Swinburne University of Technology. He obtained his PhD from the University of Melbourne in 1994. His areas of expertise include structural dynamics, residential construction, structural modelling and field and laboratory testing. He is a winner of the prestigious Chapman Medal in 2012. He was the Chairman of the Structural Branch of Engineers, Victoria division, Australia. He is widely consulted by the industry on various aspects of engineering.

Helen Goldsworthy is from the Department of Infrastructure Engineering at the University of Melbourne, Melbourne, Australia. She received a BEng (Hons1) from the University of Adelaide (1981), a MSc from the University of Illinois at Urbana-Champaign (1982), and a Ph.D. from the University of Melbourne (1990). She worked for Skidmore, Owings and Merrill in San Francisco in the early 1980s, designing earthquake-resistant buildings. Her research interests are centred on earthquake engineering; specifically the behaviour of steel, reinforced concrete and composite structures under seismic loading. She is a member of the national committee of the Australian Earthquake Engineering Society and conducted reconnaissance for the society in Christchurch in 2011. She received the Chapman medal in 2013 for a paper written by her to convey the lessons learned from that earthquake and the relevance of those to Australian engineers.

Ged Griffin is a PhD Candidate at the University of Melbourne where he is researching the next generation of emergency management. He has been a police officer for 28 years where he has performed duties in general duties, emergency management, marine policing, criminal investigations, intelligence and counter terrorism operations. He has also performed a wide range of roles in East Timor including duties at the UN Serious Crimes Unit, Victoria Police Contingent Commander and as Liaison Officer supporting the former Victorian Premier the Hon Mr Steve Bracks during his work assisting Xanana Guasmao and the new government of East Timor. He is a member of the Australian Civil Corps and has been appointed as the Team Leader for the Post Disaster Response Team. He is currently an Inspector in the State Emergency Response Coordination Division at the Victoria Police Force.

Assaf Klar is a professor of geotechnical engineering and soil mechanics. After completing his bachelor's in 1999 and his doctorate (in the special direct program) in 2003 in civil engineering at the Technion – Israel Institute of Technology, he joined the engineering department of Cambridge University, UK, as an associate researcher. In 2005 he returned to the Technion, and was appointed a senior lecturer in 2005 and associate professor in 2011. In 2018, Assaf was appointed professor in the section of geotechnics and geology at DTU. Assaf's research activity focuses on infrastructure geotechnics, covering topics such as tunneling and tunneling effects on buildings and buried pipelines, underground sensing with a focus on application of distributed fiber optic sensing, soil structure interaction and foundation systems. He is also involved in petroleum related research and development, dealing with THMC modeling of gas hydrate bearing sediments.

Shashank Kothari is a post-graduate in Structural Engineering from Graphic Era (Deemed to be University). He worked as a Junior Research Fellow (JRF) in a sponsored research project. Presently, he is working as a Assistant Professor in Graphic Era (Deemed to be University), Dehradun, India.

A. Murali Krishna is presently an Associate Professor in Department of Civil Engineering at Indian Institute of Technology Guwahati. He obtained his Ph. D degree from IISc Bangalore; M.Tech degree from IIT Kanpur and B.Tech from SV University, Tirupati. His research interests include: Earthquake Geotechnics, Geosynthetics and Ground Improvement, Site characterization and Numerical and Physical modelling of geotechnical structures. He has guided three Doctoral students and twenty Masters students. Dr. Murali Krishna has nearly 140 publications of technical papers in International/National Journals and Conference/Seminar Proceedings including book chapters. He is a recipient of BRNS Young Scientist Research award, BOYSCAST fellowship and HERTAGE fellowship. Dr. Murali Krishna is currently serving as an 'Executive Member' of the IGS and Member of TC 203 of ISSMGE, since 2011. Dr. Murali Krishna organized national and international workshops and short courses. He is a reviewer for several national and international journals.

Krishna Kumar work involves developing massively parallel micro-/macro-scale numerical methods: Finite Element Method, Material Point Method, Lattice Boltzmann - Discrete Element coupling and Lattice Element method.

Ritesh Kumar is a Civil Engineering graduate from Uttarakhand Technical University of India. He did his post-graduation from Earthquake Engineering Department of IIT Roorkee. After working as an Assistant Professor in Graphic Era (Deemed to be University), he is presently in Tokyo Insititute of Technology as a doctoral scholar .

S. D. Anitha Kumari did her Master of Science (By Research) from Indian Institute of Technlogy, Chennai and completed her PhD from Indian Institute of Science, Bangalore. Currently she is working as an Associate Professor at M.S.Ramaiah University of Applied Sciences.

Nelson Lam is a Reader in Civil Engineering at The University of Melbourne, is an internationally recognized expert in structural dynamics, earthquake engineering and protective technology. In the past 20 years, he has been researching and consulting widely in this field and has published some 200 technical articles which include some 80 journal articles. His achievement in research and knowledge transfer in this field was recognized by the award of the Chapman Medal (1999), the Warren Medal (2006) by Engineers Australia, Best Paper Award (2004-2007) by the ISET Journal of Earthquake Technology and Chapman Medal (2010). He is chief investigator of numerous research programs that have won ARC Discovery and Linkage Funding Awards. He was awarded the degree of PhD in structural engineering at the University of Melbourne in 1993.

B. P. Naveen is Associated Professor at Department of Civil Engineering, Amity School of Engineering and Technology, Amity University Haryana. His major research activities include Geotechnical and Geoenvironmental Engineering. He is the author of about 16 papers in National Conference, 12 papers in International conference,10 paper in Journal and 1 textbook.He is a member of many Professional bodies.

Abbas Rajabifard is the Director of the Centre for Spatial Data Infrastructures and Land Administration at the Department of Infrastructure Engineering, the University of Melbourne. He is immediate Past President and current Executive Board Member of the GSDI Association, and was Vice Chair of Working Group 3 of the United Nations supported Permanent Committee on GIS Infrastructure for Asia and the Pacific (PCGIAP), a member of ICA-Spatial Data Standards Commission, and a member of the Victorian Spatial Council. He has been an Executive Board member and national representative to the PCGIAP (1994-1998), member of International Steering Committee for Global Mapping Project (1997-2001) and a member of the UN-ESCAP Group of Experts to develop Guidelines on GIS Standardization for Asia-Pacific (1995). He had published broadly on SDI, land administration, spatial enablement and spatial data management.

Janaki Rama Raju graduated from Osmania University,Hyderabad in Civil Engineering and Post graduated from Andhra University in Structural Engineering. He worked as an engineer in various constructional projects. Since last two decades he is teaching under-graduate Civil Engineering students. He is a member of Indian Society for Technical Education (ISTE),Institute of Engineers (IOE) and Indian Society of Remote Sensing (ISRS) India.

M. L. Sharma carries out teaching, research and consultancy at Department of Earthquake Engineering, IIT Roorkee since 1985. President of Indian Society of Earthquake Technology -2015-17 and 2017-19.

P. V. Sivapullaiah is currently Pro Vice-Chancellor, Bengaluru Campus, GITAM University. He is a member of many Professional bodies. His primary research activities include Environmental Geotechnics; Soil Mineralogy; Ground Improvement. Published about 270 papers in International and National Journals and Conferences and contributed chapters in a few books. Guided many students for Ph.D. and MSc (Engg). Member of Board of Studies of several Universities/ Institution. Delivered Invited Lectures/Keynote addresses at Many Conferences Symposia/Courses. Reviewer for many Ph.D. theses and international journals. Handled many consultancy projects. Received several best paper awards.

Srikanth Venkatesan is a Senior Lecturer at Victoria University. His research interests are centered on the broader areas of natural hazards research and disaster management. In particular he focusses on modelling resilience and has reasonable expertise in earthquake engineering and geotechnical engineering. Professor Abbas Rajabifard is the Director of the Centre for Spatial Data Infrastructures and Land Administration at the Department of Infrastructure Engineering, the University of Melbourne. He is immediate Past President and current Executive Board Member of the GSDI Association, and was Vice Chair of Working Group 3 of the United Nations supported Permanent Committee on GIS Infrastructure for Asia and the Pacific (PCGIAP), a member of ICA-Spatial Data Standards Commission, and a member of the Victorian Spatial Council. He has been an Executive Board member and national representative to the PCGIAP (1994-1998), member of International Steering Committee for Global Mapping Project (1997-2001) and a member of the UN-ESCAP Group of Experts to develop Guidelines on GIS Standardization for Asia-Pacific (1995). He had published broadly on SDI, land administration, spatial enablement and spatial data management. Associate Professor Nelson Lam is a Reader in Civil Engineering at The University of Melbourne, is an internationally recognized expert in structural dynamics, earthquake engineering and protective technology. In the past 20 years, he has been researching and consulting widely in this field and has published some 200 technical articles which include some 80 journal articles. His achievement in research and knowledge transfer in this field was recognized by the award of the Chapman Medal (1999), the Warren Medal (2006) by Engineers Australia, Best Paper Award (2004-2007) by the ISET Journal of Earthquake Technology and Chapman Medal (2010). He is chief investigator of numerous research programs that have won ARC Discovery and Linkage Funding Awards. He was awarded the degree of PhD in structural engineering at the University of Melbourne in 1993. Professor Emad Gad is the Head of Civil Engineering Discipline in the Faculty of Engineering and Industrial Sciences at Swinburne University of Technology. He obtained his PhD from the University of Melbourne in 1994. His areas of expertise include structural dynamics, residential construction, structural modelling and field and laboratory testing. He is a winner of the prestigious Chapman Medal in 2012. He was the Chairman of the Structural Branch of Engineers, Victoria division, Australia. He is widely consulted by the industry on various aspects of engineering. Associate Professor Helen Goldsworthy is from the Department of Infrastructure Engineering at the University of Melbourne, Melbourne, Australia. She received a BEng (Hons1) from the University of Adelaide (1981), a MSc from the University of Illinois at Urbana-Champaign (1982), and a Ph.D. from the University of Melbourne (1990). She worked for Skidmore, Owings and Merrill in San Francisco in the early 1980s, designing earthquake-resistant buildings. Her research interests are centred on earthquake engineering; specifically the behaviour of steel, reinforced concrete and composite structures under seismic loading. She is a member of the national committee of the Australian Earthquake Engineering Society and conducted reconnaissance for the society in Christchurch in 2011. She received the Chapman medal in 2013 for a paper written by her to convey the lessons learned from that earthquake and the relevance of those to Australian engineers. Mr. Ged Griffin is a PhD Candidate at the University of Melbourne where he is researching the next generation of emergency management. He has been a police officer for 28 years where he has performed duties in general duties, emergency management, marine policing, criminal investigations, intelligence and counter terrorism operations. He has also performed a wide range of roles in East Timor including duties at the UN Serious Crimes Unit, Victoria Police Contingent Commander and as Liaison Officer supporting the former Victorian Premier the Hon Mr Steve Bracks during his work assisting Xanana Guasmao and the new government of East Timor. He is a member of the Australian Civil Corps and has been appointed as the Team Leader for the Post Disaster Response Team. He is currently an Inspector in the State Emergency Response Coordination Division at the Victoria Police Force.

R. Vijaya is a Doctoral Research Scholar at IIT Madras, Chennai. She completed her BTech in Civil Engineering from A.C.Tech Karaikudi, TamilNadu and MTech in Soil Mechanics and Foundation Engineering from NIT Surat, Gujarat. She did her MTech project at Stuctural Engineering Research Center-CSIR, Chennai.

Index

www.ingramcontent.com/pod-product-compliance
Lightning Source LLC
LaVergne TN
LVHW080900300926
849162LV00003B/488
9781522569480